PLANNING IN THE UK

Planning in the UK

Agendas for the new millennium

Edited by

YVONNE RYDIN and ANDY THORNLEY
The London School of Economics and Political Science, UK

Routledge
Taylor & Francis Group
LONDON AND NEW YORK

First published 2002 by Ashgate Publishing

Reissued 2018 by Routledge
2 Park Square, Milton Park, Abingdon, Oxon OX14 4RN
711 Third Avenue, New York, NY 10017, USA

Routledge is an imprint of the Taylor & Francis Group, an informa business

Publisher's Note
The publisher has gone to great lengths to ensure the quality of this reprint but points out that some imperfections in the original copies may be apparent.

Disclaimer
The publisher has made every effort to trace copyright holders and welcomes correspondence from those they have been unable to contact.

ISBN 13: 978-1-138-73692-4 (hbk)
ISBN 13: 978-1-138-73687-0 (pbk)
ISBN 13: 978-1-315-18565-1 (ebk)

Contents

List of Tables

List of Figures

List of Contributors

David Adams European Urban and Regional Research Centre, Department of Land Economy, University of Aberdeen.

Nick Bailey Department of Development and Urban Regeneration, University of Westminster.

Nia Blank Department of City and Regional Planning, University of Wales.

Caroline Brown Centre for the Built Environment and Planning, University of the West of England, Bristol.

Heather Campbell Department of Town and Regional Planning, University of Sheffield.

Stephen Connelly Department of Town and Regional Planning, University of Sheffield.

Alan Disberry Previously Research Fellow, Department of Land Economy, University of Aberdeen.

Stefanie Dühr Centre for Environment and Planning, University of the West of England, Bristol.

Keith Fernie Geddes Centre for Planning Research, School of Town and Regional Planning, University of Dundee.

Frank Gaffikin Director, Urban Institute, The University of Ulster.

Catherine Hammond PhD Researcher, Department of Architecture, Planning and Landscape, University of Newcastle upon Tyne.

Victoria Hands Centre for Local Environmental Policies and Strategies, South Bank University.

Angela Hull Centre for Research in European Urban Environments, School of Architecture, Planning and Landscape, University of Newcastle.

Norman Hutchison European Urban and Regional Research Centre, Department of Land Economy, University of Aberdeen.

Barbara M. Illsley Geddes Centre for Planning Research, School of Town and Regional Planning, University of Dundee.

Glyn Jones Environmental Project Manager, Urban Mines, Norwood Green.

Victoria Joy Environmental Consultant, Addleshaw, Booth & Co, Solicitors, Leeds.

Peter J. Larkham Birmingham School of Planning, University of Central England.

M. G. Lloyd Geddes Centre for Planning Research, School of Town and Regional Planning, University of Dundee.

Robert Marshall Department of Town and Regional Planning, University of Sheffield.

J. McCarthy Geddes Centre for Planning Research, School of Town and Regional Planning, University of Dundee.

Malachy McEldowney Head of School of Environmental Planning, The Queen's University of Belfast.

Nicola Morrison Department of Land Economy, University of Cambridge.

Mike Morrissey Co-Director of the Urban Institute in Belfast. He has worked in the Northern Ireland Polytechnic and the University of Ulster since 1975.

Thomas Munjoma European Urban and Regional Research Centre, Department of Land Economy, University of Aberdeen.

Deborah Peel Department of Development and Urban Regeneration, University of Westminster.

John Pendlebury Department of Town and Country Planning, University of Newcastle.

Susan Percy Centre for Local Environmental Policies and Strategies, South Bank University.

Jenny Poxon East Midlands Rural Action Group Development Officer and East Midlands ACRE Network Secretary.

Elaine Quinn Researcher in Environmental Planning, Environmental Impact Assessment Centre, University of Manchester.

Peter Roberts Geddes Centre for Planning Research, School of Town and Regional Planning, University of Dundee.

Yvonne Rydin Reader in Environmental Planning, Director of MSc Environmental Assessment and Evaluation, London School of Economics.

Martyn Senior Department of City and Regional Planning, University of Wales.

Liz Sharp Lecturer in Environmental Management, Course Leader of MSc in Local Sustainable Development, Department of Town and Regional Planning, University of Sheffield.

Ken Sterrett Lecturer in Environmental Planning, The Queen's University of Belfast.

Andy Thornley Senior Lecturer, Director of MSc Regional and Urban Planning Studies, London School of Economics.

Tim Townshend Department of Town and Country Planning, University of Newcastle.

Chris Webster Department of City and Regional Planning, University of Wales.

Christine M. E. Whitehead Professor of Housing, London School of Economics.

1 An Agenda for the New Millennium

YVONNE RYDIN AND ANDY THORNLEY

The year 2000 was seen in with world-wide fireworks; the year 2001 rather more quietly. Either way we are now into the third millennium. For most planners and planning work, life continues much as normal. But that could be deceptive. For any period of profound change, always carries with it continuities from the past. For UK planning there are changes (and continuities) to deal with from a number of different directions. There are the cultural, political and economic changes associated with globalisation, that we, along with many other researchers explore in the sister-volume *Planning in a Global Era* (Thornley and Rydin, 2001). For the UK there is the related aspect of learning to live with and within Europe. There is the current heightening of the environmental agenda, with a recurrent sense of crisis, political protest, repeated scientific warnings and attempts at international action. There are changes in social trends, with new patterns of household formation and ways of everyday life. Many of these, at home and work, are associated with the rise of information technology, that seems to be all around us these days: PCs, mobile phones, smart equipment of all kinds. And yet, in a very real sense, the problems remain unchanged: poor environmental quality, health and safety threats to everyday life, unemployment and economic recession, social deprivation, and spatial inequality.

So from where might the major discontinuities arise that could shape a new agenda for planning? First, there is an increasing awareness that, although there are many continuities, the pace of change has so increased as to constitute a qualitative as well as quantitative shift. This appears to have occurred with more entrenched processes of globalisation. Also many environmentalists argue that working on the basis of simple linear trends in environmental change will underestimate the actual significance of that change. Threshold effects, aggregate effects of multiple events and the chaotic nature of complex environmental systems all mean that the environment can be precipitated into a dramatically worsened situation quite quickly. A similar point may be made about complex economic systems and their tendency to cyclical patterns despite active governmental "management" of the economy. Complacency arising from the experience of living with past trends is usually not a good model for future behaviour.

But the main discontinuity in the context for planning practice will arise when there is a shift in political power, as occurred in Britain in 1997 and is now occurring in the United States. For British planners, the new agenda of the Blair "New Labour" Government has been a major pre-occupation for the last four years and, if opinion polls at the time of writing prove correct, is likely to remain so for the next four to five years. In the introduction to this volume of research into contemporary planning practice, we examine the nature of this political shift and examine its implication for planning. First, we explore the extent to which the New Labour government did represent a political discontinuity. We look at the differences with preceding governments and the nature of its political ideology, and then explore its agenda for key areas of planning practice (see also Thornley, 1999), concluding with a review of how this impacts on the field of planning theory.

New Labour: How Big a Shift?

A major change in the political landscape of the UK took place with the advent of the Labour government in 1997. After eighteen year of Conservative Party rule, the government of Tony Blair promised a new philosophy and new approach under the banner of New Labour. Priorities changed and new themes moved to the top of the agenda such as local democracy, community, transparency, sustainability and co-ordinated or "joined-up" thinking. Over the following years these themes became embodied in legislation, some of which involved significant constitutional change. As a result the context for planning also changed. Many of the new priorities, such as community empowerment, involved a reappraisal of the purpose and procedures of planning, while others changed the legislative and institutional frame within which planning operated. This volume brings together the work that has been done in recent years to trace and analyse the implications for planning of this political paradigm shift.

But how big a shift was it? We can briefly explore this question through comparing the New Labour approach to that of the preceding Conservative governments. After the initial euphoria of winning the election the Labour Party in power started to feel rather uncomfortable with its pragmatic election-oriented stance and started to look around for a "big idea". It was therefore very attracted by the concept of the "Third Way" particularly as propounded by Anthony Giddens. So what has been the influence of this concept and how far has it resulted in a political approach that is significantly different from that of the earlier period? We briefly address this issue as a context for the detailed research contributions in rest of the book.

The strategy in winning the election was to keep everyone happy and to refrain from saying anything bold or significant that might lose votes. Reliance was placed on the youth and charisma of the leader and the demonstration of organisational efficiency that was evident from the reform of the Party. However some indications of an embryonic ideological position can be detected in the election manifesto and speeches. Mr Blair made it very clear that the Labour Party would continue to support the business community and that the imperatives of the market would be paramount in the new government's thinking (Blair 1996). New Labour was keen to establish its credentials with the private sector and Blair talked about a "new era opening up in relations between today's Labour Party and the business community" (Blair, 1996: 107). For example Blair held many meetings with business leaders, including representatives of the Corporation of London. The need to compete internationally was seen as an important influence on government and Blair said that "since it is inconceivable that the UK would want to withdraw unilaterally from this global market-place, we must instead adjust our policies to its existence" (Blair 1996: 86). Such strong statements, stressing the need to respond to economic forces and to accommodate to economic interests, suggest a continuity with the ideology of Thatcherism.

However in two important respects New Labour can be contrasted sharply with Thatcherism. While Mrs Thatcher took a conflictual approach, Mr Blair placed consensus and conciliation at the top of his agenda. He presented an image of a social healer who wanted to bring all elements of society together. There were strong references to Christian beliefs in a caring society. This stands in stark contrast to Mrs. Thatcher's statement that there is no such thing as society, only individuals and their families. New Labour propounded a "belief in society, working together, co-operation, solidarity and partnership" (1996: 38). Blair called upon his Christianity to criticise Conservatives because they had "too selfish a definition of self-interest. They fail to look beyond, to the community and individual's relationship with the community". In his speech to the Labour Party Conference in the autumn of 1997 he evoked the idea of a "giving society". He said, in a rather biblical style, "believe in us as much as we believe in you. Give just as much to our country as we intend to give. Give your all. Make this the giving age... Now make the good that is in the heart of each of us serve the good of all" (Guardian, Oct 1st 1997: 8).

The second contrast with Thatcherism was the emphasis given to local democracy and decentralisation. Major institutional reforms were initiated to give more autonomy to Scotland and Wales, to reintroduce regional policy, and to reinvigorate local authorities with new forms of democracy and local involvement. The reform of the House of Lords also began. There was a general commitment to increased democracy, "New Labour wants to give power to the people - to be a government working in partnership with the people, which gives them choice and responsibility" (1996: 321).

As already mentioned, once in office the new government sought to develop an ideological framework in which to place the principles it had expounded during the election campaign. One of the major influences on Blair was Anthony Giddens, whose book *Beyond Left and Right* (1994) set out a new approach to politics. This was further developed in *The Third Way* published in 1998. These books suggest that the previous political divisions of neo-liberalism and socialism are now irrelevant and a new perspective is required that responds to contemporary trends, such as globalisation, increasing uncertainty and social diversification. The features of the approach include rethinking the welfare state using the principle of empowerment, more transparency in government utilising dialogue, and reconciling autonomy and interdependence in social life.

These ideas influenced the new government. During 1998 Blair and senior Ministers started to present the concept of the "Third Way". For example in April 1998 the Foreign Minister, Robin Cook expounded his six principles of the Third Way as: strong communities, inclusive societies, open politics, rights and also responsibilities, interdependency in a global economy, and modernisation linked to social changes (Guardian, April 23, 1998). The attempt to build this philosophical basis for the new government culminated later in the year with the Prime Minister's Fabian pamphlet entitled *The Third Way: New Politics for the New Century* (Blair, 1998). This claims to unite the two streams of thought - democratic socialism and liberalism. One of the significant features of the Prime Minister's presentation of the approach is his claim that it can "reconciling themes which in the past have wrongly been regarded as antagonistic - patriotism *and* internationalism; rights *and* responsibilities; the promotion of enterprise *and* the attack on poverty and discrimination" (Blair, 1998).

The reconciliation of this last antagonism reflects the view expounded during the election campaign that it is possible to pursue economic objectives in an era of globalisation while also achieving social goals. Giddens extends this argument to the achievement of environmental goals when he describes the "Third Way" as rejecting the assumption that there is a trade-off between economic development and the protection of the environment. Instead "environmental protection is seen as a source of economic growth rather than its opposite" (Giddens 1998: 19). This fits within an approach developed elsewhere in Europe and known as "ecological modernisation" (Gouldson and Murphy, 1998). Such an emphasis on the reconciliation of views that appear conflictual leads to a desire to build up consensus around a more co-ordinated policy approach.

The exposition of the "Third Way", as with the earlier Manifesto statements, has little directly to say on urban planning. However many of the themes discussed potentially have strong implications for planning. There is a belief that the state may need to intervene in the market to protect certain social goals, such as avoiding social exclusion. The renewed emphasis on the importance of community is of particular importance to planning as it could support the value of planning intervention in achieving community interests. According to Giddens "the theme of community is fundamental to the new politics" (1998: 79). Globalisation is seen as providing new opportunities to develop local identities while also providing a focus for collective responsibilities. Giddens does not deny that the concept of community also has its difficulties; for example it can contain differing views, is often difficult to define geographically and can create antagonism to change through NIMBY attitudes. Nevertheless the shift from the individualism of the Thatcher era to the acceptance of community cannot fail to give planning greater legitimacy. Community responsibility is also the foundation of a more environmentally aware society – another aim strongly propounded by New Labour.

The emphasis on the openness and transparency of government can also be expected to affect the planning process. The importance of local democracy and the desire to explore new forms of democracy should potentially reinvigorate participation in planning. Another of the false antitheses that Blair identifies and seeks to overcome is that between representative and participatory democracy (1998: 15). Planning participation has traditionally been circumscribed by the tensions between these two forms of democracy and so new experiments that achieve a better relationship between them could open up new opportunities in the planning process.

The rhetoric of New Labour gives greater credence to the long term view, for example in environmental and education policy. It is also stated that government must pursue policies that reinforce social solidarity. The "Third Way" claims that the pursuit of economic prosperity can be sought at the same time as achieving social and environmental goals. The details of how this can be done have yet to be spelt out. How can economic policies adapt to the imperatives of the global economy with its push to competition, minimising social costs, and short-term results while not compromising the longer term aims? As far as planning is concerned, the emphasis on consensus and co-ordination in the New Labour approach could lead to the resurrection of its role in balancing different interests. The purpose of planning could once more encompass economic efficiency, environmental sustainability and social needs, in contrast to the Thatcherite emphasis on economic interests at all costs. The interesting question will be how the inherent contradictions in pursuing all three objectives will be resolved. Will Blair's belief in the need to react to international economic competition over-ride the other objectives? The indications are that planning will be used to contribute to the aim of co-ordinating different interests. It will be do this within an ideological framework that believes that all interests can be brought together in a way that can make everyone happy. Solutions can be found that create a win-win situation for all. This seems to be a principle message of the "Third Way".

However there is an alternative view. This states that some interests will always be in mutual opposition and one person's gain will be another's loss. This is likely to apply particularly in a situation where both interests seek a scarce commodity - such as development land. To give just one simple example, the expansion of a city's international airport is necessary to ensure it retains a competitive position in the globalised economy. This is a goal which Blair would support as necessary. However this will also create noise for local residents and decrease their quality of life. Now it may be argued that a win-win situation can be found through locating the airport expansion in a place that has no adverse impact on any residents. In practice such a solution is not likely to be found in many cities and so there are bound to be some losers. In the post-war period such conflicts of interests where masked under the disguise of consensus. Thatcherism exposed the true nature of the conflicts of interest in society. However, in this period, the resolution of such conflicts, through the free operation of the market, benefited certain powerful economic interests. Under the "Third Way" we may be returning to a position in which conflicts are dismissed because of the belief that a "Third Way" can always be found to overcome antagonisms. Maybe too much faith is being place in such a possibility and in reality these conflicts need to be accepted as inevitable and a political choice made over whose interests should be given priority.

The New Labour Planning Policy Agenda

The main body of the book addresses the key planning policy issues that have arisen during the New Labour government's first term of office. Four major areas are identified: the new institutional context; ensuring social inclusion and participation; promoting sustainability; and the debate over building at higher densities on brownfield sites, an issue which encompasses both social and environmental concerns.

Since it won office in 1997 the new government has been heavily preoccupied with **institutional changes**. These stem from a concern for greater devolution and encompass the new Parliament for Scotland and Assembly for Wales, the new government for Greater London, an emphasis on regional development and major reforms to local government. Three chapters focus in particular on how such institutional changes impinge on planning. The first contribution by Lloyd and McCarthy highlights the different institutional arrangements that have been implemented across the UK and explores how these are leading to different forms of strategic planning. The second chapter by McEldowney et al. is a case study of the new kinds of regional strategy being developed in Northern Ireland with an emphasis on the consultation processes. The last paper by Hammond discusses, in a historical context, the changes taking place in regional planning as a result of pressures from the EU and the new government. The message from these contributions is that the regional level is gaining in importance again but that a variety of approaches will coexist across the UK. This will put some pressures on planning as practitioners learn to live with complex multi-tier systems (again).

Social inclusion and more participatory government are key political priorities of the New Labour government. The next section explores these themes and their implications for planning over five chapters. The first chapter by Bailey and Peel takes the government programme of "modernising" local government as its starting point and explores the issue of community empowerment. It reviews the theoretical material and applies this to a detailed case study of participation practice. The second paper by Fernie explores the government's social inclusion priority by looking at its impact on one city, Dundee. Then Hull explores the same theme through an evaluation of the New Labour approach to urban regeneration. Using another local case study, this time of the innovative policy tool of "Good Neighbour Agreement", Illsley examines its participatory potential. Finally in this section, Larkham et al. return to a traditional area of planning practice, conservation of the built heritage, and consider how public participation occurs in this context. Taken together these five contributions provide a good critical account of how the government's rhetoric of "community" is already being, and might further be, translated into practice and how this might lead to greater local empowerment.

Sustainability has been a major issue over the last decade and has been given a boost by New Labour's emphasis on environmental issues. The next three chapters address aspects of this issue. The first by Brown and Duhr gives a good overview of the way sustainability is currently being incorporated into planning at the national, regional and local levels, identifying gaps and inconsistencies in coverage. The next contribution by Percy and Hands focuses on the local level, through a case study of a sustainability initiative in the London Borough of Southwark, and raises issues of implementation. The third study by Quinn is based upon a large survey of Environmental Impact Statements and detailed case studies and focuses on the question of monitoring the effectiveness of the measures to reduce the adverse ecological and visual impacts of development. Between them these three research projects provide coverage of the way sustainability is being approached at different levels, and the various constraints that operate at different stages in the process, from political commitment through to evaluation of effectiveness.

The final section addresses probably the most controversial planning topic of the early years of New Labour in office. The government has had difficulty in coming to terms with the conflicting pressures for retaining greenfield land, especially in Green Belts, and the increase in projected demand for more housing, principally arising from demographic and social change. The Urban Task Force was asked to address the issue of stimulating an "urban renaissance" and the focus of attention has been on **developing brownfield sites**. This, it was argued, would focus demand back into the city and provide the economic and policy basis for rebuilding and reinvigorating inner city areas, as well as providing an opportunity for innovative architecture in higher density housing. This argument has meshed with one based in sustainability concerns, which argued that redirecting development back into the urban areas would reduce travel by road, contributing to a policy of reduced greenhouse gas emissions.

Much of the debate remains at a fairly general level but the final contributions to this volume attempt to give the debate greater substance. The first by Adams focuses on the question of the ownership of the land in the brownfield sites. The research indicates that the extent to which such sites get developed will depend to a significant degree on how well the owners are linked into local development networks. The second paper by Webster et al. questions, through an extensive market research exercise, the extent to which people are likely to break the preference of the last century for low density living, and choose to return to the cities. Then Roberts presents the results of recent research into the definition, identification and evaluation of brownfield sites, while Whitehead concludes the section with an answer to the question "who needs housing", focussing on the contentious south-east region.

Conceptualising Planning for the New Millennium

This body of research provides an early opportunity to assess empirically the progress of the New Labour agenda for planning. But, as emphasised above, New Labour was as much an attempt at an ideological project as a set of policy prescriptions. How does this body of research interface with the ideological project of planning and its self-reflection through excursions into planning theory? In order to explore this, the sections reporting research on particular policy areas are framed by a section that discusses recent developments in the theoretical and professional concerns of planning. (For more discussion of developments in planning theory in an international context, the reader is referred to *Planning in a Global Era*.)

Four chapters discuss these issues. This first by Campbell and Marshall explores whether the current dominance of "pragmatism" in the US theoretical debate has relevance for current conditions in the UK. This is an interesting new direction in planning theory that has spent the last two decades engaged with grand theory, largely deriving from theorists from continental Europe and reacting to the immediate post-war dominance of a systems approach. Could pragmatism prove a fruitful theoretical frame for New Labour planning practice, with its emphasis on the achievable and the compromise? The following chapter by Sharp and Connelly then considers, in theoretical terms, one key aspect of the New Labour ideology: greater involvement and participation This has particular resonance for planning and, in practice, has been a matter of interest to planners since the 1960s. Remaining at the theoretical level, the chapter addresses models of participation and concludes that a new approach is needed to relate to the current revival of interest in community involvement. The next overarching theme, covered in the following chapter by Poxon, is that of planning education, asking whether rethinking is needed in the light of changing needs as the UK moves into the next millennium. The final chapter in this section, by Morrison, focuses on the effectiveness of the land use planning system. Performance indicators have been a dominant tool in recent public policy and this chapter exposes some of the problems in applying the approach to planning.

These more conceptual discussions raise interesting questions about the New Labour agenda. As outlined above, there are inherent inconsistencies and ambiguities within the Third Way ideology that informs the Blair government. These four more conceptual accounts of planning suggest the benefits of considering an area of policy activity in detail and teasing out the problems and difficulties associated with conceptualisation, education and implementation. Such details add a little rigour to the examination of the New Labour project and suggest that, in their next term, more effort will have to be given to developing a firmer conceptual basis if the contradictions and ambiguities are not to undermine the approach.

Acknowledgements

We would like to thank all those associated with the preparation of this volume and with the running of the original *Planning Research 2000* conference, on which this volume is based. *Planning Research 2000* took place at the London School of Economics in March 2000. We wish to acknowledge the support of this event by the Royal Town Planning Institute, which was much appreciated and the particular contribution made by Robert Upton. Many at the LSE contributed to the efficient running of this event and we would particularly like to thank Yonn Dierwechter for all his hard work, and his cohort of fellow research students. We would also like to thank all the participants to *Planning Research 2000* for their lively contributions, which made the event so stimulating. Of course, publishing this selection of contributions would not be possible without a publisher and thanks are due to Valerie Rose of Ashgate Publications for facilitating this. Finally, editing a volume of conference proceedings can be a thankless task but, in this case, it has been made a pleasant and easy one due to the speed within which contributors responded to deadlines and complied with Ashgate's guidelines. Thanks to them all. And a special thanks to Michael Lehman for taking on the role of compiling and formatting the final text of the volume. We recommend him to editors everywhere.

References

Blair, T. (1996), *New Britain*, Fourth Estate, London.
Blair, T. (1998), *The Third Way*, Fabian Society, London.
Giddens, A. (1994), *Beyond Left and Right*, Polity Press, Cambridge.
Giddens, A. (1998), *The Third Way*, Polity Press, Cambridge.
Gouldson, A. and J. Murphy (1998), *Regulatory Realities: the implementation and impact of industrial environmental regulation*, Earthscan, London.
Thornley, A. (1999), 'Is Thatcherism Dead? The Impact of Political Ideology on British Planning', *Journal of Planning Education and Research*, vol. 19, pp. 183-191.
Thornley, A. and Rydin, Y. (eds.) (2002), *Planning in a Global Era*, Ashgate, Aldershot.

2 Instrumental Rationality, Intelligent Action and Planning: American Pragmatism Revisited

HEATHER CAMPBELL AND ROBERT MARSHALL

Introduction

The last two decades have not only witnessed profound changes to the practice of planning but the theories underpinning the activity have been in a state of flux. A significant influence has been the critique of planning emanating from postmodernism which seemingly has undermined central tenets of planning as traditionally conceived including professionalism, technical rationality, political neutrality and a unitary public interest. As a consequence of the apparent collapse of the modernist planning project, planning theorists have been searching for a new foundation for planning (Beauregard, 1989, 1991; Milroy, 1989, 1991; Healey, 1997; Fainstein, 1999). In this respect, the arguments associated with what has come to be known as the communicative turn in planning thought, more particularly collaborative planning, have occupied a dominant position in the British planning academy over the past decade. The "communicative turn" has also occupied a central position in the North American planning theory literature over the same period but, in this case, the emphasis on communicative rationality has often been espoused by those who embrace the broader philosophical tradition of pragmatism. The latter has indeed been proffered as a normative theory of planning which can overcome the destabilising effects of the postmodern critique restoring faith in the possibility of effecting socially constructed human progress (Harper and Stein, 1995, 1996a). Although British academics have fruitfully drawn upon the work of the North American pragmatists, pragmatism, as a philosophical underpinning of planning has made little ground. The purpose of this paper is to explore the claim that pragmatism "accepts the critique of modernism without the pessimism and relativism of the postmodern abyss" (Harper and Stein, 1995).

We offer, first, a brief description of the origins and development of pragmatism seeking to arrive at a summary of its more salient characteristics at least in respect of our own immediate academic purpose. Secondly, we examine more closely the claim that it offers a way of bridging the postmodern abyss. Thirdly, we provide a critique of pragmatism conscious that we do so from a non-American perspective which may prove either illuminating or distorted but which, either way, will add a somewhat different voice from those which have contributed to academic debate about its contemporary significance for planning practice and theory. Finally, we offer some tentative conclusions arising from an excursion into territory which hitherto has appeared to be the domain of our transatlantic colleagues.

Before we embark upon this journey, there is a cautionary note to be made. Pragmatism is a philosophical tradition which spans more than a century and, inevitably, there are difficulties in discerning, unambiguously, a set of common characteristics. As Festenstein (1997: 2) observes even "the canonical trinity of Pierce, James and Dewey splits at central points" and when we come to the present day the varieties of pragmatism are numerous. Linked to this is the plasticity of some of pragmatism's concepts with the result that it means many different (and sometimes contradictory) things. The charge of vagueness has been made by both its defenders and its detractors (Bernstein, 1985; Festenstein, 1997; Dickstein, 1998; Fish, 1998). The vagueness, indeed, derives from what are seen as some of its enduring strengths including its distrust of foundations and of dualisms and its resulting contingency or situatedness. Thus Bernstein, in reflecting upon Dewey's vagueness writes: "In his desire to soften all dichotomies, distractions, and dualisms, at times, he seems to deprive us of the analytical tools for advancing and understanding" (Bernstein, 1985: 57). Because of its vagueness and plasticity, pragmatism is not easy target for critical evaluation (Fish, 1998).

The Origins of Pragmatism

Without question pragmatism is deeply ingrained in American society, but the expression of this and the reasons behind it are a matter of contention. Although its philosophical origins go back to European roots including British empiricism, neo-Kantian idealism and romanticism (Thayer, 1981), this inheritance had to be adjusted and adapted to the distinctive conditions of American culture and especially its liberal ideology and values. It was forged in the particular circumstances of that country's history.

To equate pragmatism with an ethic of ruthless competitive enterprise and rugged individualism as some have done is, however, to miss the point. As Thayer (1981) reminds us the architects of pragmatic thinking, Pierce, James and Dewey, were engaged in a task which in essence was highly critical of some of these traits. Their project was seeking to establish a method of philosophical inquiry in which science and morals would be connected and which would result in a theory of action which could enrich the quality of experience. The mainsprings, at least of Dewey's pragmatism were the paradoxical conflicts of American life and culture and the conviction that the separate strands in American life could be brought together through an instrumental process of social engineering (Thayer, 1981).

The reductive method of historical empiricism gave way in American pragmatism to an evolutionary view of testing and evaluating the significance of claims to knowledge. Experience, for the pragmatists, was not "what we know" but became imbued with a temporal quality in which past, present and future are involved in the process of evaluating ideas and putting them to work. The pragmatic method of inquiry was both experimental and experiential and meaning and value were interlinked. Pragmatism was searching for a theory of intelligent action. The futurity aspect of their thinking is particularly important in this respect. Ideas or concepts were viewed instrumentally - as a means of guiding action in order to achieve desired future experience. Pragmatism rejected the "spectator" theory of knowledge. Mind and knowledge are not detached from the world but part of it. Knowledge and truth are transient and experiential (Light and Katz, 1996).

It was John Dewey who foremost among the classical pragmatists developed these ideas into a practical ideology of social reconstruction. In this task, Novack (1975) argues, he was fulfilling a need of middle class Americans who were dismayed and disheartened by some of the consequences of unfettered liberalism and who were searching for a reformist philosophy which would promote progressive change within a gradualist framework. Dewey, in particular, rejected the dominance of rugged individualism and argued for a new conception of individualism as "a social achievement" (Festenstein, 1997: 66). It was a hopeful philosophy based upon the conviction that the future can be made better than the present but without the need for revolution or radical changes in society. West (1993: 111) argues that the emphasis on "the ethical significance of the future" is the key to pragmatism, "maybe its unique

American character". By "ethical significance" West means that human agency can make a difference for human purpose.

In his theory of intelligent action put to use for social purpose, Dewey made an important distinction between a *planned society* and a *continually planned society* (Hoch, 1984). Prerequisite for the former were fixed blue-prints imposed from above and the use of coercive power or persuasion to make them achievable. A continually planned society, on the other hand, required the "wisest form of cooperative give-and-take" which he saw as an operational method of effecting social progress rather than a predetermined set of truths. These distinctions were also apparent in his consideration of ideals and means: "the process of growth, of improvement and progress rather than the static outcome and result, becomes the significant thing... Growth itself is the only moral *end*" (Dewey, 1957: 177). Social betterment required attention to means rather than ideals "... the expiring ghosts of a once significant kingdom of divine reality whose rule penetrated to every detail of life" (Dewey, 1929: 222-223). These ideas were to have an important impact upon ideas in policy analysis in the years following the second world war. They connect with Simon's concept of satisficing behaviour in organisations (Simon, 1976) and the disjointed incrementalism of Lindblom (Braybrooke and Lindblom, 1970). Dewey stood back from making a commitment to social purpose expressed in a tangible expression of aims. Instrumentalism provided the tools, but no firm conception of what these tools might fashion or make. Dewey made a virtue out of the instrumentalism of intelligent action but it was, in Thayer's view, a flaw which weakened its influence (Thayer, 1981).

If Dewey's reformist ideas continued to exert an influence on American public affairs in the 1940s and 1950s its place within American academic philosophy declined during the 1930s as the doctrines of logical positivism took hold (Mounce, 1997). Revival of interest, however, occurred in the 1980s with the realisation that Pierce, James and Dewey had long since anticipated many of the postmodern criticisms of the Enlightenment project. This interest has led to a re-evaluation of the work of the classical pragmatists (see, for example, Ryan, 1995; Light and Katz, 1996; Festenstein, 1997; Mounce, 1997; Westbrook, 1991) but it has also resulted in postmodernist redescriptions of pragmatism (West, 1985; Dickstein, 1998).

The revival of interest in pragmatic thinking in the 1980s occurred because it seemed to be open to the insights of postmodernism but without its cynicism and nilhism (Kloppenberg, 1998). The postmodern critique of the Enlightenment project destabilised accepted ideas on the way knowledge is produced, validated and expressed. The positivist inheritance, upon which professionalism had been founded, could no longer be sustained (Schön, 1983; West; 1993). Pragmatism appealed to those wishing to maintain a faith in liberal democracy and in West's ethical significance of the future.

In this resurgence of interest in pragmatism, the influence of Richard Rorty "the trojan horse of analytical philosophy who attacked the citadel from within" (Kloppenberg, 1998), has been enormous. Rorty is anti-foundationalist through and through. Truth is not found but is made and, says Rorty, it is made through the ways we describe the world. It is "a property of linguistic entities, of sentences" (Rorty, 1989: 7). Cultural change involves redescriptions of the world or the substitution of one "language game" for another. Paradigm shifts occur when a new vocabulary is created which supersedes the old. The new description of the world serves certain of our purposes better than the old but, paradoxically, we do not know what these purposes are until the new language, the new way of describing the world, is in place. Change and progress, if that is a word which can be properly used in such a context, is contingent. Rorty offers a line of thought which suggests that "... we try to get to a point where we no longer worship *anything,* where we treat *nothing* as a quasi divinity, where we treat *everything* - our language, our conscience, our community - as a product of time and chance" (Rorty, 1989: 22).

Pragmatism is dismissive of any claim that there is any intrinsic human nature or universal moral principles which bind all members of the human race together. Rorty remains, however, an upholder of liberalism. Indeed his task, in part, is to construct a postmodern liberal utopia. The absence of foundations, rather than undermining liberalism, strengthens and reaffirms it by extending freedom through greater critical discourse and the contingency of social practice (Gray, 1995). "An ideal liberal society", according to Rorty (1989: 61) "has no purpose except freedom".

Rorty claims to be a disciple of Dewey, but as Westbrook (1991) points out, Rorty's appropriation of Dewey is a controversial one. Rorty privileges language over experience. This is not the case with Bernstein and Putnam who sustain the mainstream of Dewyan pragmatism through an emphasis on deliberative democracy and an affirmation of action through experience. Bernstein (1998) argues that pragmatists must get along with contingency and uncertainty without descending into despair and wild relativism. His thesis is that we need to go beyond objectism and relativism (Bernstein, 1983: 223) in order to exorcise what he calls the "Cartesian Anxiety" - the feeling that, if there are no fixed points, no certainties, then we are left with nothing to which we can secure our lives. There is just chaos, darkness and insanity. In an elegant interrogation of the writings of

Gadamer, Habermas, Rorty and Arendt he draws out a series of common themes which provide a basis for challenging the orthodoxy of the dichotomy between objectivism and relativism. These central themes are "dialogue, conversation, undistorted communication, communal judgment, and the type of rational wooing that can take place when individuals confront each other as equals and participants". The move beyond objectivism and relativism he sees not as a theoretical problem but as a "practical task" of creating dialogical communities in which we can reclaim a sense of practical rationality.

Neo-Pragmatism and Planning

The pragmatist planning theorists have, like Rorty, attempted to embrace parts of the postmodernist critique of the Enlightenment project within a liberalist framework but in their case with the specific intention of affirming the possibility of a continuing role for the practice of planning. Before attempting to summarise the tenets upon which a neopragmatic theory of planning is based it is important to note that the influence of pragmatism on North American planning theory is by no means new. As previously mentioned, we claim that its influence is discernible in the disjointed incrementalism of Charles Lindblom (Braybrooke and Lindblom, 1970), a claim which is affirmed by Hoch (1997) and by Anderson (1987: 26) who, indeed, claims that Braybrooke and Lindblom's *A Strategy of Decision* "is a remarkably articulate and coherent statement of the pragmatic philosophy of Pierce, James, Royce, and Dewey formulated as a doctrine of decision making". Blanco (1994), while seeing pragmatic elements in incrementalism, casts it outside the core of pragmatism. Simon's satisficing behaviour is similarly dealt with. By introducing limits to rationality, she argues, they established "blocks" to inquiry. Hoch (1984) has identified pragmatic conceptions in other "mainstream" North American planning theories including Meyerson's (1956) middle-range bridge conception of comprehensive planning, Davidoff's (1965) advocacy planning, Friedmann's (1973) theory of transactive planning and Grabow and Heskin's (1973) radical concept of planning. They are, says Hoch, united by an attempt to satisfy within a single theory the requirements of *doing good and being right* and, in doing so, resonate with Dewey's concept of pragmatic action. Pragmatic reasoning was also central to Schön's "reflection-in-action", the process through which experience allows practitioners to deal effectively with situations of uncertainty, instability and value conflict. In his 1984 paper,

Hoch identified a significant flaw in Dewey's pragmatism and in the mainstream American theories of planning and this was their failure to take effective account of the uneven distribution of power in society which undermines democratic participation and, in turn, pragmatic action. He anticipated, however, the possibility of "a reconstructed pragmatic theory of planning that is based on human experience, practical activity, and democratic community participation, but without the naturalistic bias and liberal pluralism incorporated in mainstream planning theories" (Hoch, 1984: 343). Pragmatic planning theorists have been engaged in this process of "reconstruction" in the period since then.

What then are the defining characteristics of the new pragmatism as applied to planning? First and foremost, it is defined by what it isn't. Pragmatism rejects *foundationalism and* with it *positivism, scientism and absolutism*. Planning in these circumstances rests upon rationality which is *contingent* rather than absolute. It acknowledges that our beliefs are *fallibilistic* but, notwithstanding their possible transience, we can make judgments on the basis of what we believe to be true at the time. Fallibilism does not mean that we have to doubt everything, only that we "are prepared to doubt *anything*" (Putnam, 1995: 2). Planning is *dialogic* (practice through communication/discourse), rejects the notion of *incommensurability* but recognises the existence of many voices which should be encouraged. Pragmatism rejects the dualism of objectivity/ subjectivity and the claims of undisputed authority (professional expertise). Despite this, there appears to be a role for professionals although there appears to be some divergence in the way different theorists conceive the nature of practice in a postmodern culture. Innes (1995: 184) conceptualises planners in "the emerging paradigm of communicative action and interactive practice" as "actors in the world rather than ... observers or neutral experts". They are, in this sense, emancipated from the shackles of rationalism and embedded power relations and become free to embrace an ethical stance based on their own experience of the world. We would place Forester's work in this mould. Planning practice, in his conception, becomes partisan in that planners become "practical organisers (or disorganisers as the case may be) of attention" (Forester, 1993: 27). Because of structural power imbalances in society he persuasively argues that planners should actively intervene "seeking social justice" and "interventions that go beyond liberalist-pluralist bargaining and incrementalist strategies" (Forester, 1989: 62). In this respect Forester stands apart from many theorists who have been influenced by pragmatism. Other communicative conceptions of planning suggest, for example, a different and inherently more ambiguous role for professionals.

Sometimes they are seen as neutral mediators or facilitators in communicative discourse (Webber, 1978; Blanco, 1994). In other conceptions planners are engaged in shifting involvements so that their roles change as they adapt to the different professional situations in which they find themselves. Because of professional pluralism Hoch (1994: 316) sees their roles variously as gatekeepers, path-finders, visionaries, counsellors, power-brokers, public servants and teachers.

Despite this variability in the way practice is conceived there does appear to be broad agreement that pragmatic planning is *incremental, liberal, situational* and *critical.* There is no separation of theory and practice or knowledge and action.

Critique

As previously mentioned, there is not just one pragmatism but many. The pragmatism of planning theorists draws, therefore, upon a rich diversity of thought and there are major inconsistencies and ambiguities in the picture presented. In the review which follows we concentrate on the "problems" as we see them and these are set out in a number of inter-connected themes. These examine; the proposition that pragmatism is inherently a procedural planning theory; the arguments concerning relativism and incommensurability; the nature of the planning procedures invoked by pragmatism's "method of inquiry"; and, communicative rationality.

Progress Without Purpose?

Dewey, as we have seen, has been criticised for being vague about the social purposes of his instrumental rationality. There are some who would contest this interpretation of Dewey's thinking (Festenstein, 1997). After all, one of the hall-marks of Dewey's moral and political philosophy was its emphasis on futurity and on the possibility of creating a better future through human agency (West, 1993). Pragmatism places emphasis on consequent and not antecedent phenomena (West, 1993; Blanco, 1994). Thus in his essay "What Pragmatism means by Practical" Dewey (1916: 330) states that "the term *pragmatic* means only the rule of referring all thinking, all reflective considerations to consequences for final meaning and test". Given this emphasis on the future, it is defensible, as Putnam (1992) has done, to describe Dewey as a consequentialist. Nevertheless, the very nature of Dewey's instrumental rationality means that we cannot be firm about what is "right" or "best" until we have subjected our claims to the process of verification and that process involves experimentation which is active and reflective (Dewey, 1916). Dewey's view, writes Putnam (1992: 186) "is that we don't know what our interests are or what we are capable of until we actually engage in politics. A corollary of this view is that there can be no final answer to the question of how we should live, and therefore we should always leave it open to future discussion and experimentation".

It is, we contend, the focus on method of inquiry buttressed by pragmatism's emphasis on deliberative rationality that gives credence to the view that normative conceptions of pragmatism are inherently procedural. In a prescient paper published in 1978 Webber described planning as a cognitive style and not a substantive field and it is the cognitive relexive skills, rather than substantive questions concerning the object or purpose of planning, which have come to assume a dominant position in the theories of planning pragmatists. Communicative rationality, in particular, has come to occupy a canonical place in those theories. We turn our attention more directly to communicative rationality later in the paper but for present purposes we note that under this style of planning the most common perception of the planner's role is that of facilitator in the process of mediating stakeholder's views in order to arrive at a consensus or agreement. Purposes are not known or given but are to be communicatively discovered. Planning, in these circumstances becomes "future seeking" rather than "future defining" (Healey, 1992). As Fainstein (1999) aptly notes:

> The communicative theorists display a narcissism that makes the role of the planner the central element of discussion. Both the context in which planners work and the outcome of planning fade from view. Unlike the rational modellers the communicative theorists have found a subject, but like them they lack an object ... Instead of asking what is to be done about cities and regions, communicative planners typically ask what planners should be doing and the answer is that they should be good ...

Relativism or Moral Pluralism?

By "relativism" we mean the rejection of the notion that there is an intrinsic human nature or of the idea that there are universalising principles of moral responsibility. Relativism particularises values and ideas about moral agency either to individuals or, in the case of communitarians, to definable communities. If values become particularised in this way then it becomes necessary to confront the problem of "incommensurability", the idea that, in a world of difference, we cannot make moral judgements across those cultural divides which separate individuals or communities.

It is difficult to discern a mainstream pragmatic viewpoint on these questions. Rorty (1989: 177), on the face of it, rejects entirely the notion of a universalising human nature. "There is", he asserts, "nothing deep inside each of us, no common human nature, no built-in human solidarity, to use as a reference point". This seemingly atomistic view gives way elsewhere in his writing, however, to a different view in which he acknowledges a sense of belonging which we might have but one which eschews the idea of solidarity with all other human beings. Solidarity only exists, he argues, in the context of the separation of "us" from others: "...*we* or *one of us* always means something smaller and more local than the human race" (Rorty, 1989: 191). In fact, it transpires that the sense of belonging which Rorty himself acknowledges is with "his fellow Americans" and, as others have noted (Eagleton, 1996; Geras, 1995) there are rather a lot of Americans and feeling solidarity with all other Americans is not so very different from expressing a solidarity with all humanity.

Bernstein (1983), as noted above, has argued the need to go beyond objectivism and relativism. In his discussion of incommensurability, he observes that some have seen the new varieties of relativism as a liberating force allowing us to acknowledge that our view of the world is not transcendental and enabling a new respect for difference. For others, however, incommensurability is a dangerous doctrine which opens the way for subjectivism, nilhism and irrationality (p. 79). Bernstein's own view is that although different forms of life or traditions may be incommensurable they can, nevertheless, be rationally compared and that this hermeneutic process of seeking to understand and to interpret may allow us to develop a more critical knowledge of our own traditions and way of life leading to an opening up rather than a closure of experience.

In general, the pragmatic planning theorists appear to reject the incommensurability thesis. Hoch (1996), for example, examines the ways in which pragmatism can help planners "bridge the social differences within and between communities" (p. 37) and that shared vulnerabilities can allow us to "experience solidarity with members of alien and distant communities" (p. 41). Forester (1999a) strongly rejects "moral relativism - the refusal to make evaluative arguments - that a facile liberalism, or a cynical postmodernism, makes so seductive" (p. 181). Harper and Stein (1996a, 1996b) have rather more directly addressed the problems of relativism and incommensurability. Thus they write: "there are no conceptual barriers between individuals, communities or cultures. We do not live in different incommensurate worlds". On the other hand, they also argue that planning must recognise the existence of different voices and the importance of community. Their "critical liberal paradigm" for planning is built upon three basic principles. First, understanding, they argue, must arise out of one's own community and traditions. Secondly, there is no Archimedean point providing a universal reference point. But, thirdly, it is still possible to critique and to dialogue with and between communities (1996b: 416). There is, therefore, some correspondence between their position and that espoused by Bernstein. Despite their rejection of a meta-narrative they nevertheless, like Rorty, appear to fall back on "universalising" principles which in essence are the over-arching and taken-for-granted values of liberalism. In fact, the moral vocabulary which they employ is, as they recognise, a situated one "... a moral vocabulary which is part and parcel of a tradition ... our tradition. We cannot escape our culture, our tradition, our language" (Harper and Stein, 1996a: 23).

Relativising differences in moral values is, some pragmatists would argue, a different notion than that of moral pluralism. What is invoked here is the idea that there is no one set of ethical principles which can serve in every situation. We can only respond to problems as they present themselves. There are no bed-rock universal principles which we can fall back on: "...no list of virtues, no list of rights and duties, no table of laws, no account of the good should be expected to serve in every possible situation that we confront" (Parker, 1996: 26). Ethical considerations are situated and contingent. How then does the planner or policy-maker deal with problems of value judgement? Some pragmatists appear to evade the question by arguing that the principles which should guide action "emerge" out of the consideration of the particular problem under consideration (Anderson, 1989). Hoch (1997) sees it in a similar vein when he writes:

> Pragmatism does not tell us what ends to pursue, but offers a kind of inquiry that compares the value of different courses of action alternately weighing means and ends - facts and values. It binds together what dualistic thinking keeps apart - knowledge and action (or perhaps a bit more precisely) theoretical reflection and common sense.

Parker (1996: 27) claims that the aim of pragmatic ethics "is not perfect rightness, since there is no absolute standard for reference, but rather the creative mediation of conflicting claims to value". But this begs the question of who is to carry out the mediation and how it is to be practised. It is to these questions that we now turn.

New Planning Practices

Central to the pragmatic thesis is the argument that in a postmodern culture planning needs new practices. The old foundations of positivism and scientific rationalisation no longer provide a legitimising framework for the planning activity. The old rationality must give way to a new based upon dialogic democratic principles. The issues around which we focus our discussion are principally those which arise from this espousal of communicative rationality and its implications for planners in their professional roles. Before, however, we embark upon a discussion of these issues there is a prior task which we should like to briefly address and that is to question the implied assumption in debates concerning the contemporary crisis of planning that technical rationality maintained a hegemonic status within planning practice until it was seriously challenged by postmodernism in the 1980s. In looking at UK experience, at least, we would assert two things. First, that technical rationality held sway for a comparatively short period of time. Among planning theorists it enjoyed widespread but not unqualified support for a brief period from the mid-1960s through to the early 1970s. In the case of practice it entered the rhetoric of the profession but its direct influence was, we would contend, somewhat limited. Even during its hey-day the approach was seriously challenged. In a period of economic growth and rapid change it became manifestly evident that planning inevitably involved conflicting claims over values and was embedded in a context in which matters of political judgement were centrally involved. Planning was self-evidently not purely a technical process in which planners fitted efficient means for the achievement of externally generated goals. Practitioners, in staking out a claim to offer appropriate expertise in the cause of remedying urban problems would not, we conjecture, seek their legitimising frame of reference in something called "technical rationality" but in that much more elusive but equally powerful notion of professional judgment. This latter concept we will return to in a moment but first it is necessary to consider what, communicative rationality apart, pragmatic planning theorists offer in place of technical rationality.

In fact, technical rationality, in the guise of the comprehensive rational model of the planning process, is not entirely banished. This is especially so in the case of Blanco (1994) who claims that there is a close relationship between the rational model and Dewey's concept of inquiry, the main difference being that Dewey begins with problem recognition whereas this is not usually identified as a separate stage in the rational model. A second difference is that whereas the rational model assumes that a comparison of alternatives is made, Dewey focused on one solution at a time. In effect, Blanco appears to offer a re-interpretation of the rational model as "pragmatic reason".

Verma (1996) offers "pragmatic rationality" as a viable model of planning and his thesis is that practitioners in fact already employ this method rather than the rational model. As with Hoch's (1997) description of the pragmatic method of inquiry cited earlier, it is a process where there is no separation of stages, nor of ends and means, facts and values. It is he says "a dialectical method in which the different stages in the rational-comprehensive model do not exist in practice. It is also a method which does not deny the politics, history and psychology of idea generation" (p. 11).

Verma focuses on the role of the planner in everyday practice rather than seeking to conceptualise a planning process as the rational modellers did. This concern to link practice and theory and vice versa has become something of a *leit-motif* of pragmatist writings on planning. Forester (1999b), for example, pleads with "academic *theorists* to take practice more seriously, to recognise sensitively and to analyse powerfully what insightful practitioners do well in the most challenging moments of their work" (p. 8).

Interrogating the practices of professionals provided the basis upon which Schön (1983) arrived at his alternative to technical rationality in the form of "reflection-in-action" - "an epistemology of practice implicit in the artistic, intuitive processes which some practitioners do bring to situations of uncertainty, instability, uniqueness, and value conflict" (p. 49). Framing the problem to be addressed is important as are the intuitive, spontaneous processes which derive from a critical awareness of previous experience. In Schön's words: "It is a kind of knowing which does not derive from a prior intellectual operation but from artistic, intuitive processes" (p. 51).

MacIntyre (quoted by Bernstein,1983: 57) offers further insight into what shapes Schön's "kind of knowing" as exemplified in the actions of skilled practitioners. "Objective rationality", he writes,

> is ... to be found not in rule-following, but in rule-transcending, in knowing how and when to put rules and principles to work and when not to. Consider how practical reasoning of this kind is taught, whether it is the practical reasoning of generals, of judges in a common law tradition, of surgeons or of natural scientists. Because there is no set of rules specifying necessary and sufficient conditions for large areas of such practices, the skills of practical reasoning are communicated only partly by precepts but much more by case-histories and precedents ... The teaching of methods is nothing more than the teaching of a certain kind of history.

Schön and MacIntyre, in their different ways, articulate what have traditionally been seen as one of the defining characteristics of professionalism - the ability to apply skills not routinely but to those categories of problems which require careful, wise and creative judgment. It is, we contend, these kinds of skills which practitioners allude to when they talk about professional expertise and judgment. But, of course, the citadel of professionalism has been under attack from the early 1980s and is now a flawed concept. If, as many pragmatists argue, the old kind of rationality should give way to communicative rationality, it is not at all certain that there is a place within the planning arena for the kinds of skills to which Schön and MacIntyre refer. As previously noted there is ambiguity in the way which different pragmatists see the planner's role, but many communicative theorists see planners principally acting as facilitators or mediators in the process of debate.

Communicative Rationality

John Dewey saw "public communication" as a necessary ingredient in the process of resolving conflicts between private interests. Democratic participation should, he argued, promote individual autonomy and informed public deliberation producing "a community of critical inquiry" (Festenstein, 1997: 79). "The essential need ...", he wrote,

> is the improvement of the methods and conditions of debate, discussion and persuasion. That is *the* problem of the public. We have asserted that this improvement depends essentially upon freeing and perfecting the process of inquiry and of dissemination of their conclusions. Inquiry, indeed, is a work which devolves upon experts. But their expertness is not shown in framing and executing policies, but in discovering and making known the facts upon which the former depend (Dewey, 1927: 365).

Participatory democracy has continued to be a cornerstone of much normative pragmatic discourse and, in the recent past, has been considerably influenced by Habermas's critical theory (Beauregard, 1996; Fainstein, 1999; Forester, 1993, 1999b; Harper and Stein, 1995, 1996b; Hoch, 1992; Innes, 1995).

Habermas's principles of radical democracy are mainly deontological. His is a theory of justice but one which does not, as in Rawls' case, depend upon individuals making their own self-centred contracts behind a veil of ignorance. Instead, Habermas's communicative discourse requires all participants to imagine themselves in others' places in the process of deciding whether a proposed norm is fair. Moreover, it is essential that this process is conducted in public by real social agents. The purpose is to arrive at norms of justice by agreement which is uncoerced and concluded willingly by all those who will be subject to the norms. Habermas is concerned not with a way of deciding what constitutes the good life but with giving shape to "the structural aspects of the common good" (McCarthy, 1990: xi).

The insistence that the dialogic process is carried out by all those who will be affected by the norms being debated has led some to propose that it is more appropriate to regard communicative discourse as a means of achieving democratic legitimisation rather than a way of validating moral claims (Chambers, 1995). It is in this sense of communicative discourse as a process of democratic will formation that it has influenced planning theory. However, there are major problems in operationalising democratic discourse. Some are generic in nature (see Chambers, 1995) while others are more specifically related to the nature of the planning activity (Lauria and Whelan, 1995; Feldman, 1995, 1997; Tewdwr-Jones and Allmendinger, 1997; Fainstein, 1999). Of crucial significance in evaluating the possibility of effecting communicative discourse are the stringent requirements that must be met. All relevant voices must be included, everyone must have the right to speak and there should be no coercion or other unfair means of enforcing or extracting agreement. Argumentation is the only process allowed and in engaging in conversations participants must show equal respect and impartiality towards all others. Thus the purpose of communicative dialogue is to arrive at understanding which is brought about by putting oneself in others' shoes in the effort to see things as they do. Finally, there should be no forced closure of debate in order to secure agreement or a decision. The dialogue must continue until everyone agrees.

Of course, in bringing these *ideals* down to an operational level these conditions are inevitably relaxed. Dryzek (1990), for example, argues that it is not only the *quantity* of participation that is important but its *quality* is also of paramount significance. Universal participation is not required and, indeed, he emphasises the need for debate to involve "communicatively competent individuals". Generally, therefore, the participants "represent" major stakeholders or interests and there is acceptance that discussion may not lead to an agreement. Innes and Booher (1999) note in their evaluation of consensus building processes that some of their examples arrived at agreements but, more often, they did not. They stress, indeed, the effects of the process rather than the outcomes in that the effects were often transformative on those involved leading to changes in their knowledge and attitudes. Even accepting, however, the need to relax some of the pre-requisites for communicative discourse, there are still major problems to be resolved to make it feasible.

There is, first, the difficulty of institutionalising discourse. What are the arenas within which discourse is to take place? Who decides when discourse should be initiated and what decisions should be resolved discursively? Who should be involved and who sets the agenda? There is also the paradox that the questions we have just posed should themselves ideally be determined through discourse.

Secondly, there is the nature of the problems which planning addresses. It is an activity which is rarely concerned with single issues but characteristically is concerned with complex bundles of interconnected issues. Indeed, the distinguishing hall-mark of planning is its concern for strategically linking multiple policy arenas and decision fields across space and through time in arriving at decisions concerning land and property use and development. Given the unequal distribution of rights, resources of all kinds and, hence, power, negotiation, bargaining, trade-offs, strategic argumentation and other competitive practices, which do not meet the requirements of discursive argumentation, are an intrinsic part of the process. In any case, how is it possible to distinguish between authentic as opposed to distorted communication or between strategic and communicative interaction? (Dryzek, 1995).

Thirdly, there are the problems associated with the tensions involved in arriving at a determinative outcome (decisions or agreements on policy choices) whilst fulfilling the requirements that the process should be inclusionary and that there should be no forced closure in order to secure agreement. If the rule is that debate must continue until everyone agrees then the probability is that the process is more likely to succeed with small homogeneous groups. Localism or communitarianism would, we suspect, favour conservative and exclusionary (of others not in the group) agreements. But in any case, planning operates through a hierarchy of scales from the nation state downwards.

Communicative discourse, we conclude therefore, is unlikely to be helpful as a policy or decision-making process although it may be of value as an opinion and consensus forming process which is not seeking determinative decisions on concrete and specific matters.

Conclusion

Pragmatism has appealed to intellectuals in the United States in the last two decades because it has proffered a home-grown philosophy which rejects Enlightenment rationalism but appears to be resistant to the "anything goes" nilhism of postmodernism. Its anti-foundationalism coupled with its continuing commitment to the ideals of democracy and social hope (Rorty, 1999) within the broader framework of a humanistic liberalism are appealing. However, when we endeavour to distil what it is that pragmatism offers in place of the old rationalities of the modernist planning project, the picture becomes less clear and rather more problematic. That distillation, moreover, is made difficult by the heterodoxy and ambiguity of pragmatic thinking especially in its normative manifestations. Despite the diversity of thought which pragmatism embraces, our reading of the literature suggests that the pragmatism of the planning theorists gives emphasis to process over substantive questions of *what* is to be done, encourages a strategy of incremental and serial adjustments in effecting change and commonly views the planner as a neutral mediator who stands aloof from the values to be adjudicated in the process of reaching consent or agreement.

It is neither necessary to be a pragmatist nor a postmodernist to reject foundationalism. Scepticism concerning the status of knowledge and truth is not new and, indeed, the seeds of it were present in the Enlightenment at the very beginning. Moreover, the break with foundationalism in philosophical thought occurred in the mid- to late-nineteenth century (Giddens, 1990) and, as we have seen, Pierce, James and Dewey played a significant part in effecting the change. The more recent past has produced deeper, more penetrating and, consequently, profoundly unsettling changes concerning the way we see ourselves and the world in which we live but we do not have to concede that we have moved into a new era of postmodernity. Giddens (1990: 51) offers, we believe, a more helpful and hopeful interpretation:

> The disjunctions which have taken place should ... be seen as resulting from self-clarification of modern thought, as the remnants of tradition and providential outlooks are cleared away. We have not moved beyond modernity but we are living precisely through a phase of its radicalisation.

Within the continuing but radicalised phase of modernity, planning can have a place provided it can demonstrate that it can effect material change within a determinative framework which recognises clear social purposes. We believe that planning should not just concern itself with the preconditions of the good life but should play a part in shaping shared visions of what it might consist of and how it might be secured.

References

Anderson, C.W. (1987), 'Political Philosophy, Practical Reason and Policy Analysis' in F. Fischer and J. Forester (eds), *Confronting Values in Policy Analysis: The Politics of Criteria*, Sage Publications, Newbury Park, California, pp. 22-44.

Beauregard, R.A. (1989), 'Between Modernity and Postmodernity: the Ambiguous Position of U.S. Planning', *Environment and Planning D*, vol. 7, no. 4, pp. 381-385.

Beauregard, R.A. (1991), 'Without a Net: Modernist Planning and the Postmodern Abyss', *Journal of Planning Education and Research*, vol. 10, no. 3, pp. 189-194.

Beauregard, R.A. (1996), 'Advocating Preeminence: Anthologies as Politics', in Mandelbaum, S.J., Mazza, L. and Burchell, R.W. (eds), *Explorations in Planning Theory*, Rutgers, the State University of New Jersey, New Brunswick, New Jersey, pp. 105-110.

Bernstein, R.J. (1983), *Beyond Objectivism and Relativism: Science, Hermeneutics, and Praxis*, Blackwell, Oxford.

Bernstein, R.J. (1985), 'Dewey, Democracy: the task ahead of us' in Rajchman, J. and West, C. (eds), *Post-Analytical Philosophy*, Columbia University Press, New York, Guildford.

Bernstein, R.J. (1998), 'Community in the Pragmatic Tradition' in Dickstein, M. (ed), *The Revival of Pragmatism: New Essays on Social Thought, Law and Culture*, Duke University Press, Durham and London, pp. 141-156.

Blanco, H. (1994), *How to Think about Social Problems: American Pragmatism and the Idea of Planning*, Greenwood Press, Westport.

Braybrooke, D. and Lindblom, C.E. (1970), *A Strategy of Decision: Policy Evaluation as a Social Process*, Free Press, New York.

Chambers, S. (1995), 'Discourse and Democratic Practices', in S.K. White (ed), *The Cambridge Guide to Habermas*, Cambridge University Press, Cambridge, pp. 233-259.

Davidoff, P. (1965), 'Advocacy and Pluralism in Planning', *Journal of the American Institute of Planners*, vol. 31, no. 4, pp. 596-615.

Dewey, J. (1916), 'What Pragmatism Means by Practical', in *Essays in Experimental Logic*, University of Chicago Press, Chicago.

Dewey, J. (1927), 'The Public and its Problems', in Dewey, J. *The Later Works, 1925-1953, Volume 2*, edited by J.A. Boydson (1984), Southern Illinois Press, Carbondale and Edwardsville.

Dewey, J. (1929), 'The Quest for Certainty: A Study of the Relation of Knowledge and Action', in Dewey, J. *The Later Works, 1925-1953, Volume 4*, edited by J.A. Boydson (1984), Southern Illinois University Press, Carbondale and Edwardsville.

Dewey, J. (1957), *Reconstruction in Philosophy*, Beacon Press, Boston.

Dickstein, M. (1998), 'Introduction: Pragmatism Then and Now', in Dickstein, M. (ed) *The Revival of Pragmatism: New Essays on Social Thought, Law and Culture*, Duke University Press, Durham and London.

Dryzek, J.S. (1990), *Discursive Democracy*, Cambridge University Press, Cambridge.

Dryzek, J.S. (1995), 'Critical Theory as a Research Programme', in White, S.K. (ed) *The Cambridge Guide to Habermas*, Cambridge University Press, Cambridge, pp. 97-119.

Eagleton, T. (1996), *The Illusions of Postmodernism*, Blackwell, Oxford.

Fainstein, S. (1999), 'New Directions in Planning Theory', paper presented at Planning Research Conference, University of Sheffield, Sheffield.

Feldman, M. (1995), 'Regime and Regulation in Substantive Planning Theory', *Planning Theory*, vol. 14, pp. 65-95.
Feldman, M. (1997), 'Interpretive Planning Theory as Comedy', *Planning Theory*, vol. 17. pp. 43-64.
Festenstein, M. (1997), *Pragmatism and Political Theory*, Polity Press, Cambridge.
Fish, S. (1998), 'Afterword: Truth and Toilets; Pragmatism and the Practices of Life', in Dickstein, M. (ed), *The Revival of Pragmatism: New Essays on Social Thought, Law and Culture*, Duke University Press, Durham and London, pp. 418-433.
Forester, J. (1989), *Planning in the Face of Power*, University of California Press, Berkeley.
Forester, J. (1993), *Critical Theory, Public Policy, and Planning Practice: Toward a Critical Pragmatism*, State University of New York Press, Albany.
Forester, J. (1999a), 'Reflections on the Understanding of Future Planning Practice', *International Planning Studies*, vol. 4, no. 2, pp. 175-193.
Forester, J. (1999b), *The Deliberative Practitioner: Encouraging Participatory Planning Processes*, MIT Press, Cambridge, Massachusetts.
Friedmann, J. (1973), *Retracking America: A Theory of Transactive Planning*, Anchor, Doubleday, New York.
Geras, N. (1995), *Solidarity in the Conversation of Humankind: the Ungroundable Liberalism of Richard Rorty*, Verso, London.
Giddens, A. (1990), *The Consequences of Modernity*, Polity Press, Cambridge.
Grabow, S. and Heskin, A. (1973), 'Foundation for a Radical Concept of Planning', *Journal of the American Institute of Planners*, vol. 39, no. 2, pp. 106-116.
Gray, J. (1995), *Enlightenment's Wake: Politics and Culture at the Close of the Modem Age*, Routledge, London.
Harper, T.L. and Stein, S.M. (1995), 'Out of the Postmodern Abyss: Preserving the Rationale for Liberal Planning', *Journal of Planning Education and Research*, vol. 14, pp. 233-244.
Harper, T.L. and Stein, S.M. (1996a), 'The Theory/Practice Dualism: The Doom of Planning?', paper presented to the Joint Congress of the Association of Collegiate Schools of Planning and the Association of European Schools of Planning, Toronto, 1996.
Harper, T.L. and Stein, S.M. (1996b), 'Postmodernist Planning Theory: The Incommensurability Premise', in Mandelbaum, S.J., Mazza, L. and Burchell, R.W. (eds), *Explorations in Planning Theory*, The State University of New Jersey, New Brunswick, New Jersey, Rutgers, pp. 414-429.
Healey, P. (1992), 'Planning Through Debate: the communicative turn in planning theory', *Town Planning Review*, vol. 63, no. 2, pp. 143-162.
Healey, P. (1997), *Collaborative Planning: Shaping Places in Fragmented Societies*, Macmillan, Basingstoke.
Hoch, C. (1984), 'Doing Good and Being Right: The Pragmatic Connection in Planning Theory', *Journal of the American Planning Association*, vol. 50, no. 3, pp. 335-345.
Hoch, C. (1992), 'The Paradox of Power in Planning Practice', *Journal of Planning Education and Research*, vol. 1, pp. 206-215.
Hoch, C. (1994), *What Planners Do: Power, Politics and Persuasion*, Planners Press, Chicago.
Hoch, C. (1996), 'A Pragmatic Inquiry about Power', in Mandelbaum, S.J., Mazza, L. and Burchell, R.W. (eds), *Explorations in Planning Theory*, The State University of New Jersey, New Brunswick, New Jersey, Rutgers, pp. 30-44.
Hoch, C. (1997), 'Planning Theorists Taking an Interpretive Turn Need Not Travel on the Political Economy Highway', *Planning Theory*, vol. 17.

Innes, J.E. (1995), 'Planning Theory's Emerging Paradigm: Communicative Action and Interactive Practice', *Journal of Planning Education and Research,* vol. 14, no. 3, pp. 183-189.

Innes, J.E. and Booher, D.E. (1999), 'Consensus Building as Role Playing and Bricolage: toward a theory of collaborative planning', *Journal of the American Planning Association*, vol. 65, no. 1, pp. 9-26.

Kloppenberg, J.T. (1998), 'Pragmatism: an Old Name for Some New Ways of Thinking', in Dickstein, M. (ed.) *The Revival of Pragmatism: New Essays on Social Thought, Law and Culture*, Duke University Press, Durham and London, pp. 83-127.

Lauria, M. and Whelan, R.K. (1995), 'Planning Theory and Political economy: the Need for Reintegration', *Planning Theory*, vol. 14, pp. 8-39.

Light, A. and Katz, E. (eds) (1996), *Environmental Pragmatism*, Routledge, London.

McCarthy, T. (1990), 'Introduction' to Habermas, J., *Moral Consciousness and Communicative Action*, Polity Press, Cambridge.

Meyerson, M. (1956), 'Building the Middle-Range Bridge for Comprehensive Planning', *Journal of the American Institute of Planners*, vol. 22, no. 2, pp. 58-64.

Milroy, B.M. (1989), 'Constructing and Deconstructing Plausibility', *Environment and Planning D*, vol. 7, no. 3, pp. 313-326.

Milroy, B.M. (1991), 'Into Postmodern Weightlessness', *Journal of Planning Education and Research*, vol. 10, no. 3, pp. 181-187.

Mounce, H.O. (1997), *The Two Pragmatisms: From Pierce to Rorty*, Routledge, London.

Novack, G. (1975), *Pragmatism Versus Marxism: An Appraisal of John Dewey's Philosophy*, Pathfinder Press, London.

Parker, K.A. (1996), 'Pragmatism and Environmental Thought', in Light, A. and Katz, E. (eds), *Environmental Pragmatism*, Routledge, London, pp. 21-37.

Putnam, H. (1992), *Renewing Philosophy*, Harvard University Press, Cambridge, Massachusetts.

Putnam, H. (1995), *Pragmatism: an Open Question*, Blackwell, Oxford.

Rorty, R. (1989), *Contingency, Irony and Solidarity*, Cambridge University Press, Cambridge.

Rorty, R. (1999), *Philosophy and Social Hope*, Penguin Books, Harmondsworth.

Ryan, A. (1995), *John Dewey and the High Tide of American Liberalism*, W.W. Norton and Co., New York.

Schön, D.A. (1983), *The Reflective Practitioner: How Professionals Think in Action*, Maurice Temple Smith, London.

Simon, H.A. (1976), *Administrative Behaviour. A Study of Decision-Making Processes in Administrative Organisations*, Free Press, New York.

Tewdwr-Jones, M. and Allmendinger, P. (1997), 'Deconstructing Communicative Rationality: A Critique of Habermasian Collaborative Planning', *Environment and Planning A*, vol. 30, pp. 1975-1989.

Thayer, H.S. (1981), *Meaning and Action: A Critical History of Pragmatism*, Hackett Publishing Co., Indianapolis.

Verma, N. (1996), 'Pragmatic Rationality and Planning Theory', *Journal of Planning Education and Research*, vol. 16, no. 1, pp. 5-14.

Webber, M. (1978), 'A Difference Paradigm for Planning', in Burchell, R.W. and Sternlieb, G. (eds), *Planning Theory in the 1980s: a Search for Future Directions*, Rutgers University, New Brunswick, New Jersey, pp. 151-162.

West, C. (1985), 'The Politics of American Neo-Pragmatism', in Rajchman, J. and West, C. (eds), *Post-Analytical Philosophy*, Guildford, Columbia University Press, New York.

West, C. (1993), *Keeping Faith: Philosophy and Race in America*, Routledge, New York, London.

Westbrook, R.B. (1991), *John Dewey and American Democracy*, Cornell University Press, Ithaca New York, London.

3 Theorising Participation: Pulling Down the Ladder

LIZ SHARP AND STEPHEN CONNELLY

Introduction

The XYZ community group is developing a small area of derelict ground into a nature reserve and children's play area. It operates under the umbrella of a community forum, whose management board is a partnership of community representatives, local business interests, the local authority and the TEC and which is funded through EU structural funds managed by the regional Government Office. The project is also supported by the council as part of its Local Agenda 21 programme, which was drawn up on the basis of widespread public consultation.

The XYZ group is imaginary, but the situation described is a fairly typical example of public involvement around the turn of the century[1].

Involving the public is - once again - high on the agenda of central and local government and is an increasingly mandatory component of programmes aimed at improving service provision, regeneration, revitalising democracy and achieving sustainability. However, making sense of what is going on is not straightforward. Not only is the concept of "public involvement" or "public participation" itself notoriously ambiguous and contested (Alterman, 1982; Thomas, 1996), but the current institutional settings in which it takes place further confuse an analysis of the nature of the public's involvement, and in particular of the distribution of power between the various actors. However, such an analysis is important for planners and other practitioners who have to work with the public in such settings if they are to formulate and achieve their own goals (Thomas op. cit.), and essential for those who see their role as assisting the empowerment of citizens.

In this paper we examine the theory currently available to practitioners to assist this analysis and suggest that it is inadequate from the perspective of those interested in promoting social justice. We then explore how insights from participation, community development and policy literature can help us make some tentative steps towards a more useful theory about public involvement.

What is "Useful Theory"?

In developing this paper, our perspective on theory is informed by the planning theorist, John Forester. He argues that the value-laden nature of planning imposes particular requirements on theory:

> Not only must an adequate account of planning practice be *empirically fitting*, it must also be both *practically appropriate* and *ethically illuminating*, helping planners and citizens understand and assess the ethical and political consequences of various possibilities for action, policy or intervention (Forester, 1993: 16, emphasis added).

The requirement that theory should be *empirically fitting* is one which has always been required in science. As Forester writes: "data must be available - in principle if it is not already collected - that could substantiate or weaken the account [given by theory]" (p. 16). The requirement that theory be *practically appropriate* means that it must "speak to the working interpretations that planners have of the practical situations and problems they face" (p. 17). It must therefore provide a way to understand current practice which "makes sense" to practitioners. Finally, to be *ethically illuminating* theory must "provide explicit guides for action and explicit justifications of those guides" (p. 18). We interpret this in a slightly different way: at this stage in developing theory our aim is not to provide a guide for action which is generally applicable, but to assist people to ask the right questions, to give them the theoretical understanding which can be the basis for ethical judgements. Given our own values, this leads to a focus on understanding the distribution of power in a complex setting, not for its own sake but sharing with Gramsci the belief that:

> ...such analyses [of the relations of force] cannot and must not be ends in themselves (unless the intention is merely to write a chapter of past history) but acquire significance only if they serve to justify a particular practical activity or initiative of will. They reveal the points of least resistance, at which the force of will can most fruitfully be applied; they suggest immediate tactical operations... (Gramsci, 1971: 185).

To what extent can Forester's three criteria be used to make an assessment of theoretical approaches to participation? There are two relevant points here. Firstly, writing from an academic perspective, it is not appropriate for us to make a sweeping judgement about the extent to which theories are *practically appropriate*. In consequence we focus here on the extent to which theories fulfil the empirical and ethical criteria. The second point concerns the theoretical field to which the criteria can be applied. Forester explicitly develops these criteria to address the nature of planning theory. However, when writing about participation it is vital to recognise that public involvement in policy making and implementation goes far beyond the realms of statutory planning - as the authors' research on participation in Local Agenda 21 processes aptly demonstrates (Sharp, 1999). Forester's criteria for useful theory appear as applicable to LA21 as to other areas of traditional planning. In this article we venture further, to suggest that the same criteria can aid assessment of different theoretical approaches to the wider realm of all public participation.

Our fictional example of XYZ community group provides an empirical basis through which to test theories against Forester's criteria. The first section below begins by describing the theories which are most commonly drawn upon in current accounts of public involvement. The theories are then applied to XYZ community group, and Forester's criteria enable a critique to be generated.

Existing Theory on Participation

Paralleling the resurgence in participatory processes has been a growth in literature, principally government policy documents and guidance for practitioners (e.g. LGMB, 1994; DETR, 1997, 1998a), descriptions of programmes by local authorities and - to a far lesser extent - analytic academic writing. Within this material, however, there is very little explicit reference to, or development of, theory. Government policy from the Skeffington Report (HMSO, 1969) to the present day (e.g. DETR, 1998b; Audit Commission, 2000) typically contains a mixed and unstructured bag of reasons for involving the public, presented collectively as consistent, unproblematic and unquestionably desirable. The same is true of Agenda 21 (UNCED, 1992), the principal international policy driving increased public participation.

Where theory is drawn upon it is almost always from a single source or derivatives of this: Arnstein's paper *A Ladder of Citizen Participation* (Arnstein, 1969). Arnstein's seminal contribution was to identify a single axis along which the multiple and competing meanings of "participation" could be arranged and to present it using a simple, visual metaphor. This axis is power, and the nature of participation is defined by the balance of power between the people and the combination of state and economic interests. She saw this balance as a ground for conflict, and her analysis is explicitly normative: "citizen participation is citizen power" (p. 216). Lower levels are dismissed as tokenistic or manipulation of the people by the state. This paper is repeatedly cited and the basic idea of participation taking place at different "levels" still forms the basis of most theorising about participation.

Several authors have developed this concept - in particular to make it appear more relevant to the UK policy context. In his discussion of participation in planning in the 1990s Thomas (1996) keeps Arnstein's theory in its original form while drawing on Alterman (1982) to develop in more detail an analysis of various aspects of participation strategies (of which more below). In describing local authority LA21 programmes Young (1996) simplifies the ladder and distinguishes four types of strategy defined by the relative balance of local authority and community control. He maintains Arnstein's basic structure, however, using the typology to characterise entire strategies across a local authority area and, like her, sees genuine and desirable participation in the transfer of at least some control over their local area to local communities. Wilcox (1994) produced an influential reworking of the ladder in his *Guide to effective participation* - described by DETR as "the key text" on participation and community involvement (DETR, 1998a). More recently Davidson (1998) maintained the basic analytic concept - that "participation" includes a range of different types of interaction distinguished by level of community control - but abandoned the ladder in favour of an endless wheel, a development which is examined more closely below.

These are theories of participation on the ground, attempting to make sense of the complex of processes covered by the term, and it is this field that we are principally concerned with here. At a higher, more abstract theoretical level there have been accounts of participation which analyse its rationale in relation to broader political and social theories. Thornley (1977) provided a useful summary and analysis of these ideas, and more recently Stoker (1997) has reviewed a wider range of political perspectives from which different normative approaches to participation can be derived. These higher-level theories do not appear explicitly in non-academic writing and their absence is perhaps reflected in the rather incoherent jumble of reasons often given for encouraging participation. They can, however, provide a framework within which more grounded theories and rationales for participation can be located. It is in this sense that we draw on these theories again later in this paper.

Analysing XYZ through Arnstein

Arnstein's Ladder and its derivatives purport to provide an accurate description of what is happening in a given "participation process" through locating it on one of eight - or four (Young op. cit.) or five (Wilcox op. cit.) - possible "rungs" of the ladder. If this theoretical construct is to have utility to practitioners it must be possible to identify which rung a given participatory exercise or setting occupies. The application of Arnstein's typology to XYZ immediately reveals that this is not well-defined and that several different responses are possible:

- One answer would suggest that the XYZ group enjoy a very high level of participation, amounting to Arnstein's eighth and highest rung - "citizen control". This analysis would stress the complete control that the group enjoys over the design and management of the nature reserve.
- A second analysis would emphasise how this level of control can only be enjoyed because of funding which has been secured through the partnership of the Community Forum. This approach would highlight how the agreement and support of all parties is needed before funding applications can be made. Hence, while a non-controversial nature reserve may be the subject of a funding application, a reduction in road capacity for a new bicycle lane would fail to gain the same unanimous support. This second analysis would suggest that the example amounts to "partnership" (the fifth rung).
- The third analysis would interrogate the nature of the participating community. Several different groups of people are involved in the example. In XYZ as in most community groups, members are self selected local residents or business people. The individuals' very involvement in the group may emphasise their "atypical" nature in the local area - where most individuals are too busy for such community action. Beyond the official members of the community group, the local people's involvement in the development of plans

will at best be at the level of informal consultation - according with Arnstein's fourth (consultation) or third (informing) rung.

- A fourth analysis recognises XYZ as part of the wider LA21 process within the local area. This analysis stresses how XYZ have achieved their position in the context of priorities which have been set through a consultation process over a wider area than their own locality. This is again likely to have involved a self-selected group of environmentally-minded people, whose involvement is one of consultation - level four. The remainder of the area's inhabitants, who are plausibly affected through the direction of resources to these particular initiatives and not to others, are involved, if at all, purely through being informed of the outcome of the LA21 visioning process: level three.
- The final analysis would emphasise how real decisions about what is constructed in the XYZ district depend not just on the applicants for funding, but also on the criteria used by the funding body to determine applications. What matters here then, is the group's ability to understand these criteria, and to demonstrate that their outcomes achieve this agenda. In the current funding climate, a project such as a nature reserve is more likely to receive funding than a proposal for an ice rink or a pub. This despite the fact that the latter proposals might be far more popular with the residents of XYZ district than a nature reserve. In these wider terms of overall governance, the community is manipulated to produce the detailed plans which implement the funding authority's priorities. The considerable time and effort which is contributed by the community at both the planning and implementation stages is expropriated free of charge. This analysis would suggest that the example amounted to "manipulation" - the first and lowest rung on Arnstein's Ladder.

Although Arnstein's typology was used above it seems clear that any of the ladders - or wheels - would have yielded a similar range of possible analyses. No one of these answers provides a full understanding of the situation enjoyed by the XYZ community environment group: rather, together they demonstrate the complexities of modern participation. This suggests that there are some fundamental difficulties with the most widely-used theories of participation.

In the following sections we ask *how well do these models live up to Forester's expectations?* and extract from this some concrete shortcomings in them, for which solutions will then be sought.

Evaluating Arnstein I - Empirical Fittingness

This analysis indicates that there is a problem with *empirical validation* of Arnstein's Ladder and similar models. Although the data suggest that the notion of different levels is well-founded, the theory does not allow us to identify unambiguously which rung of the ladder a given case occupies. There are three different factors which appear to have contributed to this. One of the causes is the ambiguous boundaries of "the project", in spatial terms and in relation to the participants and the issues involved. A linked empirical issue concerns who is perceived as holding power, and who are perceived as being the appropriate citizens or "public" who benefit from the transfer of power. A final empirical issue concerns the question of what the public is participating in. The latter may be unfamiliar, and is worthy of a little extra consideration.

Arnstein appears very focused on the level of citizen power in decision making - as realised in budgetary control. More contemporary frameworks recognise that as well as participating in decision-making the public also participate in the implementation of policy (Wilcox's "acting together"). This emphasis is a realistic reflection of the inseparability of policy-making and implementation (Barrett and Hill, 1984) and the current emphasis on organisations working together on both design and implementation of projects.

In summary, we can conclude that these theories based on Arnstein's Ladder face some issues in relation to empirical validation, in part through an apparently oversimplified conceptualisation of participatory processes. Whilst the simplicity of the ladder metaphor is also one of its strengths, if this is paid for in empirical applicability it seems a high price to pay.

Evaluating Arnstein II - Ethical Illumination

Arnstein's original ladder clearly fulfils Forester's requirement that theory should be ethically illuminating. By identifying power distribution as the key to understanding participation she opened practice up for ethical inspection. Under the term "public participation" could be found state sponsored programmes in which public influence was negligible and which concealed various ways in which the poor and disempowered were kept in that state - and this is ethically unacceptable. The corollary is her strong and simple normative stance: participation that involves real transfer of control to the community is a good that should be striven for.

This analysis is powerful and enduring. Yet herein also lies the danger: such normative content without adequate practical and empirical grounding is potentially unsafe and likely to lead to inappropriate decision-making.

Young and Thomas, and some of the more radical NGOs such as the New Economics Foundation (NEF, undated) essentially subscribe to Arnstein's ethical stance, albeit in a more muted form appropriate to late twentieth century Britain as opposed to late 1960s urban America. In contrast is a theoretical development which, while rooted explicitly in Arnstein's analysis, is apparently underpinned by a very different political and philosophical position. Authors such as Wilcox (1994) and Davidson (1998) explicitly disavow Arnstein's normative stance. While recognising that participation is about relative levels of power, the distribution of this between different actors is apparently seen as unproblematic and not as a potential source of conflict. Which level is appropriate is to be determined by the practitioner initiating participation, possibly in consensus with other participants, and is a matter of "horses for courses" (Wilcox op. cit.: 8). Davidson's adoption of the wheel is a deliberate move away from the normative implications of the ladder towards a more open choice of approach to participation. Such thinking is clearly present in key central government papers laying out the modernisation agenda (e.g. DETR, 1998b) and is explicit in guidance on participation (e.g. DETR, 1997).

In this approach there is still a normative stress on participation *per se* but the suggestion is that all forms of participation -all rungs on the ladder- are ethically equally acceptable. The need for "effective" participation means that methods must be consistent with objectives. This analysis thus provides practitioners with little ethical basis through which to make a choice between alternative objectives and so runs counter to Forester's third requirement.

These normative differences between Arnstein's original conception and some more recent developments of it -typified by Wilcox and hereafter collectively referred to as "Wilcox"- can be seen as linked to underlying political and philosophical values. Thornley's study, *Theoretical Perspectives on Planning Participation* (Thornley, 1977) gives some conceptual language through which to understand the normative differences between these two approaches. He distinguishes between a *consensus* view of society (in which the current structure of society is viewed as stable and desirable), a *conflict* view of society (in which the current structure of society is seen as riven by cleavages and power differentials), and a *bargaining* view of society (which recognises cleavages in society but seeks to find a compromise between existing groups, and perhaps is underpinned ultimately by a belief in limited consensus). He suggests that participation processes will reflect the view of society held by their designers, since these views will define for them the acceptable goals and role of public involvement in governance. Hence, consensus designers will mould participation to develop support for existing elite plans, conflict designers will organise participation so decision-making is taken over by the powerless group, and bargaining designers will seek to bring about compromise through negotiation between conflicting parties.

This analysis can be extended to theorists. Arnstein and other "political" theorists present a normative ideal in which decision making is taken over by the powerless. Their underlying view of society can thus be argued to approximate to Thornley's "conflict" position. In contrast, Wilcox's apparently "value-free" position supports the existing political elite in making their own choices about which form of participation is most appropriate to their situation. In addition, the explicit emphasis on the possibility and desirability of consensus and the creation of win-win situations in which all actors' power to achieve is increased denies the existence of structural and irreconcilable conflicts. Despite his avowed and undoubted interest and commitment to empowerment of society's disadvantaged, Wilcox's analysis is underpinned by a consensus view of society, which is ultimately conservative. Given this analysis it is unsurprising that it is this strand of participation theory which is given prominence in government policy and guidance, since it defuses the radicalism inherent in Arnstein's original paper and instead orients participation towards strengthening social cohesion.

Problems with Existing Theory - a Synopsis and a Way Forward

Thus far we have considered the extent to which two of Forester's requirements for useful theory – empirical validation and ethical illumination – are met by current theoretical conceptualisations of public participation which derive from Arnstein's Ladder. It has suggested that all such theories raise issues around empirical validation, and that Arnstein's approach achieves ethical illumination in a way which explicitly non-prescriptive analyses such as Wilcox's fail to do.

On the empirical side, the central problem appears to be that a typical participation situation is more complex than the theories take account of. This was manifested in the attempted analysis of XYZ by the undefined nature of the boundaries of the process and the way that boundaries could be conceived of at different locations in a number of different dimensions – of physical scale, participants, issues and so on. We suggest that current theory oversimplifies participation in two ways.

Firstly it views the ambiguity, the multiple possible forms of participation, as separable along a single axis of what can be loosely thought of as power or influence. Secondly it treats participation as a binary event or process, involving a community and some "other participant". Arnstein's analysis essentially conflates participation to a struggle between two groups - the community and a combine of state and business interests. While she acknowledges the existence of fractures within the latter this is subordinated to the basic binary structure on which the strength and attractiveness of the analysis rests. The consensual win-win approach to participation adopted by Wilcox and Davidson, which implicitly underlies government policy, also adopts this binary model to a large extent.

The various forms of the theories achieve these simplifications in different ways – through *conflation*, both of a number of analytically distinct variables onto a single axis and of the institutional complexity of the policy making environment into two monolithic and undifferentiated blocs and through *exclusion* – focusing attention solely on a participating public and an initiating body (typically a local authority department) and ignoring the wider social and institutional context in which these actors operate. A view of participation as isolated from these wider policy processes is further reinforced by the more abstract theoretical literature, which locates participation within democratic theory – but makes no reference to the extensive literature on communities and nor on policy implementation (Stoker, 1997). A more empirically adequate theory would thus be one which conceptualises participation both as located in a broader context and as more internally differentiated.

Tested against the criterion of generating ethical illumination, consensus-based analyses arguably fail because they do not provide a way to challenge the manifest inequalities of power and material resources of our society. Simplistic conflict-based analyses perhaps go too far the other way. Their analysis of the situation coupled with the strong normative emphasis on "climbing the ladder" may promote what is, to our minds, an ethically unacceptable outcome. Citizen power, taken to its limits could result in exclusion and conflict between groups, and disempowerment through disconnection between citizen groups and the state. Ethical illumination is perhaps achieved not through such strongly prescriptive theorising, but through providing a framework for analysis informed by explicit values. This is a point we return to at the end of the paper.

Bringing these two general problems with the existing theory together, we suggest that the challenge is to develop a realistic theoretical understanding of participation which recognises its internal complexity and its context, yet at the same time preserves the ethically illuminating quality of Arnstein's approach. It will thus need to not only describe complexity, but suggest why certain patterns within it are more likely than others, by maintaining Arnstein's emphasis on power. Finally, these improvements should not be bought at too high a price in practical appropriateness, though nothing is likely to approach the simplicity and accessibility of the Ladder.

Unit of Analysis: What is the Nature of the Beast?

Above, it was suggested that Arnstein's Ladder and its derivatives excluded the context of policy making from their analysis. Here we contend that in order to understand the role of participation in a more grounded and less normative way it should be seen as part of the process of *policy* generation and implementation. This establishes it in its context, as a process which forms one part of a dynamic and institutionally complex field of activity. We understand the policy making process to be one of "communication and argument" (Mazza and Rydin, 1997), ambiguous in terms of goals and continuously contested by participants with differing interests, values and strategic and tactical motivations for action (Barrett and Hill, 1984). Public participation not only shares these characteristics, but is seamlessly connected into the wider policy making process of the state, both in its devising and implementation and in terms of what happens to the "product" of public involvement.

Within this field of activity public involvement has a natural definition as a subsection in which the public interact directly with the state: "taking part in any of the processes of formulation, passage and implementation of public policies" to quote Stoker's definition again

(Stoker, 1997: 157). The boundaries of this are not clearly delineated, since they are to some extent a line drawn across a continuous field. Nevertheless, in order to achieve an analysis of participation, some boundaries – however arbitrary – must be drawn.

Boundaries are crucial if an analysis is to understand what is occurring within a participation process. Questions about how the participation is structured, what it considers, and the style in which it takes place all concern the internal complexities of the process. Unless what is "internal" to the process is clearly distinguished from that which is external, the analysis risks facing the same analytical confusion as occurs when Arnstein's Ladder is applied. Existing frameworks to understand these "internal" factors work exclusively on a binary split between an institution and the public. If these frameworks are not to be abandoned, analysis must start by isolating the two binary parties that will be studied.

The boundaries are also crucial if the context is to be understood. The context is about the role that the participation process plays in wider policy and decision processes (within institutions, between institutions and within communities). Describing the context means telling a narrative about the role that participation has played within this set of institutions/individuals. This narrative is likely to acknowledge how the binary relationship studied is just one of a plethora of other interactions within and between agencies, organisations and individuals which make up the complex policy field. Viewing these factors as external and contextual to the participation process builds in a recognition that they cannot be isolated and treated as separate "decision variables" by the designers of a participation process.

What, then, are the boundaries of the participation process? A starting point will be a policy process that is to be the focus of the analysis, and the two parties involved. One party is likely to be some officers of an institution or partnership (the initiators of the participation), the other some (more or less representative) representatives of a "community". A second part of the boundary setting process will involve identification of the geographical scale which is primarily of interest. Thus, the complex interactions between different scales observable in the XYZ example can be avoided. Finally, isolating the boundaries also involves the specification of a period of time over which the participation process is being considered - whether this be the "natural" timespan of a one-off participation process or an arbitrary timespan imposed for the purposes of analysis. By specifying this period the analysis helps to identify precisely which interactions are being considered and how they are located in the (considerably longer) period over which a narrative account of the context might be reported.

This reframing of public involvement leads to different questions being asked, compared with the existing participation theory. Arnstein and derived work allow us to ask "who is in control?" and wonder "who should be in control?" - but fails to specify control "over what?", "where?" and "by whom?" in a useful way. Taking a discursive-policy-process approach, the need for both a broader, more contextualised scope and a more detailed analysis become apparent: the pertinent questions become "how is control/power distributed?" and "who should control what?" This analysis begins by considering how the matters "internal" to the participation process can be unpacked. It then considers how the wider "context" might be analysed.

Unpacking the Participation Process

Several authors have recognised the need to unpack "participation" further than the single axis of power and expose the complexity that characterises participation programmes or strategies. Alterman (1982) usefully shows how participation strategies can be analysed in terms of six "decision variables" which need to be considered in devising a strategy. In addition to the "level" of participation in the Arnstein sense, she proposes that strategy designers have to be make decisions about:

- the goals of the programme,
- the stage in the planning process at which the public will be involved,
- the type of issue that is to be covered,
- who will be participating, and,
- the resources which will be committed to the process.

These are clearly inter-related (Barnes, 1999) - Alterman's point being that they are separable issues about which a planner has to make a decision, that an analyst must take into account in describing participation and that each of the participants may contest.

In our framework the goals and timing of the process will be considered as part of the wider context, while the resources variable is seen as peripheral to the wider issues discussed[2]. The remaining variables in

Alterman's list are level of involvement, type of issues and who to involve. Thomas (1996) makes the important addition that the substantive nature of the issues is an important decision variable. While this is fairly tightly defined within the statutory planning process, in the more "open" planning processes of LA21 and regeneration initiatives there is considerable latitude for choice - will the public be involved in determining budget priorities, at one end of the scale, or, at the other, merely in choosing the species to be planted in a nature reserve?

A further variable which seems to be missing from current analyses is geographical scale. Public involvement is necessarily related to policy covering a specific location, and the size of the area and how it relates to other areas - both as neighbours and as larger, encompassing areas - would seem to have a bearing on the nature of participation.

While these lists of "decision variables" give a much-needed breakdown of the nature of participation in practice, they do not - in current formulations - give us much assistance in explaining their significance for different actors. In our understanding they are all political variables which different actors from the various contexts will seek to influence to achieve their interests. Moreover, the way these influences are played out during the participation process will have a profound affect on its outcomes (whether these be specific and tangible outcomes such as policy document or more intangible outcomes such as a shift in council thinking on a specific issue). These "external" processes will be considered in the "context" section which follows. The concern here is to consider how the different variables can be understood in a way which is both ethically illuminating and empirically fitting.

Arnstein's Ladder provides an ethically illuminating overview which can continue to structure our thinking about participation. Once clear boundaries have removed the complexities of the wider context, a series of different levels ("rungs") can capture an overall sense of the extent to which the participation process provides the potential to challenge existing practices. However, rather than suggesting that this "degree of danger" can be directly read from casual observation of a participation process - in our view it hinges on the combination of all this wider set of political variables:

- geographical scale,
- extent of action,
- agenda,

- participants, and,
- style of engagement.

Each variable can be seen to take a range of values, some of which are "safe" - which pose little challenge to existing practices and others are dangerous - posing a challenge to existing practices. The first two can be seen as continua, while the latter three are less easily arranged on a scale. In particular what issues and participants involve a high "degree of danger" is very context dependent. These variables are discussed in turn below.

Geographical Scale

The first continuum enables a classification of participation exercises according to the geographical scale over which they operate. At its two extremes, this axis differentiates between neighbourhood comments on street-scale traffic calming proposals, and the comments of the European public on EU transport planning proposals. Given the same degree of danger in other respects, the former is a safe extreme and the latter a dangerous extreme.

Table 3.1: Geographical scale continuum

	Large			Small
Scale	City	Estate	Neighbour-hood	Open space
Participation	LA21 consultation	Community forum - negotiating funding	Planning local initiatives	Managing nature reserve

Extent Of Action

This variable characterises what aspect of the policy process it is that the public is actually involved *in*. Most literature on participation treats it as essentially about the influence the public can exert over policy making - this is clearly reflected in Arnstein and derived theory. Yet much current practice which is referred to as "participation" is more concerned with practical action on the part of the public. This difference has emerged through studies of Local Agenda 21 practices (Sharp, 1999), and is not widely acknowledged in the literature.

At one extreme of this continuum the public are requested to carry out individual actions which are perceived as furthering the public good (for example, recycling their waste). In successive positions along the continuum the public may be asked to participate in "self-help" type action, to initiate and plan their own actions (community organising), and to plan actions which will be implemented by a council or other institutions (policy participation). At a given "degree of danger" in other respects, a high "extent of action" can be seen to be politically safe - it poses little challenge to existing practices because it merely encourage voluntary changes in behaviour. In contrast, a low "extent of action" challenges existing practices. This arises because it has the potential to lead to changes in policy: if this potential was realised then individuals might be forced to change their behaviour.

This continuum can be effectively illustrated through considering a range of examples of public participation on one spatial scale. Thus, within one council estate, "self-help" action might consist of participation in a tree-planting exercise. The tree planting exercise may have been planned and organised by a group of keen local naturalists - an example of community organising. In contrast, an estate liaison committee influences the council's housing policies provides an example of a body which has influence over the actions of another institution (institutional participation). At the low-action extreme might be an "estate visioning exercise" (policy participation, or even "pre-policy" participation).

Table 3.2: Extent of action continuum

	High Action			Low Action
Category	Individual action	Communal action	Community organising	Institutional participation
Estate example	Recycling	Tree planting	Planning tree planting	Estate liaison committee

Agenda

Alterman has stressed the importance of considering the scope and type of issue when studying participation. To some extent, of course, this has already been accounted for above in relation to the "extent of action" continuum. However, we contend that two participation exercises with the same extent of action can be more or less contentious depending on the type of issue. An issue is potentially *dangerous* if there are entrenched and well-articulated interest groups in opposition to current policy goals. The contemporary debate over transport is an example of just such an issue. In contrast, an issue is *safe* if there is a widespread consensus about the "best" way forward: in the current climate urban nature conservation policies provide an example of an issue over which considerable consensus exists.

The timing of public involvement is an important factor which has an influence upon the agenda. It is clear that early involvement is likely to address a broader agenda, and thus may be more "dangerous" than late involvement. (Note, however, that the policy process is seldom linear so that this clear division into "early" and "late" may not be simple to make.) Consequently, this is a contested area: while the public often demands earlier involvement (e.g. Sheffield City Council, 1999) they are often excluded by those controlling the process - typically in contrast with other participants in the policy process (Simmie and French, 1989).

Participants

In the "Context" section below we will explore how the choice of participants can be more or less representative of the population from which they are drawn. Here, we stress a different aspect of participant choice - how participants may be more or less likely to challenge existing policy norms. Anderson (1996) has emphasised how some participants may feel lacking in power. Research by one of the authors has also emphasised how participants from highly networked participant groups had access to information which enabled them to challenge existing policy more effectively (Sharp, 1999). We therefore suggest that participant choice is another important constituent of degree of danger. A safe group of participants are those whose institutional positions means that they are unlikely to express disagreement with existing policy. A dangerous group of participants are those whose institutional positions means that they are likely to express disagreement with existing policy norms. Thus a council with an entrenched policy of city centre car parking provision might "safely" engage city centre traders in a participation exercise on this topic while dangerous participants might be drawn from environmental or transport lobby groups. If a participation exercise drew participants from both groups then the extent of "danger" would depend on their relative numbers, and the extent to which they had connections to wider expertise.

Style

Finally, the "style" of engagement provides an important final factor which influences a particular participation exercise's degree of danger. For example, questionnaires with closed questions provide a classic form of a participation exercise in which the output has been controlled (Thornley, 1977). The closed form of questions mean that the questionnaires have little potential to influence public policy - they therefore represent a safe style of engagement. Some tightly structured public meetings provide a similar example of safe participation. In contrast - as many harassed planners and councillors will be aware - unstructured public meetings can constitute extremely dangerous forms of participation.

Overall the degree of danger is composed of the sum of all these components. "Dangerous" participation is participation which has the potential to challenge existing policy norms and thus shift the balance between different interest groups. It would combine the raising of risky issues affecting policy over a wide area, with the involvement of interest groups in opposition to the status quo, using a style of engagement which allowed public challenging of received wisdom. Any dimension individually can affect the overall balance and thus all are important components in the analysis. It is relatively easy for the state to portray an instance of public involvement as being "high on Arnstein's ladder" - and therefore "good" - by promoting a potentially challenging style (e.g. a visioning exercise) and yet to maintain it at a non-threatening level by controlling the scope of issues and the participants.

How then should we relate this overall characterisation of a participation process to the wider context of a policy process engaged in by - or less directly influenced by - a wide range of actors? The key here is the issue which appears in policy guidance as just one decision variable among many: the level of engagement which the participation process has with the wider context. It is this key issue which is explored in the following section.

Context

Who are the initiators of participation? What does participation achieve? These are the questions which are asked in this final contextual part of the analysis. The key difference between this analysis and that described above is that it requires a historical investigation of the issues and debates which are current before and after the participation. The central question to be considered is the extent to which the participation was used to contribute to these debates. For example, it may be used to publicise or legitimate particular policy or process outcomes. Indeed, it is our contention that participation can often be understood as a subtle and effective means of promoting changes to policies and policy-making structures.

Alterman (1982) suggests that participation programmes are to be viewed as contested objects which are developed by a range of interested parties in a political arena. These parties - individuals, groups, agencies - will have different goals which a participation programme could serve in different ways. This contest will largely determine both the nature and the effect of public involvement. Our attention is therefore drawn to power relations amongst the various actors. Both empirically and ethically it is important for us to understand which outcomes are promoted by participation and how this is achieved, in order that we can have grounds for deciding which parts of the process are ethically acceptable and identify places where change is desirable and feasible.

The focus on power relations raises a question over the theoretical model through which the analysis of context might take place. Broadly, two options can be identified. The first option draws on Lukes's analysis of power to identify actors and interests in the policy fields (Lukes, 1974). A second isolates competing policy discourses, and provides an account of the tactics and mechanisms of their contestation (Sharp and Richardson, 2000). It is beyond the scope of this paper to review the relative merits of these different options. Both approaches will lead to a narrative which will dissect the influences upon the design and operation of the participation, and way in which its outputs are then used. Both options are thus are likely to offer helpful insights to an analysis of the external power relations.

The context for participation can be seen as located in two broad arenas: that of the initiating institution or partnership, and that of the community or locality which is involved in the decision. These are discussed in turn below. Having addressed these arenas in broad terms, the discussion will then consider how their influence upon the process of participation might be understood through the literature on the "goals" of public participation.

A key part of the contextual narrative will be to tell the story of how participation emerges, how it is shaped and how it is used as part of the policy contestation within an institution or partnership. The literature stresses how the development of policy is a non-rational process (Lindblom, 1959). Within local authorities, for example, both formal and, in particular, informal structures and cultures have a significant influence on policy making (Ham and Hill, 1993; Leach *et al*, 1994). Thus different groups and individuals will have varying levels of influence on what will emerge as a participation programme. (For example, in fieldwork by one of the authors, the requirements of a local authority's Economic Development Department have severely constrained the scope of the Environment Department's Local Agenda 21). It is also clear that the process is not linear: "policy" develops along with implementation (Barrett and Hill, 1984). Actors may well have input at different stages, and thus turn a policy towards their own goals - or defuse its threat to their interests in midstream. Such internal divisions can make participation exercises appear incoherent from the "outside" and thus frustrating and alienating for the public. One section of an institution or partnership may initiate participation in good faith, only for those who are in a position to act on the outcomes to ignore or downplay the significance of the public's views (Tewdwr-Jones and Thomas, 1998; Audit Commission, 2000).

The continuing fragmentation of local governance adds further complexity to this field (Stewart and Stoker, 1995; Rhodes, 1997). In the regeneration sector in particular, participation is frequently part of a wider "partnership" process, including local and central government, business interests and the organised voluntary sector as well as some form of community representation. The power relations between these partners will have a significant influence on the way public involvement is sought (Hastings, 1999). For example Cochrane (1993) suggests that where authorities have entered partnerships to promote economic regeneration there is an observable tendency towards corporatist governance and loss of local authority independence which would be inimical to public involvement in decision-making over major policy areas. Moreover, both Central Government and European regeneration funds are increasingly requiring increased public participation within policy making (for the usual mixture of reasons). This in itself gives communities a curious power of veto over what are usually the more powerful members of the partnership - by refusing to co-operate in consultation exercises they can deny the other partners access to funds (Wilson, 1999). Partnerships can thus be arenas of bargaining and power struggles as well as, perhaps, places of consensus building and collaborative planning.

A second key part of the contextual narrative will consider how participation is shaped and shapes the community or locality from which it draws. Both in guidance on participation and in a large academic literature the concept of "community" is seen as being problematic, in that it often assumes a population who live and/or work in an area will be cohesive and homogenous - a "community of place". It is questioned whether such traditional notions of community are relevant in modern, particularly urban, settings or whether "communities of interest" are more appropriate descriptions of how people relate to each other (Lowndes, 1995; Warburton, 1999). Much of the literature suggests that the latter is the case, and initiators of participation are thus encouraged (e.g. by Wilcox 1994; Church 1999) to think in terms of "stakeholders" or "communities" and not simply of "community". This guidance puts forward what is essentially a pluralist recognition of the existence of different interest groups which need to be considered and potentially involved in policy development. Despite this, "communities of place" remain the focus of much state policy through area-defined regeneration initiatives and in planning processes.

The notion of a "community" is further disentangled by Chanan (1997). He recognises that, like the state sector, communities are dynamic arenas in which individuals and groups interact. His key insight is that active citizenship is about far more than participation in specific state-sponsored programmes: "local community groups are permanent vehicles for participation by people in their own lives" (p. 29), through which many services are delivered and less tangible community goals are met. This insight raises some important questions about participation processes and participants. Do participation programmes recognise or ignore the numerous structures already existing within the communities with which they interact? Are participants in state sponsored programmes isolated individuals, or are they active citizens with multiple local linkages?

The organisations and networks within the community will be structured by power relationships, which will give some actors more influence within the community and potentially more access to participation in state programmes. Thus one could suppose that the traditionally "hard to reach" - disaffected young men, ethnic minorities and so on - are not passive groups which more innovative techniques could perhaps reach and bring into participatory processes, but are involved in dynamic and power-laden relationships within their community which affect their ability to participate. Such structural constraints also affect the nature of individuals' participation. For example, Appleton (1999) considers how gender has structured the time, energy and roles which are played by women in local regeneration initiatives in Sheffield.

How can the different interests of (or the different discourses in) the two arenas of the institution/partnership and the community/locality be understood in relation to participation? Here, the literature on "goals" of participation can provide some insight. We should stress that the term "goals" does not necessarily mean that these are fixed objectives which are determined in advance of the participation. They rather show the different ways in which individuals or groups may see participation as supporting or working against their own interests. Two different types of goals are addressed. Firstly the discussion draws on the participation and democratic literature to discuss the political goals which may structure attitudes to participation. Later, the discussion moves on to institutional and policy theory to discuss the influence of the less substantive but equally important goals which individuals and groups may bring to participation process.

Political goals are concerned with achieving a (heavily contested) "desirable" level of public involvement in governance. The extent to which such goals can be explicit depends on what are seen as legitimate understandings of the public role: currently a wide range of such goals has been given legitimacy by the government's democratising agenda. In policy documents and guidance these are typically presented as a "mixed bag" of service improvement and more political aims encompassing social cohesion, rejuvenation of local democracy and legitimation of the public authorities through increased accountability (Barnes, 1999). Public involvement is presented as "a good thing" which can help attain these very varied ends and possible synergies or conflicts between the various goals are unexamined. This is where the political theory reviewed by Thornley (1977) and Stoker (1997) has a part to play in unpicking these relationships by showing how they derive from different conceptions of the role of the state and the purposes of democracy. The conflict-consensus dimension introduced earlier interprets politics in an instrumentalist fashion, in which the rationale for participation is the achievement of material ends - essentially control over the distribution of resources. A second dimension which generates further rationales for participation is that which gives popular participation an ethical value in itself. At the "high participation" end of a continuum Stoker identifies two traditions which value high levels of involvement: on the communitarian grounds that it expresses a concern for the collective and because it contributes to the development of the potential of the individual, and thus of their capabilities as a social being. Together these perspectives see increased participation as a way to create a more cohesive and perhaps juster society - a concept which is overt in current government policy documents (e.g. DETR, 1998b).

Stoker also identifies a relatively recent theoretical development which could provide another normative framework for participation. Advocates of deliberative democracy (e.g. Gutman and Thompson, 1996) prioritise citizen debate as the ground for good policy making - as a corollary certain forms of participation are favoured, though maximising participation *per se* is not central to this perspective. The current emphasis on innovative techniques such as citizen juries is to some extent underpinned by this view of democracy.

These perspectives all provide different rationales for participation and, as Thornley pointed out long ago (1977), they are in conflict with each other - actors holding differing conceptions of participation's role in democracy will have different and incompatible goals and rationally have different preferred methods for engaging with the public.

However, democratic theory only covers a subset of the goals which actors plausibly have for a participation strategy. Institutional and policy theory provide a further set of goals which will be pursued by individuals and groups at the same time as substantive - and probably more overt - policy goals. In the general local government context, Leach *et al* (1994) show that individuals and groups have interests and values which will affect the decisions they make, which may act either to support or resist particular policies or changes to the system. In general, people will act to further their interests and resist detrimental change to the extent of their power within the organisation. Interests are construed in a very broad sense: essentially as any factor which is perceived as bringing them benefit. Thus they include not only material benefits, but also status and power and also the diverse factors which contribute to job satisfaction. This can clearly be generalised to actors in every sector, not just within local government, and also extended to include other social and personal benefits that participation can deliver - such as meeting people and overcoming boredom (Alterman, 1982).

Within this welter of goals, one can identify a large number of different desired outcomes as well as some which are more expressly concerned with process. The latter - perhaps understandably - receive considerable attention from practitioners, especially recent theorising which focuses on developing normative standards for fair and inclusive participation (Healey, 1996). But we would suggest that most parties with an interest in public participation are motivated by a desire for particular outcomes - whether material or institutional - and that in order to understand why power is used in particular ways it is outcomes rather than process which should be the centre of our attention. Further, from the perspective of those interested in social justice, the key questions to ask are surely *Who benefits?* and *Has this process made any contribution to the redistribution of resources from the rich to the poor?* (Fainstein, 1999) - rather *than who has participated?* and *have they been fairly heard?* Thus, irrespective of process benefits, we return to an instrumental understanding.

This discussion of the context for participation has highlighted the importance of unpacking power relations in the two arenas of the initiating institution/partnership and the participating community/locality. The questions to be asked in both arenas are who/what is shaping the process to favour what outcomes? Literature about the "goals" of community participation provide a structure through which this shaping might be unpacked.

How does an analysis of participation "context" differ from the type of analysis which was developed in previous frameworks? The "context" described above certainly encompasses some of the factors which were previously addressed *as within* a participation process through criteria such as "power relationship" (Alterman, 1982), or "level of participation" (Arnstein, 1969). These criteria concerned the extent to which the initiating authorities were willing to see their practices altered by the outcomes of participation. As the discussion of XYZ showed, one of the difficulties in applying these criteria is the ambiguity over which set of practices/actors were under review. Here we take this analysis *outside* of the specific analysis of participation - we thus overcome the inappropriate assumption that it is the practices of the initiating individuals/authorities which are necessarily the subject of participation processes. Instead, it is recognised that participation is a process which can have an effect on a wide range of practices by a variety of actors. Above, we have suggested that analysis should consider both arenas - the initiating institution/partnership and the participating community/locality - if participation is to be appropriately located. In addition, however, we suggest that some researchers may have particular interests in how certain aspects of one arena both influence and are influenced by participation. When telling a narrative about participation such researchers may purposely focus on particular set practices which are of interest, and effectively evaluate key actors' changing "level of commitment" as the process develops.

Conclusion

This paper has suggested that existing theory - largely based on Arnstein's Ladder - provides an inadequate basis for analysing participation processes for it is not empirically fitting, and in some cases fails also to be ethically illuminating. The paper has suggested that a more useful theory can be generated by pulling insights on participation together with literature from the community development and policy implementation fields. A tentative framework for analysing participation processes has therefore been developed. This framework begins by defining the participation process which is of interest - specifying the binary actors, highlighting the scale, and identifying the period which is being studied. Having defined the process, its characteristics can then be highlighted. An investigation of scale, extent of action, agenda, participants and style enables a "degree of danger" to be specified. This indicates the extent to which the participation process has the potential to challenge existing practices. It seems clear that the general tendency of those in positions of authority will be to organise participation to reduce the degree of danger it poses - conversely it may well be in the interest of marginalised groups to press for involvement which, from the state's point of view, is dangerous. A full understanding of the participation process will therefore only be achieved if an analysis of the degree of danger is accompanied by a narrative about the context in which the participation occurs, and in particular, the ways in which that context interacts with the participation process. This narrative may indicate how the "level of commitment" of particular key actors varies through time, and thus demonstrate the mechanisms through which the participation process has influenced a set of practices.

How could this framework help in the understanding of XYZ? A key process would be the definition of the participation process under review. Researchers would need to select either the control of the local nature reserve, or the operation of the community forum, or the process of the Council's Local Agenda 21 participation as the process under review. In each case, slightly different actors are specified on each side of the binary divide, and the geographical scale of influence is clearly variable. For example, we might select the nature reserve as the participation process under review. In this case the participants will be the particular sub-group of the community forum who make and implement decisions about the nature reserve, while the institution/partnership would be the City Council department which carries out open space management. The researcher will also need to make a choice about the time period which is being studied.

Having specified the process to be studied, the researcher would then need to review the degree of danger. If we take as an example, the control of the local nature reserve, we can make the following analysis:

- geographical scale is small (safe),
- the extent of action is high (safe - to a large extent it is voluntary changes which are to be made),
- the agenda is all aspects of nature conservation on the one site (fairly dangerous - complete change could result, albeit on a relatively "safe" issue).

Further information would be needed to evaluate participants and style of engagement However, from the information collated already it appears that this type of participation can be seen as reasonably "safe".

Finally, the context for the participation would need to be evaluated. In the institutional arena a key objective of research would be to trace differing views about participation and sustainable development within the Council, voluntary organisations in the nature conservation sector involved in the LA21 process and the Government Office. The research would ask how these differences were played out in terms of goals for community participation in the management of local sites - and of the XYZ nature reserve in particular - and more widely in the kinds of community involvement it was thought appropriate that EU funding should support. In the community arena the research would seek to identify differing local uses and users of the open space - and thus differing preferences about how it should be developed. The research would trace how some of these preferences were better represented than others within the community forum. It might identify how alternative voices were structurally excluded or silenced in the debate about the future use of the land.

It is our contention that the resulting analysis would be both empirically fitting and ethically illuminating. Empirical fittingness is achieved through the clear definition of the participation process and through the specification of the range of factors which contribute to the "degree of danger". The resulting characterisation of participation processes enables them to be compared in terms of the extent to which they pose a challenge to existing practices. Ethical illumination is achieved because the analysis enables such comparisons to take place. Specific participation processes can be compared through examining the "degree of danger" they pose. However, the analysis is incomplete if this is not supplemented by an examination of the two contextual arenas in which they operate. It is only when a high degree of danger is combined with (appropriate actors') high levels of commitment that significant change to existing practices will result.

Some Final Issues...

This framework has moved beyond the isolated analysis of participation processes which is frequently advocated in the participation literature to suggest that participation has to be seen as part of wider policy processes and power relations. Drawing on insights from the policy and community development literature, it has suggested ways in which the participation process itself can be more clearly characterised as well as located in its wider context. We see this as a considerable development of previous theory. Nevertheless, some questions about the framework remain. Our intention below is not to explore these in detail, but to highlight areas where are thoughts are likely to develop in future.

Firstly, it is clear that the claim of empirical fittingness and ethical illumination remains just that - a claim. This reflects the recent development of the framework. The claim will only be verified once the framework has been applied to the analysis of specific participation processes. This is a high priority for the authors over the coming months. Similarly, the claim that the framework could be universally applied to all forms of participation (not just the planning, regeneration and Local Agenda 21 participation of which the authors have experience) also remains to be confirmed.

Secondly, we have some uncertainties about the precise nature of the ethics this framework promotes. We have discussed the framework as a means to promote social justice in decision making - but whose "social justice" is being promoted, and by whom is this defined? Linked to this we are aware that our framework, like that of Arnstein, puts a normatively high value on participation that is "dangerous" and where the "level of commitment" of key policy/community actors is high. We thus privilege participation which achieves change towards (appropriate) participants' agendas. There is an assumption here that participants' agenda are likely to coincide with our ideas about social justice, which may not always be correct. Moreover, there is a question over whether the logical extension of this (and Arnstein's) framework - some form of anarchist utopia with local control over all decisions - is actually a desirable end point (and if it is not, whether the framework remains applicable because of (which?) current circumstances.

Finally, the process of generating this paper has highlighted the way that academic analysis has constructed participation as an isolated and democracy-enhancing fragment of practice. This has occurred despite the very wide range of activities that are labelled "participation", and despite the clear part played by participatory processes in policy development. A ripe area for future consideration might consider how and why participatory activity has been constructed in this way.

Notes

1 For the purposes of this paper the most useful definition of "public involvement" is that adopted by Stoker (1997) for "political participation" (following Parry *et al*, 1992: 16): members of the public "taking part in any of the processes of formulation, passage and implementation of public policies" (Stoker, 1997: 157). "Implementation" is here taken in a very wide sense to include involvement through personal action, thus extending the traditional focus on decision-making of much politically-oriented literature on participation - a point which is elaborated below.

2 Human and financial resources are obviously crucial to the ability to carry out participation, and thus control over resources gives a significant degree of control over the process. While resource shortages are routinely cited as a constraint - and local authorities are exhorted to devote more resources to participation - this is a reflection of political priorities rather than an absolute barrier (DETR, 1998a). It is thus conceptually a secondary issue to the more substantive political issues of who to involve and over what issues etc. and will not be considered further here.

References

Alterman, R. (1982), 'Planning for Public Participation: The Design of Implementable Strategies', *Environment and Planning (B)*, vol. 9, pp. 295-313.

Anderson, J. (1996), 'Yes, but Is It Empowerment? Initiation, Implementation and Outcomes of Community Action' in Humpheries, B. (ed.), *Critical Perspectives on Empowerment*, Venture Press, Birmingham.

Appleton, Z. (1999), *The participation of women in urban regeneration: a longitudinal study in Sheffield*, PhD Thesis, Sheffield Hallam University, Sheffield.

Arnstein, S. (1969), 'A Ladder of Citizen Participation', *Journal of the American Institute of Planners*, vol. 34, pp. 216-225.

Audit Commission (2000), *Listen Up! Effective Community Consultation*, Audit Commission, London.

Barnes, M. (1999), 'Researching Public Participation', *Local Government Studies*, vol. 25, no. 4, pp. 60-75.

Barrett, S. and Hill, M. (1984), 'Policy, Bargaining and Structure in Implementation Theory', *Policy And Politics*, vol. 12, pp. 219-240.

Chanan, G. (1997), *Active Citizenship and Community Involvement*, European Foundation for the Improvement of Living and Working Conditions, Dublin and Office for Official Publications of the European Communities, Luxembourg.

Church, C. (1999), *Having Your Say: Guidelines for Effective Public Participation*, unpublished draft.

Cochrane, A. (1993), *Whatever Happened to Local Government?* Open University Press, Buckingham.

Davidson, S. (1998), 'Spinning the Wheel', *Planning*, no. 1262, 3 April 1998, pp. 14-15.

DETR (1997), *Involving Communities in Urban and Rural Regeneration: A Guide for Practitioners*, Department of the Environment, Transport and the Regions, London.

DETR (1998a), *Guidance on Enhancing Public Participation in Local Government*, Department of the Environment, Transport and the Regions, London.

DETR (1998b), *Modernising Local Government: Local Democracy and Community Leadership*, Department of the Environment, Transport and the Regions, London.

Fainstein, S. (1999), 'Opening Address' at the *Planning Research Conference*, University of Sheffield, Sheffield, March 1999.

Forester, J. (1993), *Critical Theory, Public Policy and Planning Practice*, SUNY Press, Albany, New York.

Gramsci, A. (1971), *Selections from the Prison Notebooks of Antonio Gramsci*, edited and translated by Hoare, Q. and Smith, G. N., Lawrence and Wishart, London.

Gutman, A. and Thompson, D. (1996), *Democracy and Disagreement*, Harvard University Press, Cambridge, Massachusetts.

Ham, C. and Hill, M. (1993), *The Policy Process in the Modern Capitalist State*, Harvester Wheatsheaf, Hemel Hempstead.

Hastings, A. (1999), 'Analysing Power Relations in Partnerships: Is There a Role for Discourse Analysis?' *Urban Studies*, vol. 36, pp. 91-106.

Healey, P. (1996), 'The Communicative Turn in Planning Theory and its Implications for Spatial Strategy Formation', *Environment and Planning (B)*, vol. 23, pp. 217-234.

HMSO (1969), *People and Planning*, report of the Skeffington Committee on Public Participation in Planning, HMSO, London.

Leach, S., Stewart, J. and Walsh, K. (1994), *The Changing Organisation and Management of Local Government*, Macmillan, Basingstoke.

LGMB (1994), 'Community Participation in LA21', *LA21 Roundtable Guidance No. 1*, Local Government Management Board, Luton.

Lindblom, C.E. (1959), 'The Science of Muddling Through', *Public Administration Review*, vol. 19, pp. 78-88.

Lowndes, V. (1995), 'Citizenship and Urban Politics', in Judge, D., Stoker, G. and Wolman, H., *Theories of Urban Politics*, Sage, London, pp. 160-180.

Lukes, S. (1974), *Power: A Radical View*, Macmillan, Basingstoke.

Mazza, L. and Rydin, Y. (1997), 'Urban Sustainability: Discourses, Networks and Policy Tools', *Progress in Planning*, vol. 47.

New Economics Foundation (undated), *Participation Works!* New Economics Foundation: London.

Parry, G., Moyser, G. and Day, N. (1992), *Political Participation and Democracy in Britain*, CUP, Cambridge.

Rhodes, R.A.W. (1997), *Understanding Governance*, Open University Press, Buckingham.

Sharp, L. (1999), *Contesting Sustainability: Local Policy-making for the Global Environment*, PhD Thesis, Department of Town and Regional Planning, University of Sheffield, Sheffield.

Sharp, L. and Richardson, T. (2000), 'Reflections on Analysing Discourse in Planning and Environmental Policy Research', *TRP 140, Research Papers in Town and Regional Planning*, Department of Town and Regional Planning, University of Sheffield, Sheffield.

Sheffield City Council (1999), Minutes from Sheffield 100 Forum Meeting, 1 March 1999, Sheffield City Council Directorate of Development, Environment and Leisure, Sheffield.

Simmie, J. and French, S. (1989), 'Corporation, Participation and Planning', *Progress in Planning*, vol. 31.

Stewart, J. and Stoker, G. (1995), *Local Government in the 1990s*, Macmillan, London.

Stoker, G. (1997), 'Local Political Participation' in Hambleton, R., Davis, H., Skelcher, C., Taylor, M., Young, K., Rao, N. and Stoker, G. (eds.), *New Perspectives on Local Governance: Reviewing the Research Evidence*, Joseph Rowntree Foundation, York, pp. 157-196.

Tewdwr-Jones, M. and Thomas, D. (1998), 'Collaborative Action in Local Plan-Making: Planners' Perceptions of Planning Through Debate', *Environment and Planning (B)*, vol. 25, pp. 127-44.

Thomas, H. (1996), 'Public Participation in Planning' in Tewdwr-Jones, M. (ed.), *British Planning Policy in Transition*, UCL, London, pp. 168-188.

Thornley, A (1977), 'Theoretical Perspectives on Planning', *Progress in Planning*, vol. 7.

UNCED (1992), *Agenda 21*, United Nations Conference on Environment and Development, New York.

Warburton, D. (1998), 'A Passionate Dialogue: Community and Sustainable Development' in Warburton, D. (ed.), *Community and Sustainable Development: Participation in the Future*, Earthscan, London, pp. 1-39.

Wilcox, D. (1994), *The Guide to Effective Participation*, Partnership Books, Brighton.

Wilson, N. J. (1999), Personal Communication from development worker, Netherthorpe and Upperthorpe Community Alliance, Sheffield.

Young, S. C. (1996), *Promoting Participation and Community-Based Partnerships in the Context of Local Agenda 21: A Report for Practitioners*. The Victoria University of Manchester, Manchester.

4 Shaping the Planning Profession of the Future: The Role of Planning Education

JENNY POXON

Introduction

Since the early 1900s, the development of the planning profession in practice has been inextricably linked to debates about the appropriate education which students of planning should receive:

> ...the way in which the planner is trained and educated is bound up with perceived concepts of planning and the expected role of the planner in society... it is vital to fit the man [sic] to the task he has to perform... (Cherry, 1974: 230).

The very emergence of planning as a separate profession in the UK was indeed reliant on its founders being able to demonstrate that there was a set of knowledge and skills which would be developed by "planners" which was not already possessed by "engineers", "architects" or "surveyors" (see Cherry, 1974; Hawtree, 1981). That this supposition was not welcomed by these other, long established, professions and that entry to the Town Planning Institute was initially limited to those who had qualifications in these other professions, is symptomatic of some of the problems which the planning profession has faced as it has sought to carve out a niche of its own over the last 100 years. As we enter the 21st Century, questions are being raised once more about the extent to which planning can claim to be a "learned profession with external credibility" (Grant, 1999: 1) and statements have been made to the tune that "it is widely acknowledged by many graduates and employers that graduates of UK planning courses are ill-equipped for the tasks and challenges that face them in today's planning profession" (Price and Kinghorn, 1999). Such concerns cannot be ignored. It is clear that there is now a need for the future development of the planning profession and, by implication, the development of planning education, to be fully opened up for discussion to enable a constructive debate to take place amongst all those with a concern about planning's future.

This paper aims to make a contribution to this debate. It reports on the findings of focus groups held with recent graduates of planning education and senior practitioners in the planning field from both the public and private sector. These groups were conducted as part of a review of planning education at The University of Sheffield, which is reflecting upon the appropriate mechanisms for delivery of this education as well as the contents of its professionally-accredited courses. In seeking to undertake such a review and to open up planning education to a potentially radical set of changes, it is apparent that there have been few published works on this subject in the UK in recent years (Hague, 1996). After Cherry's comprehensive commentary on the emergence and development of the profession and the role of education within this up until the 1970s (Cherry, 1974), there were some who looked at such issues in the 1980s (Healey, 1985; Reade, 1987) but more recent work has focused on the future of the planning profession without explicit discussion of the role of planning education (Evans, 1995). Papers by Hague (1996), Grant (1999) and Kitchen (1999) start to touch upon some of the key issues which need to be considered but these have not found as much exposure in the planning field as might be hoped. Other studies have been specifically commissioned to look at aspects of planning education (see Town Planning Network, 1999) or they have touched upon the need for planning education to respond to perceived shifts in emphasis within a planner's role (see RTPI, 1995).

The most recent and valuable contribution to the debate about the contemporary and future development of planning education has come from Sandercock (1999) in her "meditation" on planning education. Responses to Sandercock's propositions were invited from Healey, Kunzmann and Mazza, thus providing useful insights into the ideas on planning education held by academics in the US, UK, Germany and Italy respectively. In the US context, there have been studies undertaken which, whilst not directly transferable to the UK context, can be drawn upon in the debate which needs to be held in this country about the future of planning education. In turn, as these issues are being discussed in the US and European contexts, it has to be hoped that these debates might be informed by the UK experiences and insights such as those offered in this paper.

In order to provide some background to the issues discussed in the focus groups, the first section of this paper sets out a very brief review of planning education in the UK, culminating in an overview of the contemporary situation. This summarises how the provision of education is currently regulated by the Royal Town Planning Institute (RTPI) and what issues have been raised by some of the authors specified above. The second section of the paper then goes on to set out details of the focus group discussions which were held and the themes which emerged from these sessions. The final part of the paper seeks to draw some conclusions from this research, whilst stressing that at this stage it is difficult to see a way through the tensions and contradictions which are increasingly apparent in any arena where the futures of planning practice and planning education are being debated.

Review of Planning Education 1909-1999: Where have We Come from?

Historical Review of Planning Education

With the establishment of a Chair in Town Planning and Civic Design in 1909 at The University of Liverpool came the recognition that town planning was a "distinct and separate study in itself" (University prospectus, quoted in Cherry, 1974: 54). As Cherry makes clear, however, when the subject of town planning emerged, it was viewed very much as a "supplementary" one which needed to be added to the education of architects and engineers to enable them to be professionals who were able to be inter-disciplinary and who could co-ordinate the skills and knowledge held by the different parent professions of planning (Cherry, 1974). When the Town Planning Institute (TPI) was formally founded in 1914, the need to promote the teaching of town planning subjects and the desire to create mechanisms by which candidates' qualifications for entry to the profession could be tested were set out in its charter. Thus, it was clear that from the early days of its existence, "education [lay] at the heart of planning, as of all other professions" (Grant, 1999: 3).

The preparation of an examination syllabus in 1916 saw an early move to make the TPI a qualifying body and the first exams for entry were offered in 1920, although these exams could only be taken by those who already held qualifications in architecture, engineering and surveying. A shift in the education of planners came in the 1930s when a handful of universities were "recognised" as providing courses which could exempt students from the final examinations of the TPI. In the 1940s there was then a significant further change when a new five year course at Newcastle which did not require students to have a previous qualification in one of the parent professions was recognised (Cherry, 1974).

With the establishment of a nation-wide, comprehensive planning system in 1947 came an increased demand for qualified town planners. At this time, the Schuster Report favoured a multi-disciplinary approach to the education of planners and was thereby critical of the way in which the TPI was restricting membership to those who had come from the traditional parent professions (Schuster, 1950; Healey, 1985; Grant, 1999). The TPI was encouraged to broaden its membership but Healey notes that "planning education remained largely dominated by architectural notions of town design and management" (Healey, 1985: 497).

The expansion of planning as a state activity in the 1960s led to a further increase in the demand for qualified planners and many new planning schools were recognised. The TPI was still seen to be resisting, however, any sense that the education of planners needed to change to reflect the changing nature of the planning activity in practice, hence Cherry's remark quoted at the start of this paper. A planner's work in practice was seen to be increasingly linked to social and economic processes whilst the syllabus for planning schools set by the TPI remained focused on physical design, plan-making and implementation (Healey, 1985; Hague, 1996). This emphasis continued through into the 1970 syllabus, which was seen to reinforce a rigid and limited approach to the education of planners. This led Healey to state that "*[d]espite* the Institute, the quality of planning education improved greatly during the 1970s" (Healey, 1985: 500, author's emphasis).

The 1970s and early 1980s did then see some significant changes take place in the mode of planning education which was to be supported by the now Royal Town Planning Institute. Not without a struggle, social sciences were allowed into the planning curriculum (Healey, 1985) and more flexibility over the way in which "core" and more "specialised studies" were to be covered in planning education was permitted (see Hague, 1996, for details). Through to the mid-1980s, however, there was much internal conflict within the Institute as battles were fought over the extent to which the quality of planning education should be determined through the type of student *output* which was produced (for example, should they still be concerned with what physical plans they could produce at different scales) or whether more emphasis should be placed on measuring the *inputs* (for example, what was covered on the syllabus, how did this reflect a broader understanding of a planner's work, what were students expected to read, how were they assessed and so on). Hague's account of this period demonstrates how more emphasis came to be placed on inputs by the mid-1980s and how the RTPI were viewed as more progressive in their approach. A more flexible set of guidelines did also necessitate a shift to a more hands-on approach to accreditation through the "visiting boards" as the Institute's final examinations were phased out during this period.

The Contemporary System: Process

The educational guidelines set down by the RTPI were revised once more in the early 1990s, with slightly more emphasis placed on outputs than was the case immediately before this (Hague, 1996). Figure 1 sets out a summary of the guidelines as they currently stand (see RTPI 1996a and b for details). The extent to which a planning school is meeting these guidelines, and can therefore be given accreditation, is now established by the RTPI every five years through visits of boards, made up of officers of the Institute, practitioners and academics. There are currently twenty-three planning schools in the UK which offer accredited courses. Guidelines for this *process* are also set out by the RTPI (see RTPI, 1996c).

The Contemporary System: Issues Raised

Hague is positive about the way in which the RTPI currently polices the quality of entrants into the planning profession through this system. The style of planning education which is now provided is seen to be ideal in promoting "flexible specialisation", equipping students with the skills to take on range of tasks whilst having some specialised knowledge (Hague, 1996). It is indeed difficult to be critical of the RTPI guidelines in identifying the different components to be covered in a planning education and in suggesting that some study in depth is necessary to develop the knowledge, skills and values/attitudes of students further as they progress through their studies. In addition, the recognition that the overall quality of the planning school will impact upon the education received by the students of a course is a helpful one.

There have, however, been some criticisms raised about the manner in which the accreditation process is carried out (Grant, 1999). There have also been calls for the level of specialisation within planning courses to be increased (something supported by Hague in 1994) but the RTPI has never really been clear about how it wishes to resolve the "specialised study" debate and this has been the source of much disagreement during the

Core Knowledge Components

Knowledge about:

- Nature, purpose & method of planning
- Environment & development
- Political & institutional context of planning practice
- Key attribute being an ability to make relationships across these areas

Specialised Study

In order to ensure knowledge, skills and value awareness are developed in adequate depth, the Institute also expects students to pursue the development in depth of specialised areas of study

Core Skills Components

Competence (skills) in:

- Problem definition
- Research skills & data collection
- Quantitative & qualitative analysis
- Aesthetic dimensions & design awareness
- Strategic & synoptic dimensions & implications
- Synthesis & application of knowledge to practice
- Collaborative problem-solving
- Written, oral & graphic communication
- Information technology

Core Values/Attitudes Components

Cultivating an appreciation of & a respect for:

- issues of equal opportunity, social justice, economic welfare & efficiency in the use of resources
- role of government & public participation in a democratic society & in the balancing of individual & collective rights & interests
- diversity of cultures, views & ideologies
- conservation of natural resources & of the significant social & cultural heritages embedded in the built environment
- ethics of professional practice & behaviour

Qualities of an Effective Planning School

- organisation: management, monitoring, consultation, staff development
- acknowledged academic & research strengths in defined fields of planning
- links between the school and planning practice
- professional involvement
- resources for academic & support staff, library & IT & accommodation

Figure 4.1: RTPI guidelines for professional education programmes

accreditation process. Another source of criticism has come from those who suggest that the current system of planning education is not equipping students with the training required to "do the job" (Price and Kinghorn, 1999). Grant helpfully dismisses these jibes with the following argument:

> ...practitioners misunderstand both the function of universities and the character of their own profession if they assume that new graduates should be equipped with all the skills to be able to walk straight in to day-to-day practice. It is the duty of universities to educate their students, not to produce fully-trained planners, and not to provide free training for the professions. It is their primary duty to enhance the intellectual and reflective capacity of their students, and to develop their analytical and critical skills and to develop their capacity for further development (Grant, 1999: 7).

This is a view which is supported by the RTPI in their guidelines (RTPI, 1996a: 1-2) and is reiterated by Kitchen when he states that it is "very difficult to speculate on the future of planning education without thinking about the future of planning practice ...but this does mean that we are in sausage machine territory. It is not the job of the planning schools simply to turn a handle and produce graduates that the local government planning machine (primarily) then proceeds to consume" (Kitchen, 1999: 1).

From this all too brief review of the development of planning education, it would appear that the issues which have come to the fore throughout its history, and which continue to be apparent in the contemporary system, revolve around three main questions, namely:

- What is the nature of the planner's task for which professional education is required?
- How should planning education respond to these changing demands?
- How should this provision of a planning education be regulated?

Some contemporary writings (drawing mainly on experiences of the United States) have started to suggest how we might address the first two of these questions, but there has been little debate in recent years about how we might answer the third. The following section of this paper briefly summarises the thoughts of those who have looked at the relationship between planning practice and education most recently. From the findings of the focus groups which are then set out, it is clear that these debates need to continue and that, in addition, the question of how planning education should be regulated does now need to be raised once more.

Contemporary Reflections on Planning Education

Writing in 1994, Friedmann and Kuester state that "planners are torn between reaching out further and further into the universe of policy and planning studies ... or else defining their professional turf rather more narrowly in the expectation of greater professional security" (p. 61). As suggested above, such a dilemma lies at the heart of the question about what a planning education should provide, a question which these authors note has been "asked forever" but which does now need to be "re-examined" (p. 61). In carrying out their own re-examination of this issue, Ozawa and Seltzer suggest that "the planning curriculum needs to clearly address the communicative role that planners play in the communities for whom they work" (1999: 265). They go on to say that "we must seek a balance throughout the curriculum between our university-based desire to define the discipline as a discrete, defensible intellectual territory, and the facilitative, cross-cutting role actually played by planners in the field" (p. 265).

Within the context of such challenges, Sandercock (1999) tries to set out how a planning education might be fashioned for the twenty first century which will equip planners to work in many different realms. Crucially, she first establishes what can be identified as the "domain of planning" through highlighting six "macro-social processes", namely, urbanization processes; regional and interregional economic growth and change processes; city-building processes; cultural differentiation and change; the transformation of nature, urban politics and empowerment (p. 536-537). From this, she goes on to state that many professional accreditation bodies are too limited in their approach to education as they do not place enough emphasis upon the "kinds of qualities they would like to see in graduates" (p. 538). This leads her to identify five "literacies" with which future planners should be "armed", that is, technical, analytical, multicultural, ecological and design literacies. Through such an emphasis, Sandercock notes that planners will be appropriately "TAMED". She ends her valuable contribution to the planning education debate by stressing the need for planning to be viewed as an "ethical inquiry": "The goal of education is not how to stuff the most facts, techniques, methods and information into students' minds, but how to raise the most basic question of values" (Sandercock, 1999: 542-543).

In response to Sandercock's suggestions, Healey feels that we should be wary of generalising about experiences of planning practice and education across continents and states that "Sandercock's programme for a planning education curriculum is not so different from the programmes promoted by current policy statements in Europe and being taught in many planning schools" (Healey, 1999: 548). She does conclude, however, that agreeing on the headings used to delineate what is covered by a planning course is not the same as its "design and philosophy" and it is here that Healey feels Sandercock's challenge should be taken up:

> It is how we teach as much as what we teach that needs our creative attention if we are to show, by example and the design of educational experiences, what it means to offer practical advice about place-making concerns in a dynamic, multicultural context, where knowledge is continually contested and oppressions reinforced and challenged in the ongoing flow of urban politics and governance (Healey, 1999: 548).

The challenges are additionally reinforced in Kunzmann's response to Sandercock's paper, when he states that "planning educators [are] the most dangerous enemies of forward looking planning education" (1999: 549) and he urges schools to have the confidence to establish educational programmes which strive for "better local and regional life spaces in a globalizing world" (p. 549).

In the UK context, Kitchen set out his thoughts on the future of planning education in a plenary paper for The Planning Research Conference (1999). Here, he stresses the need for planning courses not to be tempted into cramming more and more material into their curriculum but rather to concentrate on what he terms "key skills", namely: technical; customer; personal; place; system and process; organisational, managerial and political context; and synoptic and integrative skills. These are not so far from the TAMED approach set out by Sandercock. Kitchen also emphasises the need "to take a radical look at what happens to people if they wish to stay in the world of planning after obtaining an initial planning qualification" (p. 4). This point raises once more the issue of the relationship between education and further training and demonstrates the many dimensions which there are to the debate about the future of planning education.

It is within the context of such challenges that the review of planning courses at The University of Sheffield is being undertaken and it was a desire to input graduates' experiences and those of local, senior practitioners which led to the use of focus groups as part of this review process. The remainder of this paper sets out the findings from these groups in order to add fuel to the debates which have been lit by some contemporary writers and which need to be continued beyond the scope of this paper.

Focus Group Discussions: Where are We Now?

Focus Group Methodology

The use of focus groups as a method of generating and collating information on a whole range of topics has become more fashionable in recent years, most especially due to the use of such methods by political parties and other interest groups. As a market research technique, focus groups are not new and as a tool for conducting qualitative research it is evident that they offer a number of advantages over more traditional methods of data collection. Practically, they are relatively easy to conduct and can be organised fairly quickly and cheaply. Procedurally, the focus group arena allows topics to be explored easily and for new ideas to come to the surface which might not otherwise emerge. As with any research technique, there are potential pitfalls with the use of focus groups which should be highlighted. The groups can be seen as a false social setting which could pressurise participants into responding in ways in which they would not in a one to one interview. Equally, some participants might not wish to raise controversial subjects whilst others might be easily inhibited by more dominant individuals. Bearing these potential problems in mind, the success of focus group research is largely dependent upon the ability of those co-ordinating the groups to understand the importance of maintaining the right level of moderator involvement and to manage the discussions appropriately.

The most important aspect of focus group research is, indeed, the emphasis which is placed on the value of interaction *between* group members and the largely "hands-off" role of the moderator, which enables the group to create its own agenda to a large extent and which ensures that the researcher is not dictating the topics to be covered or the manner in which they are explored. According to Kitzinger (1994), these interactions can be *complementary*, such that participants draw on common experiences and establish group norms, and *argumentative*, as it is unlikely that all members of the group will agree on all issues. Through both these types of interaction, the ways in which the similarities and differences of opinion held by group members are drawn out by the group, challenged and discussed is in itself likely to be insightful. In focus groups, the way in which the information is generated and agreements are arrived at, or disagreements are highlighted, is therefore adding value to the information. In this vein, Morgan states that "focus groups are useful when it comes to investigating *what* participants think, but excel at uncovering why participants think as they do" (Morgan, 1988).

Decisions about the number of focus groups to be used in a research project, the number of participants and the make up of the groups are to be determined in accordance with the research questions being addressed (Morgan, 1988). In this instance, the overall aims of the research were to:

- understand how graduates reflected upon the planning courses they followed in the light of their working experiences and therefore what they now perceived to be of value in a planning education;
- understand how senior practitioners, who are "employers" of planning graduates, value a formal planning education;
- feed this information into the short, medium and longer term reviews of planning education at The University of Sheffield.

With both undergraduate (3+1 years full time giving the Diploma in Town and Regional Planning qualification) and postgraduate (2 years full time giving the MA in Town and Regional Planning qualification) courses on offer at The University of Sheffield, it was decided that three focus groups were required. Thus, one was drawn from Diploma graduates, one from MA graduates and one from local senior practitioners who had experience of employing graduates from The University of Sheffield or other planning schools offering professionally-accredited courses. Within all of the groups, members were drawn from the public and private sectors, with some experience of the voluntary sector also represented. Within the graduate groups, members were drawn from cohorts who had been working in planning practice for up to three years, in order to ensure that their reflections ranged from "new" to slightly "more established" practitioners, whilst also bearing in mind that their memories of the courses needed to be relatively fresh.

The level of moderator involvement in all groups was minimal. Two co-ordinators steered the discussion when it was felt that it had lost its way and asked specific questions when it was felt that points had not been expressed very clearly. In each case, a list of topics had been set out beforehand as a minimum requirement for the groups to cover (see figures

2 and 3). It is important to note, however, that this list was not fed to the participants beforehand so as not to limit the discussion in any way. In the practitioner's focus group, some of the themes to emerge from the graduates' groups were fed into the discussion on one or two occasions. These enabled the members to respond to these issues directly and added value to the information which was gathered. In addition, a time limit of two hours was set in order to make it clear to the participants what it was that they were committing themselves to from the start. All of the discussions were taped to enable detailed analysis of the findings to be undertaken.

How do they reflect upon the course in the light of their working experiences?

How has their perception of the course changed over time?

Have their public and private working experiences led to different perspectives on the course?

How do they see the practice world as changing and how might planning education respond?

What do they perceive to be employers' attitudes to education and training?

How do they reflect upon the role of the RTPI?

How do they feel that financial pressures impacted upon their educational experience?

Figure 4.2: Questions addressed in graduate focus groups

What would they like to see offered by a planning education?

To what extent does the reality, as they are aware of it, differ from this ideal?

To what extent have these ideals changed over time?

What role do they feel they have in providing training for graduates?

How do they see the practice world as changing and how might planning education respond?

What do they see as the role of the RTPI in delivering or managing planning education?

Would a more flexible mode of delivery of planning education be something which would be welcomed and supported (financially) by practice?

Figure 4.3: Questions addressed in practitioners' focus group

Themes to Emerge From the Discussions

Whilst the explicit aims of this research were related to the review of planning education being undertaken in The University of Sheffield, it was clear from the focus group discussions that there were many issues being raised which were relevant to the wider world of planning education. It is the aim of this paper to draw these out and to suggest that, although the views of those participating in these focus groups cannot be claimed to be representative of graduates or practitioners across the planning field, there were pertinent points raised which need to be debated in a wider arena. The findings from the focus groups presented here do therefore not set out detailed comments about specific aspects of the planning courses at The University of Sheffield but they are rather set out under three themes which emerged from the analysis of the discussions. These pick up on issues related to "knowledge and skills", "teaching and learning" and "education and training".

Within these themes, some almost contradictory patterns of ideas are apparent. Some of these tensions arose from differences of opinion held by the graduate groups and that of the senior practitioners. Other differences were apparent within the groups as individuals had quite different ideas about what they valued from a planning education. In addition, during the course of the discussions, it was evident that some individuals were not clear in their own minds what it was that they wished to see given priority or how it was that they felt planning education should develop in the future. With these comments in mind, the following sections of the paper draw out the main themes to emerge from the discussions, highlighting where the tensions suggested above were most keenly evident and where differences of opinion were starkly apparent.

Knowledge and skills When the graduate focus groups were asked what reflections they had on the courses in the light of their working experiences, they seem to launch into a long list of particular knowledge components which they had felt that they had needed to know about (in addition to those which they had covered on the courses) as soon as they had started working in professional practice. This was especially the case for those who had gone into development control work in local authorities where their perception had been that they had been expected to know "everything" on day one. There were thus suggestions that they needed to be taught about the detail of environmental health, highways, and engineering and construction techniques, including, *inter alia*, the design of drainage systems. As the discussion progressed amongst the group, however, there was a noticeable shift in the emphasis which they were placing on the appropriate level of detail which was required for such knowledge components. The overriding message to actually come from both of the graduate groups was thus that it was not necessarily the detailed subject matter of planning courses that should be of concern in a planning education, although they were evidently not suggesting that this was irrelevant. They were, rather, placing an emphasis on the education which was provided in a particular way of critical thinking and evaluation and they were reflecting that this had been, and was likely to be, of most value to them in their working lives.

When the practitioners were asked what they perceived to be of value in a planning education, their emphasis from the start was very much on the need for graduates to have a "portfolio of competencies and skills" to bring into the workplace. These were to be flexible and transferable skills in management, analysis, communication, self development, information technology, spatial awareness, information gathering and evaluation and policy formulation. Graduates were seen to bring fresh, new ideas into the workplace to compensate for those who had been in planning practice for a while who had a tendency to "show an absence of connected thought" and little "serious cerebral activity". There was discussion about the need for planners to be able to "put knowledge into action" and to have vision and enthusiasm. Planning education was seen to have a role in capturing and nurturing those who had the ability to think creatively and plan positively for a sustainable future.

During the discussions with both graduates and employers, there was a strong sense that the planning world was changing in a way which had not been experienced before. Within local authorities, there was a feeling that statutory land use planning was an activity which was increasingly being placed on the "sidelines" whilst other officers within local authorities (based mainly in chief executives' departments) were taking on work which had previously been in the planners' realm. Within this context, there was a feeling that planners had to regain a role for themselves, break out of the constraints which had been imposed by the statutory system and demonstrate how planners could be a part of proactive, creative and integrated solutions to complex and messy problems. There was a fear that development control was increasingly being viewed as a negative, limited, administrative exercise which was being deskilled, whilst there was agreement that it *should* be reinvigorated as a powerful agent of change. Senior practitioners also emphasised the need for planners to fight for long termism against the political and economic short termism which was increasingly dominating decision-making.

It was equally clear that the fragmentation and diversification of the traditional role of planners *within* local authorities was reflected in the increasingly complex map of institutions involved in the planning process across the public, private and voluntary sectors. In this context, it was not seen to be the role of planning education to gear students up to working in a particular environment but rather it was felt that the emphasis should be on raising awareness of this complexity and therefore of developing the students' appreciation of the pressures placed upon the planning process in practice and the implications of this for its future development.

More specific changes in practice, to which it was felt that planning education should respond, were also identified. These related, *inter alia*, to: the increasingly integrated nature of land use and transport planning; the new regional planning processes; the impact of Europe; the "IT revolution"; new approaches to local planning which had emerged, for example, from an emphasis on urban design and the findings of the Urban Task Force (1999); the introduction of "Best Value" as a new management tool (see Boyne, 1999, for explanation); the changing nature of local and community regeneration processes and programmes; the introduction of community plans; and the increasingly popular notion of planners as "people dealers".

Teaching and learning The need to acquire knowledge and skills and to develop an appreciation of different values and attitudes through a planning education raises questions about how students are taught and how individuals learn. It is not the intention here to go into details of educational techniques but it is increasingly important for providers of planning education to be tuned into the need for "student centred learning". The graduates recalled, reflected positively upon and realised that they drew experience from those times when they were active participants in the learning process rather than passive recipients of dry material. Workshops and projects in which they were required to apply knowledge and were challenged to think critically about a case or set of circumstances were therefore an important part of their education. These findings point to the use of case studies and "live" projects as an integral part of a planning course. The graduates also reflected, however, that the use of theoretical concepts to explore, examine and evaluate practice was an important aspect of their education. Senior practitioners also agreed that it was important not to lose the value of theoretical reflections on practice as it was through such processes that students were encouraged to see issues in different ways and ponder the implications of different solutions to a problem.

The use of case studies and contemporary projects as teaching and learning tools brings into sharp focus the need for planning schools to have close links with practice. In some circumstances, senior practitioners reflected that this has become increasingly problematic as local authorities find themselves pushed for resources and individual planning officers have to justify spending time and resources on students. Planning consultancies and voluntary organisations are, of course, not exempt from such pressures. Members of the focus groups hoped, however, that the reciprocal benefits that links between planning schools and planners in practice can bring are being recognised. The involvement of students in Planning Aid cases was cited as an important development in this regard as there were clear mutual benefits to be gained from this initiative (see Terrace, 1999; RTPI, 2000 for information).

Education and training Many of the issues raised above led into discussions in the focus groups about the way in which planning courses provided a broad "education" or a more specific "training" for the job (whatever that might now be). It has already been suggested in other parts of this paper that on first finishing their courses, graduates were concerned that they did not know enough practical details of the operation of all aspects of the planning process (and that this suggested to them that they needed more specific training whilst on the course), yet a year and more into a planning job, they were aware that they had benefited from a broader educational experience. They thus perceived that they had gained from receiving an *education* in planning as once in a job they were being *trained* for specific aspects of a planner's work through their everyday experiences and, wherever possible, through training courses.

Graduates were thus keen to stress the value of education followed by training. This view was also held by senior practitioners, who wanted to be able to employ well rounded, able individuals who had the ability to learn quickly, take ideas on board and operate effectively in many different roles within the messy world of planning practice. As both groups pointed out, and as suggested in previous sections, the role of the planner is in fact so ill defined now that it is difficult to claim that a planning course is providing training for a particular job. Instead, it should be exposing students to the complexity and intricacy of a planner's role, whether in the public, private or voluntary sectors, and equipping them with the knowledge, skills and attitudes upon which further training can build.

In this regard, the members of the focus groups, and the graduates in particular, were adamant that they did not wish to specialise any further in their studies than they had done through their final year projects and dissertations. They felt that graduates could not afford to be labelled as specialising in one particular field as they had to be able to demonstrate that they could apply themselves to a whole range of tasks and that it was only really after working for a period of time in one particular area that it was possible to suggest that specialist knowledge would develop. The concept of specialised study was thus seen to be helpful in enabling the students to pursue study in depth in areas which were of particular interest to them and it was not viewed as a means by which students should be channelled down one specified, identifiable path of study. This, in the graduates and practitioners' eyes, would take away the value of a planning education and there was agreement that it was the role of further training to enable that level of development to take place.

There was not, however, agreement over the mechanisms by which such training should be delivered. The graduate groups had had a much higher expectation of what training and staff development would take place in practice than had come to fruition in reality. This was partly a reflection of their knowledge about the RTPI's requirements for continuing professional development (CPD) and their awareness that they were obliged to complete a minimum of fifty hours CPD activity in any two year period as a condition of their membership of the Institute (see RTPI, 1997 for details). One private employer had had a proactive approach to training and development which employees had found very helpful whilst other private consultancies had not wanted to send employees on courses as this was reducing the potential for them to be bringing in fees from clients. Local authorities were generally unable to support much CPD activity and those budgets which did exist were seemingly managed on a rather *ad hoc* basis. Graduates also commented that the standard of courses run for CPD purposes by the RTPI was not very high and that they had turned to University-run courses for a higher quality experience.

Some members of the senior practitioners' focus group were, in contrast, quite adamant that employees should expect to be responsible for their own personal and professional development and that they should not expect to spend allocated office time undertaking CPD activities. Whilst local authorities might be expected to pay for such training, some managers were clearly stating that it was up to each individual to commit their own time to developing their knowledge and skills for their own career development. These views did not find support amongst other practitioners, however, who felt that unless local authorities and other employers were willing to invest in the future of the profession and in encouraging the most able planners to further their development, the benefits of schemes such as Best Value and Investors in People [new managerialist practices introduced in the UK in recent years with a view to improving public and private sector service provision] were, in fact, going to be non-existent.

These discussions in turn raised questions related to the role of the RTPI in requiring, providing and monitoring CPD activity and in helping to ensure the healthy development of the planning profession in the future. In this respect, both graduates and senior practitioners viewed the RTPI's approach to enforcing CPD as a negative rather than a positive force. This was seen to result from the process by which it was carried out as it did not, on the one hand, guarantee that the CPD events which were attended were of any real value to the recipients whilst, on the other, planners who did continue their professional development but who failed to fill in the correct form in the appropriate manner faced losing their membership of the professional body. The concept of CPD from the RTPI was therefore seen to be used as a threat rather than as a positive force to encourage professionals to continue to develop their practical and intellectual knowledge and skills. It was suggested that the RTPI could itself provide more in the way of lecture courses and could be more proactive through, for example, the mentoring schemes which have recently been piloted in some areas.

The RTPI itself was not viewed in a positive light by most members of the focus groups as it was not seen to promote the planning profession in a proactive way. There was a sense that it was not clear what the Institute "did" and graduates were disheartened that not all employers seemed to care whether or not they were members of the RTPI. The value of a planning education was seen to be undermined by such attitudes and there was also a concern that planners were not being granted the same "professional status" as other officers within local authorities; some graduates felt strongly that they were not viewed as being "on a par" with other, more "established" professions within their councils. This was a problem which it was felt that the RTPI itself should be aiming to rectify. In this vein, there were suggestions that the Institute needed to be reinvigorated by the next generation of "radicals" to secure its future.

Conclusions: Where do We Go from Here?

Having drawn out the main messages to come from the focus groups under three different themes, it is clear that there is much overlap between them. Many of the issues raised by the groups are not necessarily new, as demonstrated by the brief review of planning education in the UK provided in this paper. This is not to say, however, that they are unimportant and can afford to remain unexplored in the contemporary climate. Some of the points raised have been touched upon by others, as discussed earlier in the paper, and these findings give more impetus to these debates. Here, the main points which need to be taken forward from these focus groups are summarised. It was not the intention of this paper to set out a new planning syllabus and there are no apologies offered for not doing so. Rather, the aim is to demonstrate the importance of the debate which needs to be had and to offer a structure through which this could now take place.

With regard to the different knowledge, skills and values/attitudes components which were discussed, it was evident from the focus groups that there is a need for planning education to forge a healthy balance between providing students with relevant knowledge components and challenging them to analyse, evaluate, critique and apply this knowledge in creative ways to particular planning problems. Skills in these areas will be developed and tested if students have to think laterally, communicate effectively and demonstrate how it is that they have arrived at their conclusions. Planning education therefore needs to be driven not exclusively by the need for detailed knowledge components to be covered nor for particular skills to be taught but by a desire to develop a student's ability to challenge, to think critically and to respond creatively when applying knowledge in the planning arena. It is interesting in this regard to make brief reference to Bloom's taxonomy of educational objectives, which is still highly relevant to contemporary debates about educational goals (Bloom, 1956, see figure 4).

As can be seen from figure 4, the ability to recall knowledge can be classified as lying at the bottom of the hierarchy of educational goals according to Bloom, whereas analysis and synthesis lie higher up the hierarchy, with evaluation identified as the highest level of educational achievement. This is seen, in very simple terms, as "making judgements about value" which, it is clear, lies at the heart of a planner's role. The development and nurturing of this intellectual process has to therefore be a central concern of planning educators. Unless planning education raises

Evaluation *The highest level in the taxonomy & a combination of all the other categories* *Evaluation is concerned with making judgements about value*
Synthesis *A process of logical deduction which is intimately concerned with thinking & creativity*
Analysis *A process of reasoning & thinking, determining relationships between elements & the organizing principles behind these elements*
Application *Using something in a specific manner & understanding how to create new solutions to unresolved problems*
Comprehension *Understanding or perceiving a problem and being able to process information to aid this understanding through translation, interpretation and extrapolation*
Knowledge *Can be the elementary skill of recalling or remembering information or knowing how to deal with information through classifications and methodologies or a knowledge of theories and abstractions*

Source: Based on Bloom (1956).

Figure 4.4: Major classes in Bloom's taxonomy of educational objectives

students' awareness of their own values and attitudes as well as those of others, and is concerned about developing the evaluative ability (as Sandercock, 1999, suggests it should be) then it will be difficult for students to know what it is that they think they are striving for. In this regard, the current RTPI guidelines can be seen as helpful pointers to what should be covered in a planning education. It is crucial, however, that individual planning schools are left to make decisions about how best to follow these guidelines and to fulfil their purpose by ensuring that individual students receive an appropriate education in planning.

It was also clear from the discussions that the content of planning courses and the knowledge and skills components which are being covered within them needs to be adjusted to the changing circumstances of planning practice to ensure that planners are equipped to perform the tasks which they are being asked to play in practice (as suggested by Cherry, 1974; Kitchen, 1999). It is equally clear, however, that there is not agreement over what this task might be in many respects (as pointed out by some of the US writers) and the participants in the groups were concerned that simply to respond to some of the current trends in planning practice would therefore not always be a good way forward.

With regard to development control, for example, they were aware that it is increasingly being deskilled as a process and that some in the planning field are suggesting that this should be formalised by regarding development control as an administrative process which can be carried out by those without a formal planning education. This was exemplified by the council where one of the graduates worked advertising a planning post to be filled by an individual with five GCSEs and no other qualifications. This was not something which these graduates and senior practitioners supported. They feared that this was giving in to resource and management pressures within local authorities whilst ignoring the need for development control to be valued as a potentially positive and creative process. Those involved in providing planning education should therefore not simply give a knee jerk response to changing circumstances and assume that current trends are clear and irreversible. The future of planning practice, particularly in local authorities, is unclear and is still open to influence. Without a doubt, the role which planners can play needs to be reinvigorated in many ways. Planning courses should therefore encourage students to question what is happening in practice and provide them with the tools to challenge, criticise and respond creatively to these circumstances.

With regard to teaching methods and an understanding of learning processes, it is clear that planning education needs to reflect shifts in higher education more generally to provide student centred, active, learning. As discussed earlier in this paper in relation to "knowledge and skills", the emphasis in many disciplines is now on providing transferable skills within the context of a particular knowledge arena, particularly when the emphasis is on maximising the development of educational goals by moving up the hierarchy of Bloom's taxonomy (see, for example, Allen, 1993; Gibbs, Rust, Jenkins and Jacques, 1994). In a planning education, this means using links with practice in a positive way whilst employing planning theory as a set of tools to help students to develop their ideas and enhance their understanding of planning practices in the UK and elsewhere. The climate within which students learn is crucial to their intellectual development; they need to be encouraged to experiment with ideas, explore new avenues and expand their capabilities whilst undertaking a planning education. There is then some hope that these abilities will be further developed once they find themselves embroiled in the day to day activities of planning practice. In this vein, the continued education of planners is important (as highlighted by the RTPI 1996a; Kitchen, 1999; Grant, 1999) and the current, seemingly problematic nature of the RTPI CPD requirements needs to be tackled. Planners should want to continue to learn and should be given the support required to enable them to do so throughout their working lives. If planning is important in helping communities to learn (Ozawa and Seltzer, 1999) then planners themselves need to embrace the life long learning agenda (Kitchen, 1999).

On a more specific note, in reviewing the way in which courses are delivered in the future, it was evident from the focus group findings that it should not be assumed that part time routes are to be favoured over full time ones due to the financial pressures in which students find themselves. Practitioners stressed that the two routes, and those which operate through distance learning schemes, all offer quite different educational experiences. Part time students, for example, were more likely to be aware of the political and organisational complexities affecting the planning process when they gained their qualification whilst those graduates who had followed a full time course might be more naïve of these pressures when they first started work. On the other hand, practitioners were aware that the part time students were likely only to have experience of that one authority whilst full timers had been exposed to a whole variety of planning cases and issues and were able to draw upon the high quality of output they had achieved due to having much more time to spend on assignments. In addition, practitioners suggested that local authorities were finding it increasingly difficult to fund part time students and they were aware that different ways of funding employees through a planning education would have to be found. Not all part time students are, of course, working in a planning context whilst undertaking their qualification and this raises further questions about the pressures of juggling work and study. For these reasons, both graduates and employers supported the retention of the variety in modes of delivery, with the potential for there to be a combination of full and part time delivery in any one course seen as an attractive proposition so that students could "get the best of both worlds". More flexibility in these respects is therefore something which needs to be explored both by the universities and the RTPI.

It would appear from these focus groups that the need to clarify the roles of universities, employers and the Institute with regard to education and training provision for new and more established planners is indeed great. Whilst there continue to be false expectations about the role of the planning courses currently on offer in providing specific training and whilst employers and the Institute are not proactive in the training which they provide, relationships in the planning field between these different bodies will be uncertain. The debate about the extent to which a planning education is creating graduates with identifiable, specialised, areas of knowledge and expertise is also coming to a head. All the participants in these focus groups were adamant that they did not wish to specialise further and they were content not to be labelled as planners of a particular realm. This goes against the grain of the RTPI's (and others', such as Hague's (1994 and 1996)) wishes to label schools as having specific specialisms and makes it difficult for the accreditation process to be a smooth one. Perhaps what has happened is that too much emphasis is now being placed on the *inputs* rather than the *outputs* of a planning education and the RTPI needs to reflect upon how best it can monitor and regulate the quality of those entering the profession. In addition, perhaps more flexibility is needed in the application of the guidelines to enable schools to develop their own niches in an increasingly complex planning world.

If this in turn raises questions about the continued identity of a coherent planning profession, then that is symbolic of the problem which lies at the heart of this debate. As suggested earlier on in this paper, in order to establish the future shape of planning education, we need first to establish what role it is that "planners" and "planning" are playing, and *can* play, in practice. Only when we know what we would like that role to be can we move forward. In turn, in order for that role to be forged in a positive way, those involved with planning education need to engage with practice, reflect upon experience and ensure that the RTPI wakes up to the challenges being placed at its door. The need for the issues raised here to be addressed has been discussed by, for example, Grant (1999) and Upton (1999) and they now need to be examined further so that their implications for the future of planning practice, the profession and education can be fully taken on board. The findings from these focus groups suggest that confusion, concern and uncertainty are rife in much of planning practice across the UK. In these circumstances, it is the role of planning academics and educators to stand back from these pressures, to research their sources and to then help to carve out a positive future.

Above all, we need to address the following questions:

- What is the nature of the planner's task for which professional education is required?
- How should planning education respond to these changing demands?

- How should this provision of a planning education be regulated and delivered?
- How does education and training continue in the work place and how should this be delivered?
- What is the role of the universities, practice and the RTPI in shaping these agendas?
- How do we all help to shape the future of the profession through planning education and what do we wish that future to be?

These questions are central to the future development of the planning profession not just in the UK context, which has been the explicit focus of this paper, but further afield in the US and European contexts. The answers to these questions are not yet clear but if this paper has contributed in some way to framing the nature of the debate which does now need to take place and if it has challenged some to consider the issues involved in a different light, it will have fulfilled its goal.

Acknowledgements

I am grateful to my colleagues Robert Marshall, Heather Campbell and Philip Booth for their help in moderating the focus groups and in offering their thoughts on the issues raised by the findings. Thanks also go to those who gave of their time to participate in the groups. This chapter appeared as "Shaping the planning profession of the future: the role of planning education", in *Environment and Planning B: Planning and Design*, 2001, volume 28 number 4 (forthcoming), by Pion Limited, London.

References

Allen, M. (1993), *A conceptual model of transferable personal skills*, Personal Skills Unit Booklet, Department of Employment.

Baum, H. (1997), 'Social Science, Social Work, and Surgery: Teaching What Students Need to Practice Planning', *Journal of the American Planning Association*, vol. 63, no. 2, pp. 179 – 188.

Bloom, B. (1956), *Taxonomy of Educational Objectives – The Classification of Educational Goals*, David McKay Company, New York.

Boyne, G. (1999), 'Managing Local Services: From CCT to Best Value', *Local Government Studies*, vol. 25, no. 2.

Cherry, G. (1974), *The Evolution of British Town Planning*, Leonard Hill, Leighton Buzzard.

Evans, B. (1993), 'Why we no longer need a planning profession', *Planning Practice and Research*, vol. 8, no. 1, pp. 9-15.

Friedmann, J. and Kuester, C. (1994), 'Planning Education for the Late Twentieth Century: An Initial Inquiry', *Journal of Planning Education and Research*, vol. 14, no. 1, pp. 55 – 64.

Gibbs, G., Rust, C., Jenkins, A. and Jacques, D. (1994), *Developing Students' Transferable Skills*, Oxford Centre for Staff Development, Oxford.

Grant, M. (1999), *Planning as a Learned Profession*, paper prepared for the Royal Town Planning Institute, http://www.rtpi.org.uk/discuss.

Hague, C. (1994), 'What is a planner?', *Planning Week*, vol. 2, no. 41, pp. 20-21.

Hague, C. (1996), *Transforming Planning: Transforming the Planners*, paper presented at 50th Anniversary Conference, University of Newcastle, 25th-27th October 1996, unpublished.

Hawtree, M. (1981), 'The emergence of the town planning profession', in Sutcliffe, A. (ed.), *British Town Planning: the formative years*, Leicester University Press, Leicester, pp. 64-104.

Healey, P. (1985), 'The Professionalisation of planning in Britain: its form and consequences', *Town Planning Review*, vol. 56, no. 4, pp. 492-507.

Healey, P. (1999), 'Sandercock: "Expanding the Language of Planning"', *European Planning Studies*, vol. 7, no. 5, pp. 545-548.

Kitchen, T. (1999), *The Future of Planning Education – 2020 Vision*, paper presented at The Planning Research Conference, University of Sheffield, March 29th-31st 1999, unpublished.

Kitzinger, J. (1994), 'The Methodology of Focus Groups: the importance of interaction between research participants', *Sociology of Health and Illness*, vol. 16, no. 1, pp. 103-121.

Kunzmann, K. (1999), 'Planning Education in a Globalized World', *European Planning Studies*, vol. 7, no. 5, pp. 549-555.

Mazza, L. (1999), 'The Specific Domains of Planning', *European Planning Studies*, vol. 7, no. 5, pp. 557-561.

Morgan, D. (1988), *Focus Groups as Qualitative Research*, Sage Publications, London.

Ozawa, P. and Seltzer, E. (1999), 'Taking Our Bearings: Mapping a Relationship Among Planning Practice, Theory and Education', *Journal of Planning Education and Research*, vol. 18, pp. 257-266.

Price, M. and Kinghorn, J. (1999), 'Useful Classes', *Planning*, 1342, p. 14.

Reade, E. (1987), *British Town and Country Planning*, Open University Press, Milton Keynes.

Royal Town Planning Institute (RTPI) (1995), *Planners as Managers: Shifting the Gaze*, RTPI, London.

Royal Town Planning Institute (RTPI) (1996a), *Policy Statement and General Guidance for Academic Institutions offering Initial Professional Education in Planning*, RTPI, London.

Royal Town Planning Institute (RTPI) (1996b), *Guidance Note B, Initial Professional Education Programmes in Planning: Content and Performance Criteria*, RTPI, London.

Royal Town Planning Institute (RTPI) (1996c), *Guidance Note A: Accreditation Process*, RTPI, London.

Royal Town Planning Institute (RTPI) (1997), *Continuing Professional Development*, Professional Conduct Advice Note 1, RTPI, London.

Royal Town Planning Institute (RTPI) (2000), *Planning Aid*, http://www.rtpi.org.uk/

Sandercock, L. (1999), 'Expanding the Language of Planning: A Meditation on Planning Education for the Twenty-First Century' *European Planning Studies*, vol. 7, no. 5, pp. 533-544.

Schuster, G. (1950), *Report of the Committee on Qualifications of Planners*, Cmd 8059, HMSO, London.

Terrace, S. (1999), 'Planning Aid and Education', *Planning*, 1331, p. 20.

Town Planning Network (1999), *Creativity in Town Planning*, University of Westminster Press, London.

Upton, R. (1999), *Future of the Planning Profession*, paper presented at The Planning Research Conference, The University of Sheffield, March 29th - 31st 1999, unpublished.

Urban Task Force (1999), *Towards an Urban Renaissance*, Department of Environment, Transport and the Regions, London.

5 Developing Indicators to Evaluate the Effectiveness of Land Use Planning

NICOLA MORRISON

Introduction

Over the last decade, the pressure to evaluate the performance and outcomes of the UK land use planning system has grown. These pressures have come from a number of directions. Developers have increasingly argued that the planning system imposes considerable burden upon them. A number of studies have given support to these claims, focusing on the costs imposed by the planning system, particularly in terms of increased land, housing and infrastructure costs (see e.g. Bramley et al 1995; Evans 1996; Monk and Whitehead 1999). At the same time, the Treasury has become increasingly concerned about the public expenditure implications of administering the planning system in terms of both its direct and indirect costs (HM Treasury 1994). Whilst these concerns have focused on the costs of intervention, it is equally important to evaluate the nature and scale of the benefits of having a planning system in place. Attention also needs to be given to assessing the effectiveness of the land use planning system in realising its nationally stated aims and objectives (see Pieda 1992; Morrison and Pearce 2000).

The purpose of this chapter is to focus on the difficulties in developing a methodology to assess whether the land use planning system has achieved its objectives. The chapter draws on research undertaken by the Department of Land Economy, University of Cambridge for the Department of Environment Transport and the Regions (DETR). The aim of the research was to develop a methodology that would produce an indicators package to evaluate the extent to which the land use planning system is effective in realising its objectives. The emphasis was on developing a conceptual approach and establishing the general principles that should be used to develop indicators of land use planning. Practical implementation issues with respect to obtaining and interpreting the indicators' data were a subsequent stage of research.

The following section provides some background on previous studies that have attempted to evaluate the land use planning system and the use of indicators in policy evaluation. It then outlines the general principles adopted in the research for producing indicators to evaluate the effectiveness of the planning system. The potential challenges facing policy makers are highlighted.

Evaluating the Effectiveness of Land Use Planning and the Use of Indicators

Few studies have attempted to undertake a comprehensive assessment of the planning system's performance. Until Pieda's 1992 evaluation study of the planning system, Hall's work in the early 1970s was perhaps the only example of a thematic approach to assessing the effectiveness of land use planning undertaken on a comprehensive basis (Hall 1973). In contrast, there have been a number of studies concerned with the effectiveness of particular aspects of land use planning policies and policy instruments. These include the effectiveness of Green Belts (Elson et al 1993); transport (ECOTEC 1993) and the viability and vitality of town centres (URBED 1994). Projects have also been conducted to evaluate the effectiveness of particular instruments and programmes such as the Derelict Land Grant and Urban Development Grant (DoE, 1991, Arup, 1995) and the Enterprise Zone initiative (PACEC, 1995).

There are lessons to be drawn from these previous evaluation studies. In particular, they all experienced problems in measuring policy effectiveness. Most had difficulties in distinguishing the contributions of land use planning from other influences. Only a few addressed the issue of additionality, which involves distinguishing between the actual effects of the relevant policy and the effects caused by other factors: for example, the state of the economy, the state of the property market and the development industry. The studies that have examined additionality have mainly focused on a particular policy instrument, for example Derelict Land Grants, where there is a clear and recognised "policy-on" and "policy-off" situation. Isolating the effects of land use planning policy is particularly problematic. The main difficulty lies in establishing a surrogate for the policy-off position, as the UK planning system has been imprinting on the land development process across the country since 1947.

Another feature of these previous studies is that they have attempted full post implementation evaluations of the particular policy or mechanism, whereas the purpose of the DETR research was different. Its task was to move towards a more indicative approach that could be applied across the whole system, the results of which could become part of a process by which evaluation can inform decision and implementation on a consistent basis. Developing indicators as part of the policy evaluation process was seen as a way to assess whether policy objectives have been achieved (H M Treasury 1994).

A great deal of attention has been given to developing indicators for policy purposes (see OECD 1994; DoE 1996; UNCHS 1997). This is because indicators provide a useful way of organising, simplifying and improving data that are available to help measure and track the status and progress of usually large and complicated systems, such as the land use planning system. They can provide information on the direction of change i.e. whether a situation such as land use patterns are changing or staying the same. When undertaking an evaluation exercise, it is important to distinguish between monitoring indicators (or back-cloth indicators, yardsticks or trend indicators as they are also commonly known) and effectiveness (or performance) indicators.

Monitoring information alone cannot indicate whether ultimate policy objectives have been achieved. In this respect, developing monitoring indicators is just the first step in an evaluation process. In comparison, an evaluation entails checking how far objectives have been achieved. It requires clarifying the range of objectives to be assessed; establishing a clear statement of a baseline so that any change that is witnessed can be compared alongside this baseline or benchmark. It also requires an understanding of all the different influences involved in achieving that change, so that the performance or "value-added" of particular organisations, such as the planning system, can be assessed. Through these means, effectiveness indicators can start to be established. The next section outlines the approach taken in the research to develop effectiveness indicators in relation to the land use planning system.

The Approach Adopted to Develop Indicators for Land Use Planning

The approach adopted in this research has a number of stages (see Table 5.1):

- clarification and specification of the objectives of national land use planning policies;
- identification of planning outputs; and
- identification of outcomes and land use planning impacts.

The methodology introduces a number of key terms. These are summarised in a glossary in Table 5.2.

Table 5.1: A framework for developing indicators for land use planning

Stage 1: Objectives	• type of policy objective • nature of policy objective • pursuit of policy objective • desired standard/level of policy objective
Stage 2: Outputs	• intermediate outputs (e.g. PPGs, development plans) • final outputs (development control decisions)
Stage 3: Outcomes and impacts	• intermediate outcomes ("noise" plus planning policy) • intermediate impacts (attributable to planning policy) • Final outcomes ("noise" plus planning policy) • Final impacts (attributable to planning policy)
Stage 4: The indicators package	• key indicators: prioritisation and selection • supplementary indicators • issues of practicability

Stage One: Clarification and Specification of Objectives

The number of national land use objectives that have been set down for the land use planning system is very large. Every Planning Policy Guidance, Mineral and Regional Planning Guidance note (PPGs, MPGs, RPGs) and other Ministerial statements, for example, contains land use objectives. It is therefore necessary at the outset to make a selection of key policy areas and objectives on which to concentrate. For the purposes of this research, the DETR defined a series of policy topic areas and individual objectives for which indicators were to be developed. These included Green Belts, Housing, Industrial and Commercial Development, Town Centres and Retail Development, Countryside and the Rural Economy and Transport.

Table 5.2: A glossary of terms

Indicator terms	Working definition
Planning inputs:	The human capital and financial resources used to formulate and execute national land use and development policies. Includes staff at central and local government, finance for local planning departments and land use planning divisions of the DETR.
Planning instruments/ activities:	The procedures and work tasks within the land use planning system that produce the system's outputs.
Planning objective:	A statement of what is intended for a policy object and the way that intention is to be pursued. The intention may relate to inputs, outputs, or outcomes.
Target:	An objective that seeks a desired end state for the policy object, usually though not necessarily within a specified timescale.
Intermediate outputs:	Products or services retained within the land use planning system (e.g. Development Plans).
Final outputs:	Products or services of the land use planning system that impinge on (are consumed by) those that develop land (normally applicants for planning permission). These are usually planning (e.g. development control) decisions but may also include policy documents (e.g. Development Plans, PPGs) and advice services.
Policy object:	The object of the policy objective. In land use planning these usually relate to some qualitative or quantitative characteristic or attribute of the physical environment, the use or development of land.
Intermediate outcomes:	The combined effects of the planning system and all other influences on the pattern of land use and development.
Intermediate impacts:	The effects on the pattern of land use and development solely attributable to (changes in) the planning system.
Final outcomes:	The knock-on effects of the pattern of land use and development on the socio-economic environment.
Final impacts:	The effects on the socio-economic environment solely attributable to (changes in) the planning system.

As a first step in the research, a network of objectives needs to be produced for each of the selected policy areas. The network distinguishes between key and supporting objectives. In general terms, this consists of normative statements that are increasingly more detailed and less abstract the lower down the network.

Statements of policy that contain policy objectives are often not in a form that makes it easy to isolate the key information needed to develop indicators. They are often too general for this purpose. If the wording in policy statements is found to be unclear, is imprecise or gives insufficient detail, then it is difficult to establish what effects are being aspired to and thus to gauge how any evaluation can be made. In general terms, there is agreement in the literature that, in order to produce adequate indicators for evaluating effectiveness, objectives need to be expressed so that they:

- are achievement orientated;
- give a clear declaration of what is intended/what achievement is planned;
- contain some explicit or implicit statement about the means or instruments to be used to achieve what is intended; and,
- make it possible to tell whether or not (or the degree to which) the objective has been achieved (by the appropriate means).

For each policy area, statements of land use planning policy were examined in detail by the research team in order to define clearer statements of objectives that capture the essence of what is being sought, at the level of detail needed to produce meaningful indicators. In some cases, explicit assumptions had to be made to improve clarity and specification, whilst still reflecting the objectives' original intentions. This exercise formed Stage One of the methodology and was a critical starting point to the research, as without such clarification and specification of policy objectives an evaluation of achievement cannot take place.

Stage Two: Identification of Planning Outputs

In order to evaluate whether a policy objective has been achieved, it is important to consider the mechanisms within the planning system which are used to bring about the objective's delivery, i.e. the planning outputs, and to establish whether there are likely to be any data available to measure these. This exercise formed Stage Two of the methodology.

According to the HM Treasury report on Policy Evaluation (1994), outputs are "the things or conditions produced by the organisation". Another report similarly suggests that outputs are "the products or services delivered by a particular policy generated either by an individual or an organisation". Outputs should in turn "relate directly to policy objectives...be clearly quantifiable... and ... contribute to *(the intermediate and)* final outcome*(s)* of a policy" (Jackson and Palmer, 1989: 50). An analysis of outputs is therefore a critical intermediate stage in the evaluation procedure.

In land use planning terms, outputs are those products or services from the system which directly impinge on those people or organisations who wish to use or develop land (i.e. normally the applicants for planning permission) or who are directly influenced by land development (e.g. local authorities, land users). They include, for example, local authority Development Plans, which developers may take into account when formulating their development proposals. The most important product delivered by the planning system, however, is the decision on an application for planning permission i.e. the development control decision. These decisions provide quantified or quantifiable evidence of what the planning system has contributed to the achievement of specific objectives.

Development control decisions can be categorised into refusals and approvals; by type, scale, location and the timing of development. This "output" information is likely to prove useful when developing policy effectiveness indicators because, in many instances, it helps provide a way of identifying the "counterfactual" - the situation that would result in the absence of the policy. Information on the kind of development that is *refused* planning permission is likely to be particularly useful in certain circumstances, possibly offering an indication of the kinds of development that would be built were it not for the planning system being in place. This would often enable comparison between what actually materialises on the ground, i.e. what is developed (total outcomes), and what would have occurred if land use planning had not specifically intervened (the counterfactual).

However, examining the characteristics of refused and altered planning applications does not necessarily provide a perfect counterfactual as there are many reasons why refused planning applications are not typical of what would otherwise happen without the planning system in place. The very existence of the land use planning system is likely to mean, in many instances, that developers and other applicants for planning permission do not even try to develop outside the policy framework that has been laid down. Developers are likely in many instances, to have amended the project proposals they put forward for planning permission (in comparison to their original, desired, development projects) in order to improve their chances of securing a consent. But in some circumstances, comparing refusal data (planning outputs) with what has happened on the ground (planning outcomes), along with other supplementary information, may well provide the best package of available indicators to address the effectiveness of policy that is feasible.

Examples of policy objectives and related output indicators include:

- policy objective: to make full use of land within urban areas for new housing development,
- planning outputs: the number (and %) of refused (and permitted) planning applications for new housing development on greenfield sites (by type and scale),
- policy objective: to focus retail development in town centres,
- planning outputs: the number (and %) of refused planning (and permitted) applications for out-of-town shopping developments (by type and scale).

Stage Three: Identification of Outcomes and Impacts

Outcomes can be distinguished between intermediate and final outcomes. Intermediate outcomes relate to the land use or development consequences of all influences on the pattern of land use and development. In principle, they can be measured in terms of the number, type, scale, location and timing of development. Final outcomes, on the other hand, are the consequence of land use change or development on the physical and socio-economic environment.

Almost invariably, an evaluation of the land use planning system focuses on intermediate outcomes (such as number of houses developed, areas of land reclaimed etc) because these are more tangible and hence usually measurable and relate more directly to the outputs of the planning system - see Stage Two. Evaluation systems generally fail to give sufficient attention to final (or "higher level") outcomes (for example, sustainable development, urban quality). However, conversely, it is only through developing indicators that measure intermediate outcomes that one understands their cumulative effect on the physical and socio-economic environment.

Impacts -whether final or intermediate- are those outcomes solely attributable to planning policies and instruments. On a conceptual level, the best effectiveness indicator for measuring whether or not a planning objective has been achieved is derived by deducting the effects of all other influences. It would separate out the contribution of planning from all other considerations and be a measure of impact.

In practice, the vast majority of national planning objectives are stated in terms of intermediate or final outcomes, rather than intermediate or final impacts and so assumptions frequently have to be made about intended additionality. Two main types of comparison and thus benchmarks can be used for the purposes of developing effectiveness indicators:

- A comparison between the *current state of the policy object* (e.g. the current viability of town centres) and what would have happened had the policy system (and the relevant land use planning instruments) not been in place (i.e. the counterfactual - e.g. the state of town centre viability that would otherwise have occurred).
- A comparison between the current state of the policy object and its desired state, i.e. some form of target or goal (e.g. a desired level of brownfield development) is to be set by policy makers to see if there is a movement towards this target or desired state.

There are a number of different ways of trying to identify what would have happened without planning policy being in place and hence disentangle the effects of planning policy. For example, control groups and areas can be used (e.g. by comparing what happens inside Green Belt areas with what happens outside them); the correlation between planning outputs and outcomes analysed; and the characteristics of refused and altered planning applications examined.

Another possible way of establishing the likely counterfactual is through obtaining information on those applications which either do not materialise because of the planning legislation being in place or have been changed as a result of Development Plans or negotiations with local authorities. One way in which to obtain this type of information is through undertaking a sample attitudinal survey of applicants for planning permission and other landowners. Such a survey might ascertain:

- the extent and manner by which their applications have been refused, by development type;
- the scope and nature of alterations to their applications (e.g. through negotiations with local authorities or as a result of Development Plan policies) made in order to improve their chances of gaining planning permission, or applications abandoned because of the poor likelihood of obtaining planning permission.

This information should provide some evidence of the difference that planning decisions or outputs are making, though this ultimately depends on being able to interpret the attitude surveys reliably. Only by undertaking a detailed analysis of each policy objective, one by one, is it possible to gauge the usefulness of different types of comparisons to evaluate policy effectiveness and develop appropriate indicators.

Two factors would improve the likelihood of identifying the counterfactual and hence address the issue of additionality. The first relates to whether the policy objective itself is quantifiable. Without this it would be unlikely that a counterfactual could be established. Quantification is possible, as long as acceptable assumptions can be made about the meaning of policy objectives that are phrased in broad and general terms (see Stage One). The second factor relates to whether the desired outcome is expressed in terms of intermediate outcomes (or even intermediate impacts), for example, some attribute of the development process or the physical environment. In land use planning terms, policy is hoping to influence these attributes: for example, to improve, protect, retain or achieve some target level. If the aim of the exercise is to provide a measure (or indicator) of achievement then it is critical that planning policy objectives are clearly stated and the desired outcomes are apparent and achievable.

Conclusion

The purpose of the DETR research was to establish a methodology by which a package of strategic indicators for evaluating the land use planning system can be developed. In evaluating such an approach, it is important to consider three fundamentals. First, the aim of the exercise was to provide indicators that help to evaluate, not simply to monitor, how the planning system is operating. Second, the objective was to clarify how indicators might be developed which would allow regular and consistent evaluation, not to produce full scale post implementation assessment of each particular policy. Third, the long term objective of such a package would be to evaluate the land use planning system as a whole as well as different elements within that system.

The framework adopted involved a number of distinct and important stages, namely the clarification of objectives; identification of planning outputs; and the identification of outcomes and planning impacts. Three kinds of indicators that assist in determining whether objectives can be met were specified. The first, which might be regarded as internal to the process, relate mainly to outputs and whether local planning authorities are actually carrying out specified requirements. The second group relates to whether desired outcomes have been achieved, whether because of the policy or for other reasons. The third group relates to additionality - that the objective would not have been achieved, or would have been achieved less successfully, if the policy had not been in place. These can be defined as impacts of the land use planning system.

The aim of the research was to concentrate on the challenges associated with developing indicators that are directly related to planning objectives. These challenges would need to be overcome before further work could proceed.

The first challenge relates to the clarity of the policy objectives. A particular problem was to ensure that objectives were clearly defined in a way that allows measurable indicators to be identified. Policy objectives need to quantifiable, otherwise no measure (or indicator) of achievement can be made.

The second challenge relates to the difficulty in separating the role of planning from other influences in order to identify its particular contribution to outcomes. This includes establishing what would have happened without planning intervention. It will rarely, if ever, be possible fully to quantify the impact of planning policy, separating it from the effects of other policies and the economic and social environment. However, putting together bundles of output, outcome and monitoring indicators does have the potential for identifying both qualitative and quantitative relationships and so provides an indication of the impact of specific land use planning policies.

The third challenge relates to the tension between developing a manageable number of indicators and obtaining a full coverage of policies. To evaluate the overall land use planning system requires a package of indicators. The package needs to achieve a reasonable balance across overarching objectives relating to environmental, economic and social goals, as well as more specific aims. At the same time, so that the indicators package remains manageable the inter relationship between the indicators, their complementaries and conflicts need to be identified.

The final challenge relates to the practical difficulties in developing the indicators and overcoming data problems. It was beyond the remit of the research to address detailed issues of implementation, notably those associated with gathering and analysing appropriate data. However, what is clear is that the burden of data collection and analysis must be commensurate with the potential value of the findings. Indicators must not become ends in themselves.

Acknowledgements

I would like to acknowledge the other researchers in the project, Dr Barry Pearce and Professor Christine Whitehead and thank the Department of Environment, Transport and Regions (DETR) for funding the research.

References

Arup (1995), *Derelict Land Prevention and the Planning System: the Final Report*, HMSO, London.

Audit Commission (1992), *Building in Quality: A Study of Development Control*, Audit Commission, London.

Bramley, G.; Bartlett, W. and Lambert C. (1995), *Planning and the Market*, UCL Press, London.

Cheshire, P.; Sheppard, S.; Hooper, A. and Peterson, J. (1985) *The Economic Consequences of the British Planning System: Some Empirical Results*, (Discussion Paper No. 29), Department of Economics, University of Reading, Reading.

Department of Environment (1991), *Tackling Vacant Land: An Evaluation of Policy Instruments*, HMSO, London.

Department of Environment (1996), *Indicators of Sustainable Development for the United Kingdom*, HMSO, London.

ECOTEC (1993), *Reducing Transport Emissions through Planning*, HMSO/DoE, London.

Elson, M.; Walker, S. and Macdonald, R. (1993), *The Effectiveness of the Green Belt*, HMSO, London.

Evans, A. (1996), 'The impact of land use planning and tax subsidies on the supply and price of housing in Britain, a comment', *Urban Studies*, vol. 33, pp. 581-585.

Hall, P.; Thomas, R.; Gracey, H. and Drewett, R. (1973), *The containment of Urban England: volume 2 - the planning system: objectives, operations, impacts*, Allen & Unwin, London.

HM Treasury (1994), *Policy Evaluation: A Guide for Managers (Fourth impression)*, HMSO, London.

Jackson, P. and Palmer, B. (1989), *First Steps in Measuring Performance in the Public Sector: A* Manager's *Guide*, Public Finance Foundation with Price Waterhouse, London.

LGMB (Local Government Management Board) (1995), *Indicators for Local Agenda 21*, London.

MacGillivray, A. (ed.) (1995), *Environmental Measures, Indicators for the UK*, Environmental Challenge Group, London.

Monk, S. and Whitehead, C. (1999), 'Evaluating the Economic Impact of Planning Controls in the UK: Some Implications for Housing', *Land Economics*, vol. 75, pp. 74-93.

Morrison, N. and Pearce, B. (2000), 'Developing Indicators for Evaluating the Effectiveness of the UK Land Use Planning System', *Town Planning Review*, vol. 71, no. 2, pp 191-211.

Morrison, N., Pearce, B. and Whitehead, C. (forthcoming), *Developing strategic indicators for evaluating the effectiveness of Land Use Planning*, DETR, London.

OECD (Organisation for Economic Cooperation and Development) (1994), *Environmental Indicators*, OECD.

PACEC (PA Cambridge Economic Consultants) (1995), *Final Evaluation of Enterprise Zones*, HMSO, London.

Pieda (1992), *Evaluating the Effectiveness of Land Use Planning*, HMSO, London.

Sustainable Seattle (1992), *Indicators of Sustainable Community*, Seattle.

UNCHS (United Nations Centre On Human Settlements) (1997), *Analysis of Data and Global Urban Indicator Database*, UNCHS.

6 Asymmetrical Devolution, Institutional Capacity and Spatial Planning Innovation

M.G. LLOYD AND J. McCARTHY

Introduction

The Third Way political agenda in the United Kingdom advocates an approach to public policy based on the widening and deepening of the democratic processes of an economically advanced nation (Giddens, 1998). In practice, the Third Way approach seeks a synthesis of traditional forms of state interventionism and the policy nuances of a neo-liberal, market oriented agenda. It purports to avoid the disadvantages associated with an excessively bureaucratic, top-down form of public administration and the disadvantages associated with purist, market-based ideas that seek to undermine the role of government itself (Giddens, 2000). Critics of the Third Way approach have suggested that this middle ground policy arena is not as original as it appears to be, since in reality it is essentially re-addressing the old debate about the appropriate relationship between state and market (Ryan, 1999).

Nonetheless, the Third Way argues "that the three key areas of power – government, the economy and the communities of civil society – all need to be constrained in the interests of social solidarity and social justice. A democratic order, as well as an effective market economy, depends upon a flourishing civil society. Civil Society, in turn, needs to be limited by the other two" (Giddens, 2000: 51). A practical manifestation of the Third Way philosophy is a modernisation agenda to sustain the renewal of contemporary civil society. This may be seen as a response to calls for a new way forward to replace older, interventionist ideas associated with social justice and equity (Crick, 1997). The modernisation agenda since the mid 1990s has involved devolution and decentralisation processes to combat the imbalance associated with centralised power and institutional control in the UK (Hutton, 1995). In this context, devolution involves the transfer of power (or powers) from a superior to an inferior political authority. It consists essentially of the transfer to a subordinate elected body, on a geographical basis, of functions at present exercised by Ministers and Parliament (Bogdanor, 1999). Responsibilities are defined for the devolved institutions for scale, jurisdiction and geography. Thus devolution, as presently articulated in the UK, retains the supremacy of Parliament and must be distinguished from the processes of federalism, which offer a more concrete division of powers on a sub-national basis (Bogdanor, 1999). In effect, devolution stops short of granting complete legal sovereignty to the new entity but involves the establishment of an elected body and an executive to govern that defined sub-national geographical space and jurisdiction (McNaughton, 1998).

The debate over the perceived merits and demerits of devolution, and the associated institutional form and landscape, is not a new one in the UK. There have been political movements campaigning for complete or partial independence in Wales and Scotland ever since they were incorporated into the UK State. However, the modern debate, over the appropriateness of devolved institutional arrangements, may be traced to the 1960s. The new emphasis on the devolution debate is interpreted as a response to four principal conditions (McNaughton, 1998). These factors included the relative decline in the UK national economy and its associated differential geographical impact on the regions, such as Scotland, Wales and Northern Ireland; the perceived increased centralisation of power and influence at Westminster; the perception in the regions that English culture was becoming more pervasive; and, the debate (particularly in Scotland) over the appropriate stewardship of North Sea oil and gas. As a consequence, there was a relatively marked upsurge of interest in greater autonomy for Wales and Scotland from the 1970s onwards. Mitchell (1996), for example, has traced the specific advances made towards home rule in Scotland, noting the importance of democratic politics in facilitating self-identity, awareness and self-confidence. The more recent path to devolved structures in the later 1990s has similarly not been entirely straightforward, as shown in the detailed accounts of events in Wales (Andrews, 1999; Morgan and Monaghan, 2000) and Scotland (Himsworth and Munro, 1998; Ritchie, 2000). Indeed, until recently there has been a marked resistance to devolution and regional reform on the grounds that it could lead to destabilising of government at the centre (Bradbury and Mawson, 1997).

Devolution has not been put into place in an even institutional form within the UK. In practice, it has involved the transfer of specific but differential legislative, executive and financial powers from Westminster to elected bodies in Scotland, Wales and Northern Ireland, with these elected bodies remaining subordinate to a legal framework defined by the centre (Bogdanor, 1999). This chapter considers the nature of the differential arrangements for institutional innovation and spatial planning that are evolving in order to provide a framework for accommodating interests associated with development at national, regional and local levels of decision making. It involves considerations of emerging forms of governance, comprising "the activities and relations through which we come together to manage matters of collective concern" (Healey, Khakee, Motte and Needham, 1997: 26). It also involves consideration of the role and inter-relations of institutions – "devices for achieving purposes, not just for achieving agreement" (Putnam, 1993: 8) – that are associated with devolution. Differential processes of devolution and constitutional reform are likely to result in similarly differentiated lower-order arrangements for strategy setting and policy design (Lloyd and Tewdwr-Jones, 1997). This chapter attempts to capture the diversity of institutional innovation that is taking place in the post devolution context.

Institutional Capacity and Innovation

Devolution has not taken place in a vacuum. It has been, for example, emerging in an European context in which there has been a considerable emphasis on sub-national governance and management. Institutional capacity is important in enabling the management of change. In recent times, devolution has been imposed on, and has involved new relationships with existing institutional structures and cultures. Furthermore, in order to facilitate devolution, institutional capacities may have had to change. Just as economic activity and social opportunity are not even over time and space, so too institutional provision for and capacity for the management of change is uneven.

In this context, Amin and Thrift (1995) cite four factors relating to "institutional thickness": a strong institutional presence; high levels of interaction between networks; defined structures of domination and/or patterns of coalition resulting from both the collective representation of sectional and individual interests; and, the development, amongst participants, of a mutual awareness that they are involved in a common enterprise. Arguably, the structures in place prior to devolution provided these characteristics – particularly in the context of arrangements for regional industrial policy and urban regeneration policy; the regional development agencies in Wales, Scotland and Northern Ireland; and the plethora of local bodies and partnerships concerned with economic, social and environmental change and development. Similarly, the associated cultures of networking and inter-organisational relations also arguably provided an appropriate institutional capacity.

In particular, Amin and Thrift (1995) argue that institutional thickness or capacity can constitute a framework of collective support for individual agents. This can then involve a persistence of local institutions involved in the management of change; the construction and deepening of an archive of commonly held knowledge; an ability for organisations to be flexible, to learn and change; a capacity for high innovation; an ability to extend trust and reciprocity; and a sense of a widely held common project, which is able to mobilise the local economic system as a whole with speed and efficiency. Devolution complements these features of the modern state.

Institutional innovation sits comfortably alongside the contemporary processes of devolution and decentralisation of arrangements for economic development and innovation (Cooke and Morgan, 1997). It is very much a progression beyond the traditional forms of state intervention (social democratic) and neo-liberal approaches. It also gels with the political ideas associated with the Third Way agenda. Thus, it is suggested that, in "the associational repertoire, the key issue is not the scale of intervention but its mode, not the boundary between state and the market but the framework for effective interaction" (Cooke and Morgan, 1998: 22). More traditional institutional development was focused on service delivery, such as training and technical assistance, and, later, the neo-liberal approach of service delivery at minimal cost. Now, however, there is greater emphasis on process, and arrangements for governance, empowerment and capacity development. The contemporary emphasis on institutional innovation attempts to correct the balance away from exclusively top-down, or supply-side oriented, approaches. It may be seen as a synthesis of the managerial, political and economic elements of institutional development. It sits comfortably alongside the new approaches to private and public innovation in facilitating regional development, and it stresses tangible and intangible processes of negotiation, dialogue and consensus building. Furthermore, associational economic activity of this sort requires a strong institutional infrastructure, particularly in the context of economic development and innovation. Indeed, it is suggested that a region "with a strong and cohesive cultural identity is more able to develop the associative forms of action that are nowadays deemed to be so important to innovation and economic development" (Cooke and Morgan, 1998: 163).

The creation of enhanced institutional capacity for strategic spatial planning can help to realise the full potential of participative and integrative processes involved in land use planning. Indeed, as Healey (1997: 18) has indicated, "strategic spatial planning can ... provide a transparent, fair and legitimate way of recognising and responding to the multiplicity of stakeholders, interests and value conflicts that arise in urban regions". This can constitute a process of institutional innovation which, as Motte (1997) has indicated, can occur in the context of spatial planning where new spatial planning systems are applied, where such systems are applied in a different way from in the past, or where new mechanisms, processes or organisations are created to allow planning systems to work more efficiently. The specific question for this paper is: to what extent is institutional capacity building and innovation taking place within the contextual framework of devolution?

Differential Devolution and Institutional Innovation

Devolution has not been put into place in an even institutional form in the UK. There is a powerful underlying political dimension here. It has been asserted, for example, that democratic devolution "may seem a modest institutional innovation by the standards of federal countries, but it nevertheless signals a major challenge for a London-centric political culture which, until now at least, has been impervious to campaigns to extend democracy and participation" (Morgan and Mungham, 2000: 15). The devolved arrangements have been variable, according to the devolved powers involved (McNaughton, 1998). These variable arrangements reflect the nature of the differences in the respective relationships between Wales and Scotland with Westminster prior to devolution (Morgan and Mungham, 2000). Thus, in Wales, while there has been a cultural and political renaissance associated with the creation of the Welsh Assembly, this has not yet involved an economic one (Morgan and Mungham, 2000). In Scotland, in contrast, devolution has involved the creation of a Scottish Parliament and Scottish Executive, which have assumed responsibilities for a range of administrative, political and financial functions (Himsworth and Munro, 1998). Westminster has retained legislative supremacy over reserved functions, including economic policy, foreign affairs and defence. However, there has been the contingent delegation of other responsibilities, including local government, housing, social exclusion, urban regeneration, land use planning, environmental protection and management, which have then passed to the new institutions (Himsworth and Munro, 1998). This section attempts to capture the diversity of change associated with devolution.

England

In England, the establishment of Regional Development Agencies (RDAs) is an important innovation, but it is taking place against an unpredictable constitutional position with respect to future change (Mawson, 1998). The White Paper "Building Partnerships for Prosperity" (Department of the Environment, Transport and the Regions, 1997a), for example, set out detailed proposals for the creation of RDAs in England. While the RDAs are primarily business-led, they also include people with experience and expertise from local government, further and higher education, trade unions and the voluntary sector. The intention is that they will bring greater coherence into national programmes by helping "to integrate them regionally and locally" (Department of the Environment, Transport and the Regions, 1997a: 16). The RDAs commenced operations in April 1999, and their immediate contribution is to clarify the considerable variations that have existed between individual English regions in terms of embedded institutional capacity. This situation may be characterised as one of fragmentation in which a multiplicity of agencies contested both the regional strategic arena and the responsibility for the design and delivery of economic development policies and measures (Roberts and Lloyd, 1998).

The RDA initiative in England would appear to be bringing something new to the implementation of regional policy - the promotion of development according to the determination of local or regional priorities, influenced by the social and economic needs of particular localities. Indeed, the Department of the Environment, Transport and the Regions (1997b) had earlier emphasised the need to see RDAs as "an essential first step, to provide for effective, properly co-ordinated regional economic development, to underpin wider regeneration, and to enable the English regions to improve their competitiveness" (Department of the Environment, Transport and the Regions, 1997b: 1). The intention therefore is to allow each English region, within a common framework of objectives, functions and funding arrangements, to develop a RDA that matches the circumstances and needs evident in the region. In practice, this intention will be somewhat circumscribed during the first few years due to the projection forward of various inherited expenditure commitments (Roberts and Lloyd, 1998).

The RDAs are intended to enhance the existing regional institutional capacity in terms of the roles of other stakeholders and actors in a region (Roberts and Whitney, 1991). This is particularly so in regions that already have an established structure of development organisations. Indeed, in such circumstances, a major difficulty experienced by many development organisations, and especially by new entrant bodies, is in persuading the bodies already present in a region to participate in the preparation of a regional development strategy, and, having agreed the strategy, to co-operate in its implementation. Despite the considerable variations that exist between individual English regions, a number of general problems are associated with the present system of regional development in which a multiplicity of agencies contest both the regional strategic arena and responsibility for the design and delivery of economic development and other services (Roberts and Lloyd, 1996). While the RDAs have achieved much in the initial phases of their operation, tensions remain particularly with respect to reconciling their role with established local and regional partnerships and in setting regional agendas based on consensus (Robson, Peck and Holden, 2000).

Northern Ireland

In Northern Ireland, a draft regional strategic framework has set out an agenda for action for the next 25 years. This strategic framework asserts that it "is not a fixed blueprint or master plan. Rather it is a framework, prepared in close consultation with the community, which defines a vision for the region and frames an agenda which will lead to its achievement" (Northern Ireland Office, 1999: 2). The regional strategic framework is intended to enable the strengthening of the regional economy and the reduction of social disadvantage, as well as the provision of a spatial framework for key infrastructure services and for development plans to guide public and private investment relating to land use. Strategic planning guidelines are an integral part of the framework, since it aims to provide long-term direction for the public sector, the private sector and the whole community. This would seem to conform broadly to the notion of spatial planning as a process of empowerment in which those concerned with localities and the management of sub-national space collaborate to produce strategies and plans to help guide public, private and partnership decisions. The draft regional strategic framework in Northern Ireland would appear to be a first step along that pathway of innovation.

Wales

In Wales, a new policy agenda was produced as a result of a consultation exercise defining an appropriate approach to economic development. This approach does "not set out to produce a detailed blueprint or master-plan, but to focus attention on the essential priorities which need to be tackled. These priorities require a partnership for prosperity involving all segments of both the public and private sectors, with a clear vision of the appropriate role of the public sector" (Welsh Office, 1998: 1). The new economic agenda is to provide a context to the work of the Welsh Assembly through an explicit spatial planning framework together with the reconfiguration of the principal institutions responsible for economic development. There are four economic fora in Wales: the North Wales Economic Forum, the South West Wales Economic Forum, the South East Wales Economic Forum and the Mid Wales Partnership. These economic fora have similar aims and objectives. They aim to promote economic growth and development by developing close working relationships between the primary economic agencies, and between the private and public sectors. They also aim to co-ordinate strategies for economic growth and development, and to prioritise the actions needed to take forward the strategies. The fora are intended to provide a key focus in economic regional policy delivery, whilst being independent of the National Assembly. It is therefore considered important that local authorities and other members have ownership of the fora. The fora share the same boundaries as the regional divisions of the Welsh Development Agency and the Regional Committees of the National Assembly.

Scotland

In Scotland, there is a very different approach to strategic spatial planning through the process and outcomes of community planning. It has been argued, for example, that "regulatory policies offer a potential model for the Scottish Parliament. At one level, they do not involve Government directly running services or setting up new bureaucracies and expenditures, but involve them in setting agreed rules that are then implemented by other agencies. Active government can then be linked with the politics of decentralisation. Regulatory policies are thus an appropriate vision of government in an age where on one level the state is under attack and critique, yet more and more is expected from it by the general public" (Hassan, 1999: 16).

The Scottish Parliament and the Scottish Executive has put into place a relatively broad approach to strategic planning, institutional innovation and economic development. Since the creation of the Parliament and the setting up of the Scottish Executive, there has been considerable attention paid to its intended political programme. This has tended to reflect the inherited socio-economic conditions, the policy regimes which were already in place, and the institutional capacity available to address these issues across Scotland. The latter consideration is important because of its unevenness in terms of the bodies involved in planning and development matters.

Initially, the Parliament stressed the need to promote social inclusion and social justice throughout Scotland. It has attempted to do this by integrating local actions, often involving expenditures by different government agencies and departments, and often involving wider partnerships. This approach has reflected and built on the established importance of partnership working in Scotland. It is particularly evident in the prevailing policy regime for urban regeneration, which operates principally at the local level - either in geographically defined areas or in thematically defined social groups - albeit within a competitive bidding framework for available resources (Turok and Hopkins, 1998). Running in parallel with the social justice agenda has been the development of the community planning initiative.

In Scotland, community planning involves local authorities working together with their principal public sector partners to plan for and deliver services that meet the needs of their local constituent communities. More formally, it has been described as a means by which local authorities can put forward an informed view of the challenges and opportunities facing their communities (Community Planning Working Group, 1998). This process is intended to be inclusive of the key players and interest groups involved in localities so as to facilitate the overall well being of the community at large. Clearly, the notion of partnership is writ large in the community planning process. Initially, community planning was driven from within the local authority community but now it is being increasingly drawn into the centre stage of politics. The Scottish Parliament is to introduce a statutory power of community initiative and of community planning in a forthcoming local government bill which will encourage joint working, cross-cutting initiatives, and an important role for community leadership and community planning. Community planning is an attempt to enhance existing institutional capacity in Scotland, in order to address prevailing sub-national planning issues.

Community planning has been described as a means by which local authorities can put forward an informed view of the challenges and opportunities facing their communities. This process is inclusive of the key players and interest groups involved in those localities, who contribute to the overall well being of the community at large (Sinclair, 1997). In conceptual terms, community planning seeks to focus on three aspects of local authority decision making and activity: the determining of policies and the design of plans and priorities; the setting and enforcement of service standards and specifications; and, the monitoring and review of performance (Rogers *et al*, 1999). More specifically, community planning seeks to encourage cost effective service delivery, together with the concomitant empowerment and inclusion of communities. It is based on four basic sets of ideas: a strategic vision for a whole area; the need for effective community consultation and participation; partnership between key players in policy implementation; and, the importance of community leadership (Rogers *et al*, 1999). While there is a focus on integrated service delivery there is a hint of a wider strategic role for community planning. The available research evidence would suggest that community planning is a viable responsibility for local government in managing change in localities throughout Scotland. It provides the opportunity to lay down a new agenda for local authority activism in Scotland, and provides an opportunity to secure a more integrated and cost effective delivery of services (Rogers et al, 1999).

The Scottish Parliament has also devised an economic development agenda to address the differential regional conditions and relative performance of the economy (Peat and Boyle, 1999). It is clear that the unbalanced development of the economy (urban-rural; east-west; highland–lowland) can have important impacts at the national level. The Enterprise and Lifelong Learning Committee (2000) of the Scottish Parliament examined the provision of local economic development, post-school vocational education and training, and business support services in Scotland. It concluded that there was considerable congestion in these arrangements, with evidence of confusion, overlap, duplication and inter-organisational competition for available resources (Enterprise and Lifelong Learning Committee, 2000). The Committee confirmed, however, the extent to which positive partnership working is being achieved in individual localities, with positive results. This represented a vote of confidence in the established institutional capacity for sub-national development, but the Committee advocated that the roles of the key players should be more precisely defined in order to secure a more productive deployment of available resources. Indeed, the Committee suggested that enhanced partnership working at the local level would not achieve the broader strategic objectives of national and regional economic development.

The Enterprise and Lifelong Learning Committee (2000) advocated an economic development strategy for Scotland as a whole to provide the context to local action. This would be based on economic growth through competitiveness and sustainability, but would take into account regional development and social integration within the national economy. To facilitate this national perspective, it proposed a spatial or sub-national framework of local economic forums for each Local Enterprise Company

(LEC) area of the Scottish Enterprise and Highlands and Islands Enterprise networks. Each local economic forum would be based on local major interests, including the local authorities, LEC, Chambers of Commerce and institutions of local, higher and further education. Each forum would create a local economic strategy for its area, including the unambiguous delineation of the responsibilities of each organisation involved in local economic development (Enterprise and Lifelong Learning Committee, 2000).

This was confirmed by the subsequent publication of the economic development framework for Scotland (Scottish Executive, 2000b). This set out a vision for the Scottish economy, based on attaining economic growth through international competitiveness, regional development across Scotland, social integration and the principles of sustainability. Echoing the approach of the Enterprise and Lifelong Learning Committee, the economic development framework seeks to more precisely define the roles and priorities of the bodies involved so as to enable more effective working. The economic development strategy is predicated on the need for partnership, comprehensiveness and the adoption of a longer-term perspective. Attention is drawn, for example, to the need to "consider the basic organisation of economic activity and the degree to which it promotes entrepreneurial dynamism. The environment must facilitate a range of activities including the establishment of new enterprises, the establishment of productive bases in Scotland by overseas enterprises, the expansion of small enterprises, collaboration and joint ventures between productive enterprises and centres of knowledge and research, and the development of the formal and informal networks that help to lower the costs of economic transactions" (Scottish Executive, 2000b: para 4.1). This suggests the definition of particular roles for institutions and the setting of explicit priorities within the Scottish economic environment.

Enhanced Institutional Capacity for Strategic Spatial Planning?

To what extent is a regional dividend emerging in terms of strategic spatial planning and institutional innovation associated with the processes of constitutional reform? Institutional capital includes the components of intellectual capital, social capital and political capital (Healey, 1997). Together, these comprise the institutional capacity of a locality to understand the nature of regional and local dynamics and to be able to respond to change. The concept of institutional capacity "refers not merely to the formal organisational structures, procedures and policy measures that actors can draw upon in developing strategic spatial plan making activity. It also includes the collective store of relationships and alliances and the institutional loci of interactions" (Healey, 1997: 23). It can be viewed as the way in which an institutional infrastructure can manage change in the wider social or community interest. It would appear from the evidence that different forms of institutional innovation and spatial planning are emerging in the context of the different constitutional arrangements.

In Wales and Northern Ireland, for example, institutional innovation is based on the design of a parent economic development strategy that sets out a framework for economic management and that contains a spatial dimension at its core. The strategic and spatial dimension is also linked to associated institutional innovation. Thus, alongside the strategic planning innovation, in Wales there is the overhaul of the existing development bodies, and in Northern Ireland there is the introduction of strategic planning guidance. It would therefore appear that in Wales and Northern Ireland there is a concerted attempt to enhance their respective institutional capacities. In England, in contrast, there is an attempt at institutional capacity building through the introduction of the RDAs which operate within a contested and congested institutional landscape (Benneworth, 2000). These are an attempt to enhance intellectual and social capital in the individual regions. The future direction of regional government in England may further liberate a potential enhancement of political capital by the RDAs. What the current arrangements would suggest, however, is that the new institutional forms do not nest within a coherent spatial planning context. In Scotland, the approach seeks to enhance provision for the management of change through a reliance on existing institutional arrangements, particularly the development agencies and local government. There is as yet little attempt at spatial planning.

Spatial planning is described as "a social practice through which those concerned with the qualities of places and the spatial organisation of urban regions collaborate to produce strategies, policies and plans to help guide specific decisions in order to regulate and invest in development activity" (Healey, 1997: 21). It is significant that a principal focus for a spatial planning process is that of attaining balance in terms of social, economic and environmental development and change within a national or regional economy (Alden, 1996). In this respect, the potential of enhanced institutional capacity is important in establishing the structural capacity for, and the cultural means of, securing effective and accountable decision making at a time of change. The capacity of the new constitutional arrangements is also significant in terms of the need for appropriate economic, social and environmental strategies. Specifically, an enhancement of institutional capacity in relation to spatial planning may allow more effective engagement with the geopolitical divisions and variations within national economies.

References

Alden, J. (1996), 'Regional development strategies in the EU: Europe 2000+', in Alden, J. and Boland, P. (eds.), *Regional Development Strategies: A European Perspective*, Jessica Kingsley Publishers, London, pp. 1-13.

Alexander, A. (1997), 'Scotland's Parliament and Scottish Local Government: Conditions for a stable relationship', *Scottish Affairs*, vol. 19, pp. 22 - 28.

Amin, A. and Thrift, N. (1995), 'Globalisation, institutional thickness and the local economy', in Healey, P.; Cameron, S.; Davoudi, S.; Graham, S. and Madanipour, A. (eds.), *Managing Cities*, John Wiley, London, pp. 91 – 108.

Andrews, L. (1999), *Wales Says Yes. The Inside Story of the Yes for Wales Referendum Campaign*, Seren, Bridgend.

Bachtler, J. and Turok, I. (eds.) (1997), *The Coherence of EU Regional Policy*, Jessica Kingsley, London.

Benneworth, P. (2000), 'An initial assessment of the eight final regional economic strategies', *Regions*, no. 225, pp. 9-21.

Bogdanor, V. (1999), *Devolution in the United Kingdom*, Oxford University Press, Oxford.

Bradbury, J. and Mawson, J. (1997), *British Regionalism and Devolution. The Challenges of State Reform and European Integration*, Jessica Kingsley, London.

Community Planning Working Group (1998), *Report of the Community Planning Group*, Scottish Office, Edinburgh.

Cooke, P. and Morgan, K. (1998), *The Associational Economy. Firms, regions and innovations*, Oxford University Press, Oxford.

Crick, B. (1997), 'Still missing – a public philosophy', *Political Quarterly*, vol. 68, no. 4, pp. 344 – 351.

Danson, M.W.; Lloyd, G. and Newlands, D. (1993), 'The Role of Development Agencies in Regional Economic Regeneration', in Hart, M. and Harrison, R. (eds.), *Spatial Policy in a Divided Nation*, Jessica Kingsley Publishers Ltd. and The Regional Studies Association, London, pp. 162 - 175.

Department of the Environment, Transport and the Regions (1997a), *Regional Development Agencies. Issues for Discussion*, DETR, London.

Department of the Environment, Transport and the Regions (1997b), *Building Partnerships for Prosperity*, DETR, London.

Enterprise and Lifelong Learning Committee (2000), *Inquiry into the delivery of local economic development services in Scotland, First Report 2000*, Scottish Executive, Edinburgh.

Fordham, G.; Evans, R.; Fordham, R.; Harding, A.; Parkinson, M. and Harrison A. (1998), *Building Partnerships in the English regions*, DETR, London.

Giddens, A. (1998), *The Third Way. The renewal of social democracy*, Polity Press, Cambridge.

Giddens, A. (2000), *The Third Way and its Critics*, Polity Press, Cambridge.

Glasson, J. (1992), 'The fall and rise of regional planning in the economically advanced nations', *Urban Studies*, vol. 29, pp. 505 - 531.

Halkier, H. and Danson, M. (1998), 'Regional Development Agencies in Western Europe. A survey of key characteristics and trends', in Halkier, H.; Danson, M. and Damborg,

C. (eds.), *Regional Development Agencies in Europe*, Jessica Kingsley, London, pp. 26 - 44.

Halkier, H.; Danson, M. and Damborg, C. (1998), 'Regional Development Agencies in the new Europe', in Halkier, H.; Danson, M. and Damborg, C. (eds.), *Regional Development Agencies in Europe*, Jessica Kingsley, London, pp. 343 - 358.

Hassan, G. (1999), 'The new Scottish politics', in Hassan, G. (ed.) *A Guide to the Scottish Parliament*, The Stationery Office, Edinburgh.

Healey, P. (1997), *Collaborative planning. Shaping places in fragmented societies*, Macmillan, London.

Himsworth, C. M. G. and Munro, C. R. (1998), *Devolution and the Scotland Bill*, W. Green, Edinburgh.

Hutton, W. (1995), *The state we're in*, Jonathan Cape, London.

Illsley, B., Lloyd, M. G. and Lynch, B. (1997), 'Local government decentralisation in Scotland - an opportunity for planning?', *Town and Country Planning*, vol. 66, no. 7, pp. 206 - 208.

Lloyd M. G. and Tewdwr-Jones, M. (1997), 'Unfinished business? Planning for devolution and constitutional reform', *Town and Country Planning*, vol. 66, no. 11, pp. 302 - 304.

Lloyd, M. G. and Illsley, B. M. (1999), 'An idea for its time? Community planning and reticulism in Scotland', *Regional Studies*, vol. 33, no. 2, pp. 181 – 184.

Mawson, J. (1998), 'Remaking the Union: devolution and British politics in the 1990s', *Regional and Federal Studies*, vol. 8, no. 1, pp. 158 – 175.

McNaughton, N. (1998), *Local and regional government in Britain*, Hodder and Stoughton, London.

Mitchell, J. (1996), *Strategies for self-government. The campaigns for a Scottish Parliament*, Polygon, Edinburgh.

Morgan, K. and Mungham, G. (2000), *Redesigning democracy. The making of the Welsh Assembly*, Seren, Bridgend.

Motte, A. (1997), 'The institutional relations of plan-making', in Healey, P.; Khakee, A.; Motte, A. and Needham, B. (eds.), *Making Strategic Spatial Plans*, UCL Press, London, pp. 231-254.

Northern Ireland Office (1999), *Shaping Our Future. Towards a strategy for the development of the region*, Northern Ireland Office, Belfast.

Peat, J. and Boyle, S. (1999), *An Illustrated Guide to the Scottish Economy*, Duckworth, London.

Putnam, R. (1993), *Making democracy work. Civic traditions in modern Italy*, Princeton University Press, Princeton.

Rhodes, R.A.W. (1996), 'The new governance: governing without government', *Political Studies*, vol. XLIV, pp. 652 - 667.

Ritchie, W. (2000), *Scotland Reclaimed. The inside story of Scotland's first democratic parliamentary election*, The Saltire Society, Edinburgh.

Roberts, P.W. (1997), 'Sustainability and spatial competence', in Danson, M.; Lloyd, M. G. and Hill, S. (eds.), *Regional Governance and Economic development*, Pion, London, pp. 7 - 25.

Roberts, P. W. and Lloyd, M. G. (1998), *Developing Regional Potential. Monitoring the Formation of the Regional Development Agencies. Progress Report and A Way Forward*, Nabarro Nathanson and BURA, London.

Robson, B.; Peck, J. and Holden, A. (2000), *Regional Development Agencies and local regeneration*, Joseph Rowntree Foundation, York.

Rogers, S.; Smith, M.; Sullivan, H. and Clarke, M. (1999), *Community Planning in Scotland. An evaluation of the pathfinder projects*, Report to COSLA, University of Birmingham, Birmingham.

Royal Town Planning Institute (1986), *Strategic planning for regional potential*, RTPI, London.
Ryan, A. (1999), 'Britain: recycling the third way', *Dissent*, vol. 46, no. 2, pp. 77 – 80.
Scottish Office (1994), *Local Government Reform - Structure Plan Areas*, HMSO, Edinburgh.
Sinclair, D. (1997), 'Local Government and a Scottish Parliament', *Scottish Affairs*, vol. 19, pp. 14 - 21.
Welsh Office (1998), *Pathways to prosperity. A new economic agenda for Wales*, Welsh Office, Cardiff.

7 Shaping our Future: Public Voices – a New Approach to Public Participation in the Regional Strategic Framework for Northern Ireland

MALACHY McELDOWNEY, KEN STERRETT,
FRANK GAFFIKIN AND MIKE MORRISSEY

Introduction

A key component of the recent approach to regional strategic planning in Northern Ireland has been the involvement of an independent "Research Consortium" in its extensive public participation exercise. This consortium comprises the School of Environmental Planning from Queen's University, the Urban Institute from the University of Ulster, Community Technical Aid (Northern Ireland) and the Rural Community Network (Northern Ireland) – a combination of academic and urban and rural community interests. The Consortium has published two reports at different stages of the process and has formally participated in the Public Examination. It is now involved in community response to the Public Examination Panel Report and in an evaluation of the process as a whole. A previous conference paper (McEldowney et al., 1997) on the first stage of consultation examined key aspects of the **concept** involved and the **process** adopted. This paper uses the same structure to provide a more comprehensive retrospective account of a strategy-building exercise which is now nearing adoption.

Concept

Key aspects of the initial concept were the **spatial definition,** which evolved from "city region" to "region" over time, the **national/**

international context, which embraced relations with the rest of Ireland as well as with the UK and the EU, and the fundamental objective to go **beyond land use planning** in its analysis and prescription – to attempt "joined-up" thinking on a multi-dimensional problem. These will be discussed in turn.

Spatial Definition

The initial idea was to plan for the Belfast City Region (BCR) – an area within a thirty-mile radius from the city centre incorporating thirteen District Council areas. There was considerable functional justification for this in that it linked an employment centre to its commuter hinterland – reflecting the systems planning analysis which has underlain most structure planning initiatives in the UK since the seventies. Hall and Hay (1980) in their analysis of European city regions in 1980 had defined a similar but slightly more extensive city region for Belfast and Diamond's (1982) standard metropolitan labour area in 1982 included eight of the thirteen District Council areas. But even the functional basis of this definition can now be questioned as commuting distances extend and economic activities become more diffused – Hall (1999) now claims one hundred miles as a realistic radius for the London deconcentration zone and specifically states in relation to this locality that:

> the basic geography of Northern Ireland suggests strongly that there is in reality only one city region for planning purposes… of course Londonderry is a distinct centre in its own right but it does not command a region sufficiently large to act as a logical planning unit.

Such functionalist prescription incidentally uncovers an underlying political debate of long-standing intensity - the perception of "Belfast bias" and the geopolitics of perceived neglect "west of the Bann" - which were fuelled in planning terms initially by the Matthew Plan of 1963 and a series of contested development decisions in subsequent years. The Regional Physical Development Strategy of 1975-95 offered some recognition of this problem, with a decentralised settlement policy which sought a fairer spatial development balance. Nevertheless, a major finding of the initial consultation process for the BCR (1997) was the fear in peripheral District Councils of a "return to Matthew" and evidence of unreconstructed contestation (McEldowney et al., 1998) between urban and rural interests which extended well beyond the boundaries of the designated city region.

Accordingly, following the submission of the Research Consortium's first report and the reorganisation and rethinking which followed the election of a Labour Government and a fresh ministerial team, the new Minister announced the incorporation of the BCR into a wider Regional Strategic Framework (RSF):

> I have responded to concerns that it would be more sensible to reach decisions on the future development of the Belfast City Region within the context of an overall strategy for the whole of Northern Ireland... you will see the emerging thinking which has been formulated through intensive consultations with those in the City Region but it is absolutely right that we now treat this as a regional strategy and engage the whole community across the region in the debate (Minister's Speech, 1997).

As well as a more realistic spatial definition for the strategy this contained a positive reaffirmation of the notion of "planning as debate" (see below).

National/International Context

All strategic spatial planning initiatives must now take account of the Transnational Studies (European Commission, 1994) and the European Spatial Development Perspective (EU Ministers, 1997) to attempt to achieve some coherence in policy-making. While there may be some dissonance between these - the former being Commission-led while the latter is inter-governmental - there is little doubt that, as Williams (1996) argues, "spatial policy will continue to be developed and play a role in giving coherence to and co-ordinating the actions of EU sectoral policies". He also argues that spatial policy will be important as a tool for the redistribution of wealth in the adjustment to EU enlargement – the key concern over the next few years.

This implies some form of spatial planning hierarchy, possibly with the ESDP at the top, followed by transnational studies (in this case, for the Atlantic Arc) as contexts for regional strategies (such as the RSF), urban and rural development plans and local planning regulations. Of more importance than such a hierarchy (the implementation of which would inevitably lag behind rapidly changing political and economic circumstances) is the correlation between spatial planning and EU resource allocation. In this context, the relationship between spatial planning north and south of the border within Ireland is interesting as it confronts both the above issues – relations within the Atlantic Arc transnational framework and preparations for the new round of EU structural funding up to 2006.

As part of the Atlantic Arc, key considerations are economic and infrastructural links between Belfast and Dublin (two of the "pearls" on the Atlantic "string"), the analysis of trading patterns and the reconciliation of disparate planning traditions in cross-border areas. Further considerations are the wider relationships with Scotland, Wales, Cornwall, Brittany, the Basque country and Portugal, in which communication links are an immediate priority but more fundamental shared concerns in relation to culture, political autonomy and economic peripherality need to be addressed. As regards spatial planning's contribution to the coherence of EU sectoral funding, the experience of the Republic of Ireland (RoI) is instructive – its emphasis on the production of an economic/social strategy (the National Development Plan (NDP) first, to be followed by a subsequent strategic spatial plan is an interesting model (Murray and Greer, 2000).

This opens up an important area of debate - and one which was highlighted in the RSF consultation process: the integration of economic and spatial planning strategies and of the agencies responsible for them. Northern Ireland's economic strategy (EDSR Steering Group, 1999) and the RSF have been produced more-or-less simultaneously but as essentially separate exercises, giving rise to public concerns about effective co-ordination in an era of separate Assembly Ministries. RoI's approach has been to secure EU approbation for a spatial sub-division within an economic/social strategy, to bc followed subsequently by detailed spatial strategies. This is more in accordance with EU objectives (although EU structural funding will provide less than 10% of the NDP's budget) but the time-lag between the production of the social/economic and spatial strategies will be problematic. Land-use planning is therefore relegated to a subordinate position within a wider strategic enterprise – an issue which is highly relevant for RSF and is discussed in more detail below.

Beyond Land Use Planning

In relation to strategic planning in Britain, the RSF should have more relevance for emerging regional structures (Regional Development Authorities (RDAs), Regional Chambers) than for traditional structure planning which focused mainly on land use planning for city regions. The Minister's recommendations (1998) on "modernising planning" included an emphasis on partnerships with stakeholders, the facilitation of RDAs' regional economic strategies, the inclusion of an integrated transport strategy and, most significantly, the provision of a framework for EU financing bids and consistence with EU spatial planning initiatives such as ESDP. These are analogous to the RSF's original objectives – particularly the necessity to establish a coherent regional vision and integrated strategy before giving it land use expression; hence the emphasis on getting "beyond land use planning".

This objective was clearly set out in the discussion document for the city region, but it is even more essential at the regional level as indicated above. It has also been argued (McEldowney, 1997) that there are two additional strong justifications for this approach – the universal imperative to achieve "sustainability" in planning and the local imperative to address the issue of "planning in a divided society". "Sustainability" objectives require more integrated approaches to, for example, planning and health, planning and employment, planning and education and, most obviously, planning and transportation – not to mention the politically contentious issue of achieving compact settlement patterns in a region with a dispersed settlement tradition. The "divided society" manifests itself in both sectarian and socio/economic terms, and the problems it presents cannot obviously be solved by spatial planning alone, although both these divisions have strong locational implications, previously ignored or underestimated by local planning practice. This problem was explicitly addressed in the Consortium's public consultations and produced a typically "divided" response – some (mainly urban) communities saw it as the dominant issue while other (mainly suburban) communities preferred to regard sectarianism as a non-issue in planning terms.

Process

Although the argument in this paper is that "public participation" is a more accurate description of the comprehensive process described below, the official title of the research project was "consultation", so that term is used when reference is made to the Northern Ireland case study. This section considers firstly some aspects of public participation in current theory and past practice, and secondly describes key features of the case study within that context. It concludes with an update of current progress and some general reflections on the issues identified here and above.

Theory and Practice

The concept of "public participation" has an established status within the British system of town and country planning, although there has been continuous debate about its meaning in practice. Since the Skeffington Report of the late sixties through the era of structure and local planning in the two-tier local governmental system of the seventies and eighties, there has been formalised public involvement in all stages of plan-making - from objective-setting in the initial stages through representation-making at draft stage to formal objections prior to public inquiry and adoption. This has been praised for its good intentions but frequently criticised for either its tokenistic character or its capacity for delay. Legislative changes of the early nineties, however, have introduced a notable switch of emphasis - Barlow (1995) in his comparative survey of public participation in Britain, Sweden and France, suggests that the current British "plan-led" system is moving away from its discretionary tradition and coming closer to the European zoning-based models. This implies the necessity for more effective and more formal public involvement at the early stages, as there is less discretion once the plan is adopted. The public local inquiry is now more crucial than subsequent planning appeals, so powerful interests such as landowners, developers and commercial organisations tend to dominate through professional representation, often to the exclusion of a wider public debate. The planner's responsibility in this situation of competing and unequal interests begs questions about his/her role in a situation of representative democracy (Kitchen, 1997) and gives rise to a series of difficult dilemmas – who is the customer? who is the client? who is the user? what, if anything, is the public interest?

European Union objectives also have an influence on public participation in planning, which is seen as a positive force in the reconciliation of member-states' individual planning systems to ensure that they do not establish non-tariff barriers to transnational investment. Plan-making is central to this as planning policy has the capacity to discriminate between internal and external investment sources. In relation to this, Morphet (1997) emphasises the importance of *transparency* and *openness* in the strategic planning process and the desirability of achieving *social cohesion* in the outcome – this requires a meaningful public participation process in which the technical process of plan-making and the social process of community involvement are continuously intertwined.

With regard to such a process of meaningful public participation, Healey's (1992) advocacy of a more communicative form of planning action – the importance of "story-telling" as well as statistical analysis, sensitivity to different "discourse communities" with different "systems of meaning", the necessity for "mutual learning" and "consensus-building" and the dominance of the "power of the better argument" – are particularly useful. Her admission of uncertainty as to the outcomes of the process represents commendable realism, while her central theme – "the democratic project of making sense together while living differently" – has particular resonance for a divided society which is seeking to build power-sharing governmental structures as a framework for strategic policy-making.

This thinking is developed and applied more specifically to planning systems and practices (the "hard" and "soft" infrastructures) in a later book (Healey, 1997), although some of the uncertainties referred to above remain. Healey here uses Giddens's theory of *structuration* to illuminate planning's embedded power through the modes of thought and sets of values embodied within its doctrines, but also, significantly, its potential for transformation – "in the fine grain of planning practice, planners not only bring power relations into being ... for Giddens they also have the power to change them … the practice of planning involves delicate day-to-day choices about whether to follow the rules or whether to transform the structure". These transformations also involve changes of culture, so as a consequence the management of co-existence in shared spaces becomes an exercise not merely in consensus-building but also in local culture building – in the creation of a public realm.

Habermas's theory of *communicative action* is then considered as one effective way of reconstituting this public realm – as a vehicle for transforming the abstract systems of bureaucracy and the market through a process of open, public debate. In this way the traditionally dominant position in planning of instrumental/technical reasoning can be challenged by moral reasoning and by emotive/aesthetic reasoning, which are equally valid. So, collaborative planning asserts peoples' collective capacity to transform the structures which they mould and are moulded by – offering open dialogue based on inclusive reasoning as a means of achieving this. As such, it provides valuable theoretical underpinning for the approach to public consultation undertaken in the RSF for Northern Ireland.

In relation to government and governance, Healey (1999) has represented current tendencies in urban governance as a series of processes from traditional practices to current approaches: from providing to enabling; from representing to empowering; from producer-driven to consumer-driven; from functional separation to integrated collaboration. These processes can be summarised in three models of urban governance – the traditional "service delivery" model which was bureaucratic, hierarchical and professional; the 1980s' "multiple initiative model" which was characterised by contracting out, partnership arrangements and multiple voices; the current "strategic – capacity-building" model which involves strategic collaboration, shared visions, local as well as technical knowledge and innovative learning.

It can be argued that in Northern Ireland local governance is now moving rapidly from limited experience of the first two models to become firmly established in the third. Because of limitations in the power of local representative government under direct rule (from Westminster) the "strategic capacity" of the community and voluntary sectors is already well-developed. In the terms of Healey's other categorisation of western governance systems (1997), new local "representative democracy" is seeking to replace the established "corporatist" model of direct rule in a situation where "pluralist democracy" has an established infrastructure in community institutions but as yet has little effective power. It is within this context that the public consultation process for the RSF must be considered.

The RSF Public Consultation Process

The initial part of the consultation process – for the Belfast City Region – was completed by the end of 1996, and its methodology was then revised and applied to the wider region during 1997 and 1998 with the significant agency change of the addition of the Rural Community Network to the Research Consortium to cover what was now a very extensive rural constituency. The threefold objective of exercise was **dissemination** (providing scene-setting information, basic statistics and translations of technical language for community discussion) **engagement** (helping identify and develop community organisations and encouraging their participation) and **innovation** (facilitating cross-sectoral debate and the promotion of fresh ideas). This required a deliberate diversity of modes (visioning exercises in the planning workshops and youth forums, exhibition-based discussion and debate with Councils, formal presentation and workshop discussions at the sub-regional conferences, structured interviews with statutory agencies and interest groups) to achieve the objective of "open conversation between diverse peoples".

A more difficult objective is the achievement of "consensus building" across diverse "discourse cultures", but this has to a longer-term project. The problem was addressed firstly by deliberately structuring all discussions under integrative themes such as "valuing people", "building prosperity", "caring for the environment" and "improving communications" – this helped to break down sectoral barriers and extend consideration beyond land use planning issues. More specifically, a series of regularly recurring *integrative specialist forums* under the four themes outlined above were held at a central venue, with cross-sectoral representation and a consistent membership. The focus here was on innovation, debate and small-scale consensus – building, with the opportunity for protracted open-ended discussion amongst a small number of people. Representation, attendance and time-pressure were problematic here, but valuable contributions were made. This was the only central venue, care being taken to hold workshops, interviews and even sub-regional forums as close as possible to the local "milieu" in which people felt comfortable. Over the extensive consultation period more than 500 separate organisations – from community, voluntary, professional, commercial and environmental constituencies – were involved.

The public consultation initiative took place alongside the essentially technocratic process of data-collection and analysis carried out by in-house teams within the DOE (NI). There was close co-operation between these two complementary streams of activity – regular interim reports were exchanged between the responsible agencies and personnel from each set of agencies appeared at public meetings and at internal strategy-building conferences. Nevertheless, the independence of the Consortium's position was maintained and accepted, as was the input from a third strand of activity – the advice from a "Panel of Experts" which comprised academic and professional expertise from British European and North American practice. This Panel organised study-tours for professionals and addressed conferences for the wider constituencies during the course of the strategy-building process. The culmination of this three-strand process was the publication of the Draft Regional Strategic Framework for Northern Ireland (1998) at the end of 1998.

The Draft RSF could be seen to have responded to consultation inputs in its initial "guiding principles" section which stressed the importance of "people and community-focused" and "sustainable" approaches to development; as well as setting objectives on "a more cohesive society", "competitiveness in education and quality of life", and "integration between aspects of Government and the regional institutions". It also proposed a "shared vision" derived from the consultation process which sought to achieve "an outward-looking, dynamic and liveable region with a strong sense of its place in the wider world". However, the rest of the document focused specifically on the development of a spatial development strategy which was much more familiar to land-use planners – a "hub, corridor and gateway" approach which allocated specific housing growth targets to a three-tier system of settlements, identified the major ferry-ports and airports as regional gateways and suggested five transport corridors (mainly road-based) as the key axes of development. All of these were encapsulated in thirty "strategic planning guidelines" (SPGs) which were to form the basis of discussion at the Public Examination. The difference in tone between the integrative, inclusive and visionary aspirations of the "guiding principles" and the more physically-focused prescriptions of the SPGs is striking – it reflects an acknowledgement of the necessity to transform structures and frames of reference, but also the underlying strength, perhaps even the immutability, of sectoral thinking and established procedures in planning.

Another illustration of the tension between established and innovatory thinking was evident in the rural-urban debate, which originated in the limitations of the original city-region definition and was a key part of the second-stage consultation. The initial RSF discussion paper drew attention to the disproportionate population growth in rural areas as compared to the decline at the core of the Belfast city region and outlined a strong protectionist theme as part of its environmental agenda – the conservation of agricultural land, the designation of extensive protected areas, the application of environmental capacity tests to new development and, in general, a compact rather than dispersed settlement pattern. A very strong and well-marshalled counter argument from rural interests (West Rural Region, 1999) advocated a "living and working countryside", a "diverse economy" and a settlement hierarchy which included dispersed rural communities. Its key demand was for "social and spatial equity" and its underlying principle was one which accorded totally with the original RSF vision – that social and economic factors should dictate spatial planning requirements rather than vice-versa.

The organisation of this response created new alliances within the rural region as evidenced by the insistence of five District Councils on organising themselves collectively as the West Rural Region and formally replying to the RSF as a coherent entity – subsequently being similarly represented at the Public Examination. Their agreed development themes were strengthening community, to support these goals by providing locational choice and accessibility. This argument was substantially accepted by DOE (NI) in the draft RSF (although contested by some environmental interests) – a reflection of the "power a good argument" perhaps, but more realistically a recognition of the "institutional thickness" of the rural constituency and the benefits of open dialogue as part of the consultation process.

Update and Reflection

The draft RSF was the subject of a formal consultation period in Spring 1999 during which submissions for hearing at a Public Examination were canvassed and received. The Consortium's role in this phase was the organisation of community workshops and cross-sectoral forums to test the compatibility of draft RSF policies with previous consultation feedback and to help facilitate formal representation – this helped highlight the key points of dissension and set some of the agenda for the Examination. There was general approval of the consultation process but concern about lack of integration between Governmental agencies – particularly those with spatial and economic planning responsibilities; there was approval of the general principles but little perceived evidence of them being translated into policies or measurable targets; there was support for sustainable development but concern about the Spatial Development Framework reinforcing existing spatial inequalities. The housing target of 200,000 was questioned and the public transport policy was generally perceived to be weak, but there was recognition of a more realistic cross-border dimension. Environmental interest groups challenged the commitment to prioritising "brownland" development and generally questioned the planners' understanding of real "sustainability".

The Public Examination took place between October and November 1999, involving invited participants from the range of objectors who had made written submissions, under the scrutiny of a three-person expert Panel. Fourteen separate sessions were held, at three different venues to provide regional accessibility – each focused on a set of related issues and involving round-table discussions by approximately twenty representatives. Development and commercial interests were professionally represented but community, voluntary and environmental groups tended to make their own cases. The Research Consortium attended many of the sessions to represent the findings of the consultation process and to provide independent professional and academic analysis. Discussion was structured but informal – there was little attempt at adversarial cross-examination and little time or opportunity for significant consensus-building. The weighing of the evidence and the drawing of conclusions were very much at the discretion of the Panel. The Panel's Report (2000) has recently been issued and is now the subject of formal public comment and part of an ongoing evaluation exercise by the Consortium.

The process continues and a full evaluation must await the next crucial stages; it is possible, nevertheless, to reflect briefly on the key issues identified above. The argument about **spatial definition** has come full circle with the Panel suggesting a balance of housing growth between the Belfast Travel to Work Area and the rest of the region – an indirect reintroduction of the "city region" concept in relation to a key aspect of the strategy. The **national/ international context** has been better recognised in succeeding drafts of the RSF and has been effectively linked to new British-Irish and European political structures in the Panel report. The RSF's guiding principles endorse the necessity to go **beyond land use planning**, but the narrowing of focus in the SPGs and the confined nature of the Panel discussions tended to reinforce the significance of physical issues and the dominance of technocratic thinking.

However, the public participation initiatives were successful at drawing in under-represented groups, facilitating a degree of mutual-learning and, importantly, generating some sense of ownership of a previously-unfamiliar **process** – now evidenced by the self-organisation of conferences by coalitions of interests to respond to the Panel report. They were less successful at involving land-owning and development interests, who preferred to wait for the opportunity of professional representation at the Public Examination, a forum which facilitated informal discussion but little effective consensus-building. For Government planners the process provided genuine learning benefits – alternative concepts to "predict and provide" and different perspectives on the inter-dependence of spatial, social and economic planning. It also provides some form of legitimisation for a centralised governmental system in a situation of erstwhile "democratic deficit". It is to be hoped that the lessons learnt from the process will inform future planning in the more democratically-accountable but professionally – segregated system which is now emerging.

References

Barlow, J. (1995), *Public Participation in Urban Development – the European Experience*, Policy Studies Institute, London.

DETR (1998), Modernising *Planning, a* Policy Statement by the Minister, London, January 1998.

Diamond, D. R. (1982), in House, J.W. (ed.) (3rd ed.), *The U.K. Space*, Weidenfield and Nicholson, London, pp. 426-486.

Economic Development Strategy Review Steering Group (1999), *Strategy 2010*, Belfast, March 1999.

EU Ministers for Spatial Planning (1997), *European Spatial Development Perspective*, First Draft, Noordwijk, Netherlands, June 1997.

European Commission (1994), *Europe 2000+: Cooperation for European Territorial Development*, Office for Official Publications of the European Communities, Luxembourg.

Hall, P. (1999), in Gaffikin, F. and Morrissey, M. (eds.), *City Visions*, Pluto Press, London, pp. 61-78.

Hall, P. and Hay, D. (1980), *Growth Centres in the European Urban System*, Heinemann, London.

Healey, P. (1992), 'Planning through Debate – the Communicative Turn in Planning Theory', *Town Planning Review*, vol. 63, no. 2, pp. 143-162.

Healey, P. (1997), *Collaborative Planning: Shaping Places in Fragmented Societies*, MacMillan, London.

Kitchen, T. (1997), People, *Politics, Policies and Plans*, Paul Chapman, London.

McEldowney, J. M.; Sterrett, K. and Gaffikin, F. (1997), 'Beyond Land Use Planning', Paper to XI AESOP Congress, Nijmegen, Netherlands.

McEldowney, J. M. et al. (1998), 'Strategic Planning in Northern Ireland' in *Pleanail*, the Journal of the Irish Planning Institute, no. 14, 1997/8, Dublin, pp. 110-116.

Minister's Speech (1997), 'Shaping Our Future', Conference, Belfast, 27th November 1997.

Morphet, J. (1997), *Belfast City Region: the European Dimension*, Report to Belfast Executive, January 1997.

Murray, M. and Greer, J. (2000), 'The Republic of Ireland National Development Plan 2000-2006: Some Strategic Planning Implications for Northern Ireland', *Economic Outlook*, vol. 15, no. 1, pp. 37-42.

Research Consortium (1997), *Belfast City Region: Public Voices*, The Stationery Office, Belfast.

Williams, R.H. (1996), *European Union Spatial Policy and Planning*, Paul Chapman, London.

8 New Approaches to Regional Spatial Planning?

CATHERINE HAMMOND

Introduction

Since the beginning of the 20th century regional activity, and regional land use planning in particular, has been cyclical. There have been periods of slow but steady decline, and periods of no regional planning at all, so that "the wider regional planning function, ... has not been given the opportunity to prove its worth" (Roberts, 1997). However, as Wannop said in 1995, "The need for regional spatial planning has never gone away - when ever it has been in retreat circumstances of one kind or another have always brought about a revival" so that the 1930s, the mid 1960s to the mid 1970s and the mid 1990s onwards have been periods of intense regional activity.

The current revival looks set to continue:

- The current regional revival began under a Conservative government, demonstrating that they at least appreciate that it can be a necessity. Labour Party commitment to a future of regional activity is well demonstrated by the steady increase in regional activity and institutions since their election in 1997. These include the strengthening of regional land use planning; the establishment of Regional Development Agencies and Regional Chambers; and an explosion in regional policy documents.
- Support from the public, private and voluntary sectors is evident in the bottom up pressure which was highly instrumental in instigating the regional revival.
- There is increasing UK activity in EU programmes which operate at the regional level and require regional partnerships, organisation and integrated planning (Baker, 1998; Roberts, 1998; Alden, 1998).
- There are a number of areas, such as sustainable development (CPRE, 1994; Baker, 1998; Roberts, 1997 and 1999) and strategic transport which arguably need a strategic tier that is best provided at the regional level.

This revival also provides regional spatial planning with an opportunity to develop and flourish. We already have a new style of Regional Planning Guidance in PPG 11 to complement a host of other regional strategies with some degree of spatial element. But it is still in its infancy, and has some way to go to meet the needs of changing regional organisations and activity and thus to be more effective.

This paper stems from my PhD research into a new form of regional spatial planning. Taking a very land use planning perspective it analyses past eras of regional planning; explores the impact EU policy and funding may have on regional spatial planning; and looks at what is being asked of regional planning at the moment. Through this it begins to identify a series of criteria against which regional spatial planning will need to conform if it is to overcome the mistakes of the past and meet the needs of the regional agenda into the future.

The terms "regional spatial planning" and "regional spatial strategy" are used to indicate a move away from both the separation between regional land use and regional economic planning, and also away from the strict adherence of recent RPG to development plan issues - a move towards a wider, more integrated form of regional spatial plan.

An Historical Perspective – Strengths to Build On and Mistakes to Avoid

Any foray into the future of regional spatial planning needs to be placed in its historical context. A critique of past attempts at regional planning is also important as a starting point from which to build a new system. Despite fundamental changes since the 1960s, many of the core issues for regional spatial planning remain the same. Looking at the successes and past failures of regional planning and the reasons for them is therefore valuable in guiding us to a more lasting and effective system in the future.

This section of the paper outlines and appraises England's two most recent eras of regional planning. The analysis of the strengths and weaknesses of each cycle are then fed into the criteria for a new form of regional spatial planning in the paper's conclusion.

1. The Mid 1960s to the Mid 1970s

The mid 1960s to the mid to late 1970s are generally considered to be a golden age of strategic and regional planning (Breheny and Hall, 1984; Thomas and Kimberly, 1995). Regional activity took off with the election of a Labour government in 1964, (Wannop, 1995) and the setting up of Regional Economic Planning Boards and Councils in 1965/6. The Councils consisted of government appointed representatives of different regional interests. They dealt with regional development strategy, including the preparation of studies and plans. The boards were made up of civil servants seconded from the regional offices of relevant Whitehall departments (Wannop, 1995). Their role was to provide technical expertise to the Councils (Glasson, 1978) who in return gave them some degree of accountability.

Despite their title Regional Economic Planning Councils and Boards also entered the realm of land use planning (Baker, Deas and Wong, 1999). Indeed, according to Wannop (1995), the intention behind their creation was to bring together economic and physical planning which was functionally separate at national, regional and local levels. The regional strategies of this period tended to be discussion documents rather than firm plans (Glasson, 1978), produced as a form of advice to central government.

In the early 1970s Standing Conferences of local authorities were also established. They helped prepare the regional strategies in a tripartite arrangement between central government, local authorities and the Regional Economic Planning Councils. This tended to lead to a more land use oriented approach to problem solving (Wannop, 1995), as well as bringing more accountability to the process.

Regional strategies were increasingly backed up by rigorous research, and a systems approach involving analysis, and monitoring. Through this they achieved a degree of regional specificity, with more emphasis on the economic dimension in the northern regions, and more emphasis on physical planning in the South East (Glasson, 1978; Wannop, 1995). Some regional plans also considered resource implications, and required central government to adjust national policy and spending priorities in order for them to be implemented (Wannop, 1995). Responsibility for strategy preparation from 1966 lay with the Regional Economic Councils, which were in fact only advisory and central Government was often unwilling, or unable, to make the changes necessary to implement the plans, highlighting the tension between strategy and delivery mechanisms.

Table 8.1: Key features of regional activity 1965–1979

Feature	Description – 1965–1979
Dominant direction of policy	Top down
Model of governance	Centralised control of functional devolution, i.e.: • centralised control • devolved civil servants • centrally appointed "independent" regional council
Method of approach / degree of stakeholder involvement	Centrally dominated
Type of regional strategies	Theoretically comprehensive and multi sector – in practice tended to focus either on economic or land use issues
Key strategic objective	Promoting / accommodating growth
Integration of strategies	High
Integration of governance	None
Style of planning	Hierarchical
Policy instruments	Bureaucratic regulation, financial inducement, public sector provision

Source: Adapted from Roberts and Lloyd (1998).

Strengths

Regional strategic planning during the 1960s had three key strengths. Firstly, policy was based on rigorous research and analysis. This technical approach was reinforced by the belief that "understanding was a prerequisite to action, and that action must be co-ordinated across both different spheres of activity and across different spatial scales" (Breheny, and Hall, 1984). Academics and practitioners had mutual goals and worked closely together, integrating research and practice in all spheres of land use planning.

Secondly, although the process was dominated by central government and mechanisms for participation and ownership were lacking, Breheny and Hall (1984) argue that a transparent approach did achieve a far greater degree of democratic accountability than would otherwise have been achieved.

Thirdly, while it tended to be referred to as regional economic planning, it did recognise the wider land use implications (Glasson, 1978) and tried, albeit unsuccessfully, to bring together economic and physical planning.

Weaknesses

Along with political and economic changes a number of weaknesses in regional planning were responsible for its decline from the mid 1970s.

A lack of integration of land use and economic planning The Barlow Report in 1940 favoured linking regional land use and economic planning (Roberts and Lloyd, 1998). The regional strategies of the late 1960s and early 1970s tried but failed to bridge the gap between the two, achieving at best an "uneasy ... marriage of physical and economic components" (Wannop, 1995), and at times the complete separation of economic and land use planning (Roberts and Lloyd, 1998). This was at least partly due to the fragmentation of powers and responsibilities within central government (Roberts, 1999), with economic and land use policy separated between the Departments of Trade and Industry and Environment respectively, leading to dislocation and at times conflict (James, 1968). This separation was heightened by the lack of co-ordination of regional agencies (Glasson, 1992). While fragmentation was less of a issue in times of stability, the cohesion between land use planning and economic development became divided once there was competition for resources due to economic decline (Roberts and Lloyd, 1998).

Lack of implementation mechanisms The general consensus is that the effectiveness of regional strategies was greatly hindered by a lack of implementation mechanisms, both in terms of economic and land use planning (Roberts, 1999; Roberts and Lloyd, 1998; James, 1968; Wannop, 1995; Glasson, 1992), in particular the inability to set or control budgets (James, 1968; Roberts and Lloyd 1998; Roberts 1999). This was heightened when the idea of linking regional plans with investment programmes was abandoned.

Process / technical issues There were also weaknesses with the technical preparation and process of developing regional strategies (Roberts, 1999).

Firstly there was a tendency for regional planning to be too detailed and over-complex (Glasson, 1992). Strategies took too long to prepare

(Roberts, 1997), while many research studies were one offs and lacked continuity (Glasson, 1992).

Secondly, as a tripartite organisation for the preparation of regional strategies was developed as a way of recognising the interests of central government, local authorities and other regional actors, disagreements arose over the scope and content (Minay, 1992; Roberts and Lloyd, 1998). No adequate mechanisms were developed to overcome this.

Thirdly, regional planning remained a state-led, top down activity. This led to standard policies (Roberts and Lloyd, 1998) and a failure to meet the individual needs of different regions (Minay, 1992; Roberts and Lloyd, 1998; Roberts, 1999). It also meant a lack of genuine regional participation (James, 1968; Wannop, 1995; Baker, Deas and Wong, 1999), which weakened the commitment to and implementation of regional policy. In addition over-centralisation provided insufficient flexibility to respond to changing economic and social conditions (Minay, 1992) and inter-regional inequalities.

Finally, regional strategic planning was undermined by its failure to retain relevance in a time of economic recession; by at times claiming too much for itself; and by being misunderstood (Breheny and Hall, 1984).

The late 1970s heralded the end of this era of regional planning. Weaknesses in regional policy making were coupled with deep recession and the beginning of expenditure cuts and a reluctance for sub national strategy making. This was reinforced in 1979 by the election of Margaret Thatcher's Conservative government with its free market philosophy.

2. The Late 1980s to 1996

However, by the end of the 1980s the weaknesses of a laissez faire approach to strategic planning were becoming increasingly apparent. Deregulation of service provision resulted in duplication in some areas and an absence in others (Roberts and Lloyd, 1998), while the resource driven approach was breaking down from a lack of a long term or strategic overview.

The Government was also facing pressure from a number of sources to solve problems which would benefit from some form of strategic planning, Planners needed strategic guidance to cope with large scale development proposals and increasing development pressure (Roberts, 1998). Business was demanding a more strategic framework, both to ensure sufficient land for future development, and to provide security for major investments (Baker, 1998; Roberts, 1999). NIMBYs and the wider environmental movement, especially in the South East, were putting electoral pressure on the Government to resolve the conflict between market led development and environmental quality (Roberts and Lloyd, 1998; Roberts, 1999; Baker, 1998). Changes were also needed for access to European Union Structural Funds (Thomas and Kimberly, 1995; Roberts and Lloyd, 1998; Roberts, 1999).

A reintroduction of strategic planning was becoming politically expedient and a distinctive element of the regional revival of the early 1990s was its emergence from bottom up pressure. This reawakening of regional interest led to both institutional and policy initiatives.

On the policy side the Government's recognition of the need for strategic planning culminated in the 1991 Planning and Compensation Act. This reintroduced the requirement for England to be covered by a complete set of Structure and District Plans, set within a strategic framework of Regional Planning Guidance. RPG was to be issued by the Department of the Environment, from 1994 through Government Regional Offices, on advice by groupings of county and metropolitan authorities, entitled regional planning conferences. Widespread consultation by the regional planning conferences was urged by the DoE. Details were initially set out in PPG 15, shortly followed by PPG 12 in 1992.

The institutional results were threefold. The requirement to undertake some form of regional planning and activity resulted in the revival of regional groupings of local authorities, and by 1992 coverage of England by standing regional conferences was complete. In a parallel sphere of activity other service providers and the private and voluntary sector also developed regional structures, creating a "fragmented and dispersed landscape of new and revamped organisational forms" (Roberts and Lloyd, 1998).

In 1994 the Major administration also created new Government Offices for the Regions. These brought together the regional arms of the Departments of the Environment, Transport, Employment and Trade and Industry to improve effectiveness by co-ordinating policy across central government departments (Baker, Deas and Wong, 1999). They also provided a regional voice for EU matters.

Table 8.2: Key features of regional activity 1990–1997

Feature	Description – 1990–1997
Dominant direction of policy	Top down and bottom up
Model of governance	Centralised with some functional devolution, i.e.: • centralised control • devolved civil servants • local authority regional associations
Method of approach / degree of stakeholder involvement	Public sector dominated
Key strategic objective	Guiding / accommodating growth
Type of regional Strategies	Weak with a land use focus
Integration of strategies	None
Integration of governance	Little (Government Offices for the Regions)
Style of planning	Hierarchical – national direction some local variation
Policy instruments	Deregulated. Reduced financial support. Mixed public, private and voluntary sector

Source: Adapted from Roberts and Lloyd (1998).

Strengths

In addition to providing much needed strategic planning, the period benefited greatly both from increasing skills and confidence in regional land use planning and developing regional institutional capacity. Growing confidence and experience lead to a steady improvement in regional planning as lobbying by regional planning conferences led to more coherent RPGs based on analyses and regional priorities (Thomas and Kimberly, 1995). Later RPGs broadened the vision and scope of the plans and included issues such as sustainable development and economic development (Thomas and Kimberly, 1995; Roberts, 1998; Baker, 1998).

The bottom up process of developing regional activity and partnerships was strengthened by RPG preparation and a partnership approach to accessing and managing Structural Funds (Roberts, 1997). A new style of informal regional governance, reflecting the belief systems of the organisations involved, arose from this mixture of top down and bottom up initiatives, providing a far richer institutional landscape (Roberts and Lloyd, 1998) than in the past. It also led to more flexibility into policy making (Glasson, 1992).

Weaknesses

While the revival in regional strategic planning and regional organisation was a good thing in itself it did have a number of weaknesses.

Scope Two weaknesses stemmed from Government advice in PPG 12. This determined restricted RPG to matters relevant to the preparation of development plans (DoE, 1992). Widespread criticism of this narrow approach (Wannop, 1995; Thomas and Kimberly, 1995; Baker, 1998) was summed up by Wannop (1995) as a "blinkered absurdity", failing to recognise integral economic and social concerns (Minay, 1992) and separating land use planning from economic, social and environmental planning (Roberts, 1997).

Another frequent criticism (for example: CPRE, 1994; Thomas and Kimberly, 1995; and Baker, 1998) was that RPG was hindered by its strict adherence to national planning policy. This militated against the production of regionally distinct plans and led to bland, uniform policy (Roberts, 1997) and a "cascade of platitudes" (Breheny, 1991), rather than forward looking, effective strategies (Thomas and Kimberly, 1995).

However it must be noted that both of these criticisms were overcome to some degree in later RPGs as plans improved, their value was recognised (Thomas and Kimberly, 1995; Baker, 1998) and arrangements for consultation and partnership were strengthened (Roberts, 1996; Baker, 1998).

Content There were a number of weaknesses in the content of the RPGs, although again these improved.

Criticisms included:

- the insubstantial nature of RPG documents (Baker, 1998);
- a lack of comprehensiveness between policy sectors within RPG (Roberts, 1996);
- a lack of robustness;
- a lack of regional distinctiveness (CPRE, 1994; Baker, 1998);

- a lack of vision (Roberts, 1996; Baker, 1998); and
- failure to include national environmental objectives, for example an absence of environmental appraisal (CPRE, 1994).

Integration / coterminousity The 1990s saw an emphasis on functional rather than territorial integration, so that the first round of RPG tended to be prepared in relative isolation from other areas of emerging regional activity. This sectoral segregation resulted in more than one regional strategy, often with conflicting views, which was reinforced by the continuation of individual thinking and visions pursued by different organisations (Roberts and Lloyd, 1998).

The sectoral approach to RPG was exacerbated by the separation of different areas of regional policy under different Government departments, most notably the Departments of the Environment; Transport; Trade and Industry; Employment and Education. While they were brought together to some extent through the Government Offices for the Regions these did not cover all areas of Whitehall activity with a spatial dimension (Roberts, 1997).

Integration was further undermined by a lack of coterminousity between the regional boundaries of central government departments and other regionally operating bodies (CPRE, 1994; Roberts, 1997). This problem worsened with the ever increasing number of quangos.

Centralisation and accountability A number of criticisms of the PPG 12 versions of RPG relate to an over-centralisation of power, and a lack of accountability and public participation. Firstly was the centralised influence on process and content (CPRE, 1994; Roberts, 1997). Despite RPG advice being prepared by regional associations of local authorities, the RPG itself was still prepared by the Regional Government Offices and issued by the Secretary of State. This tended to result in a watering down of regional issues, the inclusion or exclusion of policy according to central government priorities and also enabled central government to intervene and override in any conflicts between the RPG advice and government policy.

Secondly RPG had no formal means of accountability. It relied on the "widespread" consultation process required by the DoE to achieve stakeholder participation, legitimacy and regional ownership, and it is generally agreed that this was weak (CPRE, 1994; Thomas and Kimberly, 1995; Roberts, 1997; Baker, 1998).

Procedural weaknesses In addition to the lack of adequate mechanisms for public participation there were a number of weaknesses in the preparation of RPG, including:

- a process which discouraged weighing up the pros and cons of different options (Minnay, 1992);
- a "lowest common denominator" type approach;
- poor mechanisms for monitoring and review (CPRE, 1994);
- weak implementation mechanisms (Roberts, 1997); and
- a lack of any means of conflict resolution between the differing interests of the local authorities preparing the plan (Claydon, 1999). Disagreements varied from a relatively consensual view in the West Midlands, which produced the most coherent strategy, to incoherent and timid advice from regional planning conferences which avoided political and intra regional confrontations (Thomas and Kimberly, 1995).

Resources Inadequate resources for plan research and preparation tended to affect the quality of the advice and the level of inclusiveness and consultation in the RPG preparation process. (Thomas and Kimberly, 1995; Roberts, 1997). It also weakened implementation.

Conclusion

There are marked similarities in the weaknesses of the above two eras of regional planning - restricted scope, a sectoral approach to strategy preparation, over centralisation, a lack of accountability and a lack of mechanisms for conflict resolution and implementation. These weaknesses played a big part in the failure of regional planning to reach its potential. The conclusion of the paper develops these weaknesses into positive criteria for successful regional spatial planning in the future, building on the strengths from the 1960s and early 1990s which can be carried forward.

3. 1997 to 2000

At this point it is helpful to very briefly outline the development in regional organisation and activity since 1997 to see the direction in which things are moving and the degree to which past criticisms are being addressed.

Table 8.3: Key features of regional activity 1997–2000

Feature	Description – 1997–2000
Dominant direction Of policy	Top down and bottom up
Model of governance	Centralised with some functional devolution, i.e.: • centralised control • devolved civil servants • local authority regional associations • government appointed regional development agencies • regional chambers – "independent" council
Method of approach / degree of stakeholder involvement	Land use – public sector dominated partnership Economic – quango dominated partnership
Key strategic Objective	Guiding / promoting sustainable growth
Type of regional strategy / s	Superficial sectoral plans, although becoming more in depth: land use; transport; economic; sustainability; biodiversity
Integration of strategies	Some in theory. Little in practice
Integration of governance	Integration within GORs and co-ordination between GORs and Regional Planning Bodies, Regional Development Agencies and Regional Chambers. Developing between RPBs, RDAs and Chambers
Style of planning	Hierarchical – national direction some local variation
Policy instruments	Deregulated. Reduced financial support. Mixed public, private and voluntary sector. More co-ordination

Source: Adapted from Roberts and Lloyd (1998).

The new PPG 11 on Regional Planning Guidance (DETR, 2000) provides a strengthening of RPG, a process which is being undertaken in the current round of RPG preparation. It potentially widens RPG into a broader spatial strategy, extending its scope and content beyond development plan issues. Rather than simply giving advice to Government Offices for the Regions, RPG is now being prepared by regional associations of local authorities, termed Regional Planning Bodies (RPBs) with much greater involvement of regional stakeholders in the process. Key issues arising after extensive consultation are being discussed at a Public Examination. PPG 11 gives a challenging timetable for RPG preparation to overcome previous criticisms of the length of time taken to prepare guidance.

Regional Development Agencies were proposed in the White Paper "Building Partnerships for Prosperity" (DETR, 1997) and established in April 1999. They are powerful non departmental public bodies, accountable to ministers and informally to Regional Chambers. Members are appointed by the secretary of state and they have a large budget, incorporating among other things English Partnerships, some of the Rural Development Agency and the Single Regeneration Budget. They are charged with preparing Regional Economic Strategies, advising other agencies in the economic development field and administering regeneration and inward investment programmes. This means that they have limited powers of implementation.

Regional Chambers were introduced through the RDA legislation to "bring together elected representatives and other regional partners in a forum for the consideration of issues of shared interest, such as transport, land use planning, and economic development" (DETR, 1997). Membership can not be more than 70% local authority members. One of their functions is to provide some degree of democratic accountability to the RDAs. In some regions they have taken over from the local authority regional body, but in others they are remaining separate organisations. Some Chambers are also producing regional strategies, in some cases Regional Planning Guidance but in others a broader strategic framework.

Regions, either through their local authority regional body or through their Chamber, are now also required to produce wider reaching regional strategies, including a Regional Sustainability Framework and a Regional Cultural Strategy. These add to the strengthening of other bodies' regional strategies, for example the Housing Corporation's Regional Housing Statement.

The Impact of the EU on the Future of Regional Spatial Planning

Many EU objectives - for example the promotion of social and economic cohesion and reducing regional inequalities - have a spatial dimension, and many EU activities - Structural Funds, Trans-European Networks and environmental policy - impinge on regional spatial planning (Nadin, and Shaw, 1998; Roberts, Regional Studies, Vol. 31.9). The development of the European Spatial Development Perspective indicates a likely rise in the importance of a spatial dimension to EU policy. At the same time the concept of a Europe of the Regions, embedded in Europe 2000+ (CEC, 1995) advocating sub-national authorities as the most appropriate spatial level for implementing much EU policy (Roberts Regional Studies, Vol. 31.9), suggests that in the future EU initiatives will have a greater influence on spatial planning in member states. Indeed, Roberts foresees that, while the pace of change is uncertain, future EU policy will include co-ordinated and integrated spatial policy and will push member states into adopting internal structures to accommodate this. While this looks ore to the future, at present PPG 11 requires that RPG take account of inter-governmental and EU legislation, policies, programmes and funding regimes that impact on the region.

This section of the paper therefore looks at the EU initiatives with the greatest spatial dimension and begins to consider how they might influence the future development of regional spatial planning in England. The analysis feeds into the criteria for future regional spatial planning in the conclusion.

The European Spatial Development Perspective

The ESDP is an intergovernmental document, acting as a policy framework for co-operation between community sectoral policies with significant spatial impacts - an EU spatial policy. It was adopted by the Committee for Spatial Development in Potsdam in May 1999. Its objectives are environmentally sound economic development and balanced spatial development at the regional level.

It is not yet clear what impact the ESDP will have on domestic sectoral and spatial policies within England, or on their co-ordination and implementation (Nadin and Shaw, 1998). Strictly speaking the EU, at present, does not have a legislative competence in spatial planning, although there is a spatial dimension to much EU policy, including the

Maastricht and Amsterdam Treaties. Also, the ESDP is not legally binding, giving Member states the option to ignore it if they wish. However, in other policy areas this sort of influence has led to the establishment of new Community competences. It is therefore reasonable to speculate that it will be a potentially significant influence in terms of spatial policy priorities.

In the future the ESDP is likely to influence regional spatial planning in two ways.

One of the roles of the ESDP is to provide "a framework for policy thinking which will provide a stimulus at the national, regional and local level" (Williams, 1999). In this respect it is already in a position to influence regional spatial planning policy. Whether a regional spatial plan will want to take on board the ESDP will depend to a large extent on the relevance of the policy priorities and policy processes.

Secondly, although there is, as yet, no formal relationship between the ESDP and the structural funds, Williams (1999) speculates that, in time, the ESDP will play an increasing role in planning investment and infrastructure in the EU. While it is too late for the ESDP to formally influence the bidding for the 2000 - 2006 round of structural funds it may well have an informal influence. For example DG Regio are considering criteria for Single Programming Documents which accord with the ESDP. This would provide regional spatial planning bodies with a very good incentive to take the ESDP on board in regional spatial planning as a means of obtaining funding for implementation.

Structural Funds

Since 1994 the structural funds have been allocated in England according to bids though a regional Single Programming Document, in effect a form of regional plan, albeit with only a minor spatial dimension.

It is apparent that these single programming documents will to some extent be duplicated in the preparation of both Regional Planning Guidance and the Regional Development Agencies' Economic Strategies and visa versa. PPG 11 goes so far as to say that it is important that RPG, RDAs and Single Programming documents share a consistent vision. It would therefore seem valuable, as a minimum, to co-ordinate these strategies to achieve maximum benefit, avoid unnecessary duplication of effort, resolve any conflicts and to avoid weakening activity through strategies undermining one another. There are also implications for implementation. Reflecting the Single Programming Document in the regional spatial strategy could provide funds for policy implementation.

The integration and partnership approach to EU Structural Funds and ESDP is also relevant in so far as it reflects wider EU thinking which may have an increasingly direct bearing on spatial planning. It endorses the need for integration and partnership if effective structures are to be established which meet the needs of a wide range of activities and implementation methods at the regional level.

INTERREG 2000 - 2006

In October 1999 the EU set out the guidelines for INTERREG III (Guidelines for the Community Initiative INTERREG 2000 - 2006), a form of transnational structural funds for cross border spatial planning initiatives. It is Strand B, for transnational regions, which is most relevant to England and is to receive between 14% and 40% of the budget. All of the UK is eligible for funding under either the North West Metropolitan area, the North Atlantic Sea area or the Western Seaboard.

Although INTERREG funds transnational projects, these transnational territories are made up of adjacent regions, which in turn benefit from EU funding for spatial planning initiatives. There are therefore advantages in having a degree of complementarity between regional spatial planning strategies and transnational INTERREG funding bids. The regional spatial strategy can contribute to the aims of the transnational area, while the funding gained from the transnational strategy through INTERREG can aid in the implementation of regional spatial planning policy, particularly now that sufficient funds are available for infrastructure works as well as for feasibility studies.

Other Initiatives

Other EU initiatives which are also likely to impact on regional spatial planning in the future include:

- Environmental Policy;
- Trans-European Networks;
- Territorial Impact Assessment; and
- Common Agricultural Policy.

Calls for Future Changes to Regional Spatial Planning

This section of the paper analyses literature putting forward the views of planning academics and practitioners as to the direction regional spatial planning should take.

Content The content of a regional spatial strategy will very much depend on its function and its degree of integration with other regional strategies. PPG 11 on Regional Planning Guidance (DETR, 1999) now requires a comprehensive spatial strategy, broadening the scope significantly from past restrictions to development plan matters.

However, as the TCPA (1999) point out, even in conjunction with RDAs, RPG does not encompass everything with a regional spatial dimension. With the RTPI (1998b) and CPRE (1999) they believe that in the future RPG, or a regional spatial plan, should encompass most policy issues with a spatial impact. They foresee no problem in including issues which can not be delivered through the development plan system (RTPI, 1998a), although they do raise awareness of the dangers in giving too broad a brief.

It might be that, rather than a regional spatial strategy covering all topics, it should select only those of relevance to the region (RTPI, 1998a).

Sustainability There is widespread support for the principles of sustainable development to underpin future regional strategies (for example: Roberts, 1999; LGA, 2000). It is a central part of both EU and national policy, demonstrated in the requirement for sustainability criteria in structural fund programmes and a sustainability appraisal of RPG and Regional Economic Strategies. Indeed, Marshall (1998) argues that this is the new focus to regional policy and Roberts (1999) that "the sustainable development debate is a matter of immense significance and it cannot be resisted or ignored".

Integration The cry remains for better integration of spatial planning with other regional strategies, particularly since sustainability issues have highlighted the interrelationships between land use, economic, environmental and social issues (Roberts, 1998 and 1999; TCPA, 1999; CPRE, 1999). Another frequently cited reason for the greater integration of regional strategies is the need to avoid the duplication of effort inherent in preparing a tranche of independent, sectoral strategies and the benefits of strategies pulling in the same direction. The degree of integration being suggested varies.

At the bottom is the integration spectrum is the co-ordination of regional strategies with a spatial dimension. Government support for this principle is demonstrated by advice on both RPG and RDAs (Building Partnerships for Prosperity, DETR, 1997; The Future of Regional Planning Guidance, DETR, 1997). However, if co-ordination across a range of strategies is to be achieved then some form of co-ordinating element would help. Roberts (1999) suggests the preparation of an overarching vision for the region, based on the principles of sustainable development. This would then be used as a base document for all regional strategies. Shared aims and objectives or a shared sustainability or other appraisal could be used in the same way. The Regional Sustainable Development Framework will perform this sort of function.

Going one step further is the concept of an overarching strategy, with one strategy providing a framework within which other regional strategies would be prepared (Roberts, 1998; TCPA, 1999). Many advocates argue that a spatial dimension tends to be a co-ordinating theme, so that a regional spatial strategy could provide the most appropriate framework (CPRE, 1999; RTPI, 1998; TCPA, 1999). Other regional strategies, for example the RDA's economic strategy, could take on a role more of implementation of policy (CPRE, 1999). Sustainability is also a co-ordinating theme though, so that the Regional Sustainable Development Framework could equally claim such a role.

At the far extreme, regions could have one, comprehensive regional strategy. Baker (1995) suggests RPG be allowed to broaden its remit to become a more comprehensive form of regional planning, encompassing economic development and future infrastructure and public spending plans. The LGA (2000) supports a spatial development strategy that binds together key regional strategies. In this respect Roberts and Lloyd (1998) believe that "the provision of a single, overarching, regional strategy would help to clarify matters and would provide a starting point for all future strategic programmes. Indeed, there is a good case for bringing most regional level activities and functions together in a single regional programme". This would also help to bring together the different visions for a region held by different organisations, many with similar overall hopes, but coming from different perspectives.

Coterminousity One problem for the future of further co-ordination of regional strategies is the lack of coterminousity between regional boundaries. For example in the Eastern and South East regions even the RDA and RPG boundaries differ considerably. The problem worsens when other policy areas are introduced.

Regional government / regional accountability It is generally agreed that regional strategies are being prepared under varying and insufficient systems of democratic accountability (e.g. Roberts, 1997; 1998). A wider ranging and more formal system of regional governance is considered necessary to: improve accountability of regional strategies; enable regional partnerships and joint working (Roberts, 1999); implement regional proposals (Roberts, 1998) and EU policies (Roberts, 1997) and funds. Various degrees of regional accountability and control are advocated by different parties.

Several organisations, for example the RTPI (1998b), consider the preparation of RPG by regional planning bodies as a first step to improving accountability to the people and ownership by regional players. Both the LGA (2000) and CPRE (1999) would like to see the regional planning bodies extended to include regional stakeholders, as they are sceptical that a local authority / Government Office partnership will provide for public scrutiny or ownership. PPG 11 supports this to some degree by advising that the Regional Chambers contain the Regional Planning Body, but this is not a requirement. The LGA believes that the Public Examination Panel should report to this regional planning body rather than to the Secretary of State. Roberts (T. 1998) takes this view one step further and advocates the concept of a Regional Planning Authority, at least indirectly accountable to the electorate, and funded through powers of precept on local authorities.

As a means of developing regional accountability further, the RTPI (1998(b)) and CPRE (1999) among others, support the establishment of Regional Chambers as an interim measure towards statutory assemblies. The logical extension of these assemblies is democratically elected regional government, which has widespread support as the most accountable option, enabling the co-ordination of regional strategies and ensuring the implementation of plans (Roberts, 1997).

Conflict resolution Historically a criticism of regional planning has been the risk of a "lowest common denominator" approach being taken in the absence of any means of resolving conflict in an essentially collaborative (albeit restricted) process of plan preparation. Baker, Deas and Wong

(1999) point out that, in the case of RPG at least, regional plans are still prepared by a system of joint policy making so that the risks remain. This is likely to increase if participation by regional stakeholders increases.

As regions are also required to pursue the often conflicting aims of economic competitiveness and sustainable development without the powers to prioritise (Baker, Deas and Wong (1999) there is also a need for a means of conflict resolution between the differing aims and sectors of a regional spatial strategy.

Preparation / process Looking to the future there is also support for modifying the process for preparing RPG. There seems to be general support for the stringent timetable for RPG preparation and review set out in PPG 11, although there are concerns over feasibility and the implications for stakeholder participation (RTPI, 1998; CPRE, 1999).

Looking at plan preparation, a need is perceived for a comprehensive review of regional problems and opportunities (Roberts and Lloyd, 1998) which takes into account the interrelationships of the different policy sectors, particularly if RPG is broadened.

Roberts and Lloyd (1998) and Roberts (1999) highlight the imperative for regional spatial planning to reflect the full views of regional stakeholders, while Baker (1995) and the RTPI (1998) dwell on the need for effective consultation. There is general support for plan preparation by regional associations, in partnership with regional stakeholders from the beginning, to improve validity. A public examination to address outstanding issues of debate is also advocated, now enshrined in PPG 11.

There is widespread support for a sustainability appraisal of regional strategies, although CPRE (1999) is concerned that this will lead to the neglect of the environment by replacing an environmental appraisal. The TCPA (1999) suggests that this appraisal is undertaken by a Regional Planning Commission, appointed jointly by the DETR, Local Government Association and partner organisations, and consisting of a permanent committee of independent regional planning, development and academic experts. This would improve the accountability of appraisal procedures and, again, improve ownership of the process.

Finally the importance of procedures for the monitoring and review of regional plans, including quantifiable targets or performance indictors are stressed (Roberts, 1997; 1999; RTPI, 1998b; TCPA, 1999; CPRE, 1999). Applying these to all regional strategies could also help co-ordination. The TCPA believes that targets and performance indicators should be specific to the region, to ensure their relevance, although this would also make it easier for regions to move away from national policy. Some half way point between the two - perhaps a regional interpretation of national targets - would be most effective.

Resources While all regions now have a Regional Development Agency to undertake the preparation of the Regional Economic Strategy, many regions are still without permanent or experienced staff to undertake Regional Planning Guidance. This "capacity congestion" (Roberts, 1999) has hindered effective research and development of RPG in particular, and has held back the greater commitment to public participation and consultation. Roberts (1997) and the RTPI (1992, in Baker 1998) are among the authors on regional planning that are calling for an independent and permanent team of professionals at the regional level to undertake, monitor and review regional planing guidance. This call is equally valid for any broadened form of regional spatial planning and is rapidly taking place.

Implementation Concern remains that regional spatial planning remains ill equipped with the necessary powers of implementation (Roberts, 1997; TCPA, 1999). The most advocated improvement is through strengthening links between plan policy and public expenditure, which in tern would indicate to the private sector where it would be wise to invest (James, 1968; Hall, 1995; CPRE, 1994). This is not perfect as real influence over large areas of activity such as infrastructure provision would still be lacking.

Conclusion

The body of literature reviewed - on the strengths and weaknesses of past eras of regional planning, on the impact of the EU, and of writings on what academics and practitioners feel needs to change - provides a fairly clear and consensual picture of the direction regional spatial planning needs to take. Obviously it provides only a very broad picture, and no doubt opinions would vary far more widely in the detail. The conclusion brings together the findings from the three areas of literature and summarises them in the form of a set of criteria for successful regional spatial planning in the future, some of which have now been partially met by advice in PPG 11.

A regional spatial strategy needs to look beyond not only the requirements of development plans but also beyond traditional land use planning issues to consider all strategic policy issues with a spatial dimension. In particular it should be based upon the principles of sustainable development. Of central importance is that it breaks out from its sectoral mode and embraces at least some degree of functional integration with other sectoral regional strategies. This needs to be backed up by better co-ordination of the various regional organisations acting and making strategies at the regional level, and by better sectoral co-ordination at central government level.

A regional spatial strategy needs to be rigorous, robust, visionary and relevant to the individual region. An important part of achieving this is through developments in the strategy preparation process. The timetable for completion needs to allow for a review every 4 to 5 years. Reviews need ongoing, regular, background research and analysis, including a comprehensive study of the state of the region, its problems and opportunities; they need a partnership approach to preparation; an assessment of plan options, including public consultation; they need an EIP to discuss outstanding or controversial issues; targets/performance indicators, based on a regional interpretation of national indicators; and mechanisms for monitoring and review. This would be aided by better resources for plan research and preparation including a permanent, independent, regional professional "planning" team. As a part of this process a regional spatial strategy needs a means of conflict resolution between players in a collaborative process of plan preparation and a means of resolving conflict between different aims and aspects of a regional spatial strategy. A regional spatial strategy will also need effective mechanisms for implementation, perhaps even linked with investment priorities and the allocation of public and guidance of private.

To be successful a regional spatial strategy needs to be regionally led and owned, through a regionally distinct organisational structure, based on the belief systems and values of the component regional stakeholders. It requires a bottom up networks of regional organisations, and a more democratically accountable and powerful system of regional governance. Accountability is an important part of this, requiring open debate, genuine regional stakeholder participation and a partnership approach to strategy preparation at all stages.

Finally a regional spatial strategy needs to take account of its EU context and take account of the likely future role of EU spatial policy in regional spatial planning. In this respect it should incorporate the principles set out in the ESDP; co-ordinate with Single Programming Documents; be able to cater for transnational and interregional planning initiatives and include a partnership approach to policy development and implementation, including formal structures.

Current regional developments, particularly with the advent of PPG 11, are moving in the direction of many of these criteria. But a detailed assessment of the regional picture would show that we still have some way to go if we are to avoid the mistakes of the past, build on its strengths and meet the needs of regional stakeholders in regional spatial planning.

References

Alden, J. (1998), *Scenarios for the Future of the British Planning System*, Draft Report for the Royal Town Planning Institute, Cardiff University, Cardiff.

Baker, M. (1995), 'Return to the Regions?' *Town and Country Planning*, October 1995.

Baker, M. (1998), 'Planning for the English Regions: A Review of the Secretary of State's Regional Planning Guidance' *Planning Practice and Research*, vol.13, no.2.

Baker, M.; Deas, I. and Wong, C. (1999), *Obscure Ritual or Administrative Luxury? Integrating Strategic Planning and Regional Development.*

Barraclough, D. (1999), 'What is the Relationship Between RDA Strategies and Regional Guidance?' *Planning*, 19 March 1999.

Breheny, M. and Hall, P. (1984), 'Strategic Planning and the Victory of the Know-Nothing School', *Built Environment*, vol. 10, no. 2.

Brenikov, P. (1967), 'Regional Planning in Practice', *Planning Outlook*, vol. 2, Spring.

Buunk, W.; Hetsen, H. and Jansen, A. (1999), 'From Sectoral to Regional Policies: a First Step Towards Spatial Planning in the European Union?', *European Planning Studies*, vol. 7, no. 1.

Cabinet Office Performance and Innovation Unit (2000), *Reaching Out: The Role of Central Government at the Regional and Local Level*, The Cabinet Office, London, February.

Claydon, J. (1999), 'The South West Through the Looking Glass', *Town and Country Planning*, March 1998, pp. 70-80.

Cochrane, A. (1999), 'Thinking About the English Regions: Some Notes', *paper for the ESRC Oxford Brooks University Regional Research Seminar 2*, June 3rd 1999.

Committee on Spatial Development (1999) *European Spatial Development Perspective: Towards Balanced and Sustainable Development of the Territory of the EU*, EU, Potsdam, May.

CPRE (1994), *Greening the Regions*, CPRE, London.

CPRE (1998), *The Future of Regional Planning Guidance: CPRE response to the DETR consultation paper*, April.

Crookston, M. (1998), 'Regional Policy and the Great Housing Debate', *Town and County Planning*, July, pp. 213-215.

DETR (1997), Building *Partnerships for Prosperity*, Cm 3814, December, DETR, London.

DETR (1999), *Planning Policy Guidance Note 11: Regional Planning. Public Consultation Draft*, HMSO, February, DETR, London.
DETR (1999), *Regional Development Agencies', Regional Strategies*, Apr, DETR, London.
DETR (1999), *Supplementary Guidance to Regional Development Agencies*, Apr, DETR, London.
DETR (1999), *Modernising Planning; A Progress Report*, HMSO, Apr, DETR, London.
DETR (1999), *Subsidiarity and Proportionality in Spatial Planning Activities in the EU: Executive Summary*, EU, Brussels, September, DETR, London.
EU (1999), *The Structural Funds and Their Co-ordination with the Cohesion Fund: Guidelines for programmes in the period 2000 – 06*, Communication of the Commission, Brussels, 1 July.
EU (1999), *Guidelines for the Community Initiative INTERREG 2000 – 2006*, EU, Brussels, October.
Glasson, G. (1992), 'The Rise and Fall of Regional Planning in Economically Advanced Nations', *Urban Studies*, vol. 29, nos. 3 and 4, pp. 505-531.
Glasson, J. (1978), *An Introduction to Regional Planning*, Hutchinson, London.
Glasson, S. (1995), 'Regional Planning and the Environment: Time for a SEA Change', *Urban Studies*, vol. 32, nos. 4 - 5, pp. 713-731.
Hague, C. (1996), 'Spatial Planing in Europe: the Issues for Spatial Planning in Britain', *Town Planning Review*, vol. 64, no. 4.
Hall, P. (1995), 'Planning Strategies for Cities and Regions', *Town and Country Planning*, May / June 1995.
Hull, A. (1998), 'Changing Regional Governance and the Role of Integrated Spatial Planning', in *New Lifestyles, New Regions - Regional Studies Association Conference Proceedings*, November 1998.
James, J.R. (1968), 'Regions and Regional Planning', *Geography*, vol. 64, no. 2.
Local Government Association (2000), *LGA Statement on Regional Planning*, LGA, London, January.
Martin, S. and Pearce, G. (1993), 'European Regional Development Strategies: Strengthening Meso - Government in the UK?', *Regional Studies*, vol. 27, no.7, pp. 681-685.
Mawson, J. (1997), 'The English Regional Debate, Towards Government or Governance?', in Bradbury, J. and Mawson, J. (eds.), *British Regionalism and Devolution*, Jessica Kingsley and The Regional Studies Association, London.
Mawson, J. (1998), 'Transnational Spatial Planning in Europe: The Role of INTERREG II in the UK', *Regional Studies*, vol. 32, no. 3.
Mawson, J. and Spencer, K. (1997), 'The Origins and Operation of the Government Offices for the English Regions', in Bradbury, J. and Mawson, J. (eds.), *British Regionalism and Devolution*, Jessica Kingsley and The Regional Studies Association, London.
Middleton, M. (1998), 'Have Vision... Need Framework', *Town and County Planning*, July 1998, pp. 210-211.
Minay, C. L. (1992), 'Developing Regional Planning Guidance in England and Wales, A Review Symposium', *Town Planning Review*, vol. 63, no. 4.
Nadin, V. and Shaw, D. (1998), 'Transnational Spatial Planning in Europe: the Role of INTERREG IIc in the UK', *Regional Studies*, vol. 32, no. 3, pp. 281-299.
Roberts, P. (1994), 'Sustainable Regional Planning', *Regional Studies*, vol. 28, no. 8, pp. 781-787.
Roberts, P. (1996), 'Regional Planning Guidance in England and Wales: back to the future?' *Town Planning Review*, vol. 67, no. 1.

Roberts, P. (1997), 'Strategies for the Stateless Nation: Sustainable Policies for the Regions in Europe', *Regional Studies*, vol. 31, no. 9 pp. 875-882.

Roberts, P. (1999), *Finding a Way Through the Regional Maze: The Journey So Far*, TCPA / RSPB Annual Planners Conference, Birmingham, 26th January 1999.

Roberts, P. (1999), 'Incorporating the Environment: Towards a New Model for Regional Planning', paper for the *ESRC Oxford Brooks University Regional Research Seminar 2*, June 3rd 1999.

Roberts, P. (1999), 'Guiding Regional Development Strategies', *Town and County Planning*, January 1999, pp. 4-5.

Roberts, P. and Lloyd, G. (1998), *Institutional aspects of Regional Planning, Management and Development: Models and Lessons from the English Experience*, Centre for Planing Research, School of Town and Regional Planning, University of Dundee, Dundee.

Roberts, T. (1998), 'Revitalising the Statutory System - Chucking a Few Stones into the Pool', paper to *RTPI Councillor's summer school.*

RSPB (1997), *Regional Development Agencies Discussion Paper: Comments from the RSPB*, RSPB Sandy, Bedford.

RTPI (1997), *A Consultation Response to the DETR on the First Draft of the European Spatial Development Perspective*, RTPI, London.

RTPI (1998), *Regional Development Agencies: Draft Guidance on RDA's Strategies*, letter ref. R1A DJR? ADB.

RTPI (1998), *The Future of Regional Planning Guidance; memorandum of observations to the DETR on its consultation paper*, RTPI, London.

RTPI (1999), *The Future of Regional Planning Guidance; memorandum of observations to DETR on its consultation of a new Planning Policy Guidance Note*, RTPI, London.

Sharpe, L. J. (1997), 'British Regionalism and the Link with Regional Planning', in Bradbury, J. and Mawson, J. (eds.), *British Regionalism and Devolution*, Jessica Kingsley and The Regional Studies Association, London.

Shutt, J. and Colwell, A. (1998), 'Towards 2006: European Regional Policy and UK Local Government: A New Regional Agenda', *European Planning Studies*, vol. 6, no. 6.

TCPA (1999), 'Moving Forward with the RDAs, A Response from the TCPA', *Town and Country Planning*, March.

Thomas, K. and Kimberly, S. (1995), 'Rediscovering Regional Planning? Progress on Regional Planning Guidance in England', *Regional Studies*, vol. 25, no. 4.

Williams, R. (1999), 'Constructing the European Spatial Development Perspective: Consensus Without a Competence', *Regional Studies*, vol. 33, no. 8.

Williams, R. (1999), 'European Integration Through Transnational Spatial Planning', paper submitted to the conference on *The Changing Role of Boarder Regions as a Catalyst for New Policy Directions*, Faculty of European Studies, University of the West of England, Bristol, May.

9 Building Sustainable Networks: A Study of Public Participation and Social Capital

NICK BAILEY AND DEBORAH PEEL

Introduction

This chapter takes the Government's (1998a) White Paper *Modern Local Government: In Touch with the People* as its starting point and draws on the theoretical literature relating to social capital and public participation. It describes the development and delivery of a public consultation strategy instigated by Brighton and Hove Council as part of the preparation of its pre-deposit draft local development plan. It evaluates the experiences of participants involved in a number of linked participation processes to do with strategic planning issues. The objective is to investigate the links between a number of current themes which contribute to the "democratic renewal" and "active citizenship" debates by focusing on town planning and urban regeneration. The research was prompted by a series of questions:

- Are there ways of targeting and encouraging non-joiners to participate?
- At the local level, is it just a small proportion of individuals who are actively involved in voluntary and community organisations – the "usual suspects"?
- Is it possible to encourage and to sustain involvement?
- Are members of local organisations likely to widen their interest in community activities and local democracy on the basis of positive experiences of participation?
- Is there any evidence that the newer methods of community participation are likely to lead to greater community involvement and democratic renewal?

In this chapter we distinguish between the "narrow" issues of the effectiveness of methods of participation, as well as the "wider" issues of whether involvement in some local networks and organisations necessarily leads to a greater commitment to, and involvement in, local democratic processes. In doing so, we consider both individual and collective aspects of citizen involvement.

The chapter is divided into three sections. We begin by briefly reviewing some of the terminology currently dominating social policy. We then report the findings from our case study of Brighton and Hove. The final section sets out some thoughts on lessons learned and good practice.

Theoretical Overview: Social Capital and Community Capacity Building; Public Participation and Democratic Renewal

Social Capital and Community Capacity Building

Set up in 1992 by the late John Smith, the Commission on Social Justice reported its strategies for national review in 1994. Amongst its recommendations was that of "Responsibility: Making a Good Society":

> A good society depends not just on the economic success of the individual, but the "social capital" of the community (Commission on Social Justice, 1994: 10).

The Commission on Social Justice quotes Putnam, who describes social capital as the "networks, norms and trust that facilitate co-ordination and co-operation for mutual benefit" and continues that it

> consists of the institutions and relationships of a thriving civil society - from networks of neighbours to extended families, community groups to religious organisations, local businesses to local public services, youth clubs to parent-teacher associations, playgroups to police on the beat (*ibid*: 307-308).

Bottom-up regeneration, building strong communities and reviving local democracy are also key themes of the Report and are closely intertwined. "Community capacity building", linking economic, physical and social capital, now underpins the Government's urban regeneration programs. It is "a process for empowering a community - and all the stakeholders in that community - to plan effectively for its own future" (LRN/PLCRC, 1999). "It can be thought of as comprising a mix of skills, knowledge, resources, power and influence" (DoE, 1995: 128). Thake (1999) has described how the waft and weft of civil society have been worn away as recognised institutions of the nation state, such as the trade unions, have been undermined or people simply move away from neighbourhoods. He warns, however, against the patronising colonialist mindset inherent in the notion of capacity-building and which assumes a deficit model, and, instead, advocates capacity-building for growth, based upon an understanding and recognition of resources. He argues for trust between individuals and groups and for there to be an acknowledgment of the fact that capacity-building is not a one-way street, with reverse capacity building of institutional players also required. At the heart of stakeholder engagement is thus a learning process: learning about the views and skills of local communities, listening to the needs of local people and finding ways of jointly meeting these needs. Increasingly, it is the various arena for learning which have become the focus for attention:

> Social capital is about the strength of communities and their ability to make a difference to the lives of their members. At the heart of the ability to make a difference, the ability to shape the course of one's own life, is the issue of democracy (Commission on Social Justice, 1994: 350).

Social capital can, therefore, be seen as being of crucial importance in providing a foundation for participation and effective local democracy, linking community associations, private business and public agencies. How far do ideas of social capital relate to the functions of local government and public policy making? Are we talking in terms of "active communities" or looser networks of active organisations and individuals?

Public Participation and Democratic Renewal

Turnout at local elections has fallen to 40 percent, or sometimes less. Since their election in May 1997, modernising local government, devolving political power and reducing the democratic deficit have been high on the current Labour Government's agenda. The objective is to tackle the old culture of paternalistic provision of services; the risk of corruption and wrong-doing by inward-looking administrations; and the associated indifference of communities to democratic processes. The aim is to overcome apathy and encourage "active citizenship", to have "local people taking a lively interest in their council and its affairs" (DETR, 1999: para 1.21). It is a two-way street: modern local government is to be "in touch with the people" and local people are to have a "bigger say":

> ...modern councils fit for the 21st century are built on a culture of openness and ready accountability. They have clear and effective political leadership, to catch and retain local people's interest and ensure local accountability. Public participation in debate and decision making is valued with strategies in place to inform and engage local opinion (DETR, 1999: para 1.2).

Leach and Wingfield (1999) summarise the four key elements of the democratic renewal program within the White Paper as:

- a set of proposals focused on improving turnout in local elections;
- a commitment and proposed legislative framework for facilitating community leadership;
- a set of proposals for transforming the internal political management structures and processes of local authorities (centred on the idea of an executive/assembly split); and, importantly for this discussion;
- guidelines aimed at developing opportunities for citizens to participate in local government.

They state that whilst "the public participation proposals are so far the least prescriptive of the elements in the government's modernisation program [...] they are arguably the most fundamental" (1999: 46). The Government wishes to see "consultation and participation embedded into the culture of all councils [...] Every Council will have to decide which methods are the most appropriate in their own particular circumstances" (DETR, 1998a: paras 4.6 and 4.7).

It is beyond the remit of this paper to review public participation in planning. This has been done at length elsewhere (see, for example, Arnstein, 1969; Thornley, 1977; Thomas, 1996; Stewart, 1996). Suffice it to say, that there is a burgeoning literature on more direct, deliberative and innovative arrangements for public participation, offering practical guidance and/or developing new theoretical perspectives (Healey, 1997; DETR, 1998; Sanderson, 1999; Wates, 2000). In its evaluation of the advantages and disadvantages of alternative methods for consultation of development plans, the University of Westminster reviewed a diverse range of options: newspapers; press releases; newspaper advertisements; council leaflets; posters; radio; exhibitions; public meetings; surveys; focus groups; citizens' juries; Planning for Real and visioning (Brighton and Hove Council, 1998: 4-5). Since then, the DETR has published *Planning Policy Guidance Note 12: Development Plans* (1999) (PPG12) which came into effect in January 2000 and which formalises pre-inquiry procedures and is intended to speed up the plan-making process. In terms of pre-deposit consultation, PPG12 states:

> Local people and other interested bodies should have the opportunity to express their views on plan proposals before those proposals are finalised. The aim should be to encourage local people to participate actively in the preparation of plans from the earliest stages so that they can be fully involved in decisions about the pattern of development in their area. Consultation with the general public, community groups, conservation and amenity groups, business, development and infrastructure interests helps local planning authorities secure a degree of consensus over the future development and use of land in their area (DETR, 1999: para. 2.10).

With regard to the form that consultation should take, PPG12 does not specify how pre-deposit consultation should be undertaken by local authorities, but does require them to explain how their proposals and mechanisms established for comment are "sufficient to encourage all sectors of the community to be involved". Where full replacement plans are concerned, they should

> aim to consult with a range of key organisations and business interests on the strategic principles that will ultimately underpin the policies in the structure plan. But different forms of consultation may be needed for the public at large (DETR, 1999: para. 2.13).

It is the public at large with whom we are concerned here.

Although the revised PPG12 was published after the adoption of the Consultation Strategy adopted by Brighton and Hove Council, it was in draft form at the time and reinforces the approach taken. In the wider sense, however, to what extent does public participation assist democratic renewal? The next section sets out our case study findings.

Case Study: Public Participation in Brighton and Hove

In April 1997 a new unitary authority was created by amalgamating the adjoining towns of Brighton and Hove into a single tier council. Considerable rationalisation was required to amalgamate the two councils, as well as incorporating those services previously provided by East Sussex County Council. One of the first decisions taken by the new Council's Land Use and Transportation Sub-Committee was to prepare a new Local Plan while ensuring that the views of the local community were fully taken into account. Rather than simply consult on the basis of a draft plan, the Council determined first to seek the input of residents in drawing up the pre-deposit policies. In April 1998 a team from the University of Westminster was commissioned to prepare a consultation strategy and, on the basis of extensive interviews with key stakeholders, to recommend a range of methods of consultation to ensure that the full range of public opinion was taken into account. The University team was retained to monitor and observe the consultation process and to carry out a final evaluation of community involvement.

We do not discuss our evaluation of the Strategy as a whole, but instead, describe the consultation process and methods used to involve local citizens and then evaluate this by focusing on the views of the various participants, based on written sources and a series of interviews with Council officers, councilors and members of the community. This forms the basis for our conclusions about the advantages and disadvantages of the different methods employed and the wider implications for enhancing social capital and promoting democratic renewal.

The Consultation Strategy

In preparing the Consultation Strategy, the University team was aware of the Council's commitment while a shadow authority to "listen to, identify and respond to local needs" and that the first of the Council's seven *Values* was "putting people first". This included the Council's commitment to be responsive "to people's needs, giving them information and access to decision making". The team was also aware that, within its Community Planning initiative, the Council had identified the twin goals of engaging the community in the democratic process and working with local people to set priorities. In addition, both Council members and the Planning Policy Team, based in the Environmental Services Department, were fully committed to ensuring that all sections of the community should be encouraged to participate.

The team carried out an extensive literature review of the wide range of consultation methods currently in use in the UK. They were also aware that the Department of the Environment, Transport and the Regions (DETR) was carrying out a review of development plan procedures and, as part of the review of PPG12, was committed to "promote shorter, clearer plans and better targeted consultation". The preparatory stages also included a series of interviews with key officers in different Council departments in order to identify what other consultation initiatives had been completed or were anticipated, and which groups and organisations were used for consultation purposes.

In summary, the following stages were specified in the Strategy:

- identify and analyse examples of good practice in public consultation elsewhere, relevant to the Brighton and Hove context. Briefly review alternative methods and techniques of public consultation;
- identify relevant public, private and community sector interests which should be encouraged to participate. In particular, review the social, community and sectoral interest groups which do not normally comment on planning matters;
- discuss with key officers their perceptions of the purpose, expectations and likely format of the Local Plan, as well as the contribution they see public consultation making;
- prepare a strategy for public consultation; and
- monitor the public consultation process, provide additional advice where appropriate, and submit a final report to committee with an assessment of methodologies and outcomes (Brighton and Hove Council, 1998: 3).

The team identified the criteria for selecting methods of consultation as:

- involve the widest possible range of community, business and voluntary sector interests currently represented in Brighton and Hove;
- make contact with sections of the population which do not normally get involved in planning issues;
- promote a non-confrontational dialogue between individuals and interests; and
- clearly structure meetings and events so that participants are fully briefed on their purpose and are kept informed of the outcomes and subsequent stages.

The outcome was that the team recommended the use of a series of participation methods, the two primary methods being focus groups and community visioning workshops. The Consultation Strategy was adopted in full in April 1998 and a series of consultation events was held between October 1998 and January 1999. Table 9.1, below, summarises some of the main organisational aspects of the focus groups.

Table 9.1: Organisational aspects of the focus groups

Composition	Between 6 and 12 members took part each attending two separate sessions. In total, 64 organisations were represented.
Selection process	Members were selected by the facilitators and the Planning Policy Team from businesses, developers, representative organisations and associations. All meetings were attended by a Planning Officer.
Topics for discussion	7 topics were selected from the SERPLAN draft policy document A Sustainable Development Strategy for the South East considered appropriate to Brighton and Hove: housing; sport, recreation, leisure, tourism and the arts; higher density and mixed use development; retail and town centres; sustainable development; wildlife; and employment.
Facilitation	Chris Blandford Associates, planning consultants from East Sussex.
Briefing	All members were sent briefing material on the topic and purpose of the event in advance.
Feedback	All participants were asked to complete a feedback form.

Source: Brighton and Hove Council, "Focus Groups" (1999a).

One advantage of focus groups is that discussion can be guided to key questions which can be introduced in briefing material. Moreover, it is often easier to achieve a consensus between a small number of participants, whilst the facilitator can ensure all have a fair hearing. Interaction between different interests can lead to innovative proposals and, where members gel, the focus group can become sustainable. Nevertheless, there are disadvantages. Much depends on the quality of facilitation and discussion can become too wide-ranging or be "highjacked" by forceful individuals; indeed, some participants may feel excluded. It is sometimes difficult to maintain the focus, for example, on planning issues, rather than a wider agenda. Notwithstanding careful planning, absence or non-attendance can bias outcomes (Brighton and Hove Council, 1998).

In this particular case, the facilitator's comments are of particular interest because they highlight the learning which is an inherent part of such deliberative participatory practices (Forester, 1999). Facilitated dialogue can bring new understanding:

> One positive outcome of the focus groups is that the discussions seem to have enhanced understanding between different parties. In some cases participants have been pleasantly surprised by the level of agreement between "opposing" interest groups. Moreover, in many cases participants were able to contribute and air views on a much broader range of issues than they had originally thought (Brighton and Hove Council, 1999a: 5).

Moreover, in the facilitator's view, new social relationships had indeed been formed through the process:

> Particularly encouraging was the way in which the focus groups opened up opportunities for communication between participants, and it appeared that this "networking" would continue even after the meetings had finished (ibid: 6).

The second primary method used in the Strategy were community visioning workshops. These aimed to assist in establishing:

- a context and vision for the local plan;
- a set of objectives for the local plan;

and to inform:

- the preparation of land-use policies;
- the preparation of site specific land-use proposals (Brighton and Hove Council, 1999b: 5).

The role of the facilitator is to guide the discussion and focus on key issues.

The advantages of this particular consultation method are that people with similar ages, interests and outlooks can discuss their preferences and priorities in small groups which can stimulate everyone to contribute and generate confidence in expressing opinions. Moreover, small groups can give the impression that individuals are privileged in being invited and can thus feel valued. Forester (1999) suggests that such public participation is a way to re-create social identities. An organisational advantage is that participants can be selected to be representative of a sector of the total population. Yet, as with the focus group approach, there are disadvantages. There are no guarantees in practice that views will be representative and, again, small groups can be influenced by forceful individuals. Moreover, some may resent being grouped with others in the same category, for example, pensioners. Some groups, such as non-English speakers, may be hard to contact and involve. This can often be a major obstacle and, as the facilitator pointed out, although it seemed to be a very valuable process, under-representation of some sections of the population limited the effectiveness of some of the workshops. The main organisational aspects, such as the composition of the workshops and topics for discussion, are summarised in Table 9.2 below.

Table 9.2: Organisational aspects of the community visioning

Composition	8 workshops: 1. A random sample of 17 selected from voluntary and community organisations. 2. A random sample of 10. 3. 12 older people. 4. 6 people with disabilities. 5. 4 people on low incomes. 6. 10 members of gay/lesbian organisations. 7. 5 women. 8. 5 members of black/ethnic minorities. 1 follow-up feedback event involving 12 of the above.
Selection process	Participants were selected by the Council's Performance, Quality and Accountability Team. Contacts were made by letter and an advertisement in the local gay press.
Topics for discussion	• "The best thing about where I live." • Priorities for the next 10 years. • Discussion. • What the Council should do. • Next steps for participants.
Facilitation	Projects in Partnership Ltd., community facilitation specialists.
Briefing	None.
Feedback	Feedback forms were circulated after each session. All participants were invited to a follow-up feedback session and received a written report on the focus groups. Of those attending, 2 gave the session 10 out of 10; 5 gave it 8 out of 10; 4 gave it 7 out of 10. 10 each said they thought the Council was listening to individual views and the results of the workshops. Some wanted to mix different social interests groups in workshops, rather than separating them.

Source: Brighton and Hove Council, "Community Visioning Workshop Reports" (1999b).

In addition, a number of other methods were used to raise the profile of the consultation process and to seek maximum public involvement:

- A two-page insert was included in the Council's own newspaper, the *Brighton and Hove News*, which is delivered to 125,000 homes. This resulted in 303 respondents suggesting 645 issues to be included in the Local Plan.
- One-to-one meetings were held between the Planning Policy Team and local institutions and organisations, such as the University of Sussex, Friends of the Earth and the Planning for Business Group.
- In November and December 1998 five urban design visioning workshops were held in three different neighbourhood centres. In addition, one was open to all residents and one open to architects practicing in Brighton and Hove.
- A Schools event was organised in the Cardinal Newman School to encourage pupils to reflect on their likes and dislikes in the local environment and to set out their vision for the future.

A further proposal in the Consultative Strategy was that an Advisory Group should be established, composed of members of local organisations with an interest in planning. Members were drawn from disability and pensioners' groups, housing organisations, an amenity society and a local architect. This was chaired by the Chair of the Land Use and Transportation Sub-Committee with the objective of meeting four times a year in order to oversee and advise on the consultation process.

The outcome of the process was that the Planning Policy Team drew up a draft set of "pre-deposit policies" which were themselves the subject of further consultation. The next stage was the draft plan which was then put on deposit. This constitutes the formal opportunity for residents and others to submit objections or observations on the plan before the Local Plan Inquiry.

Evaluating the Consultation Process

We do not have the space here to evaluate fully the effectiveness of the whole Consultation Strategy but, instead, simply report on our findings from a series of interviews we carried out several months after the participatory exercises. Our focus is the public. We record three perspectives: those of a member of the Land Use and Transportation Sub-Committee; a planning officer; and the feedback from the citizen participants at the visioning workshops.

The Councilor's View

The Councilor who was interviewed was the lead Councilor for the Local Plan and Chair of the Local Plan Task Group. He has been heavily involved in local politics for a number of years. He took over responsibilities for the local plan in May 1999 so was not involved in approving the Consultation Strategy or the selection of methods of participation.

His general view was that the approach adopted had been successful:

> I think it was a different way of doing things and a very successful exercise. I think it's a difficult subject to get people interested in and, unless you are able to say to ordinary people that you may get a supermarket at the end of your road, the Local Plan remains very abstract. This way you get small groups of the community together and say, "This is what we want to do - so what do you think?" and it all works very well.
>
> The most important stage is coming up when it [the local plan] goes on deposit. Despite having done all this consultation, we will get accused of not having consulted and haven't listened and that will happen even though we have the evidence that we have. In a way, the consultation will count for nothing when the plan goes on deposit because it may include controversial things, such as a community stadium [for Brighton and Hove Albion football team] at Falmer and a park and ride scheme. These issues are hot topics around the town and pressure groups will latch onto them, even though some of them have already been consulted.

An important part of the Strategy was designed to encourage the Council to operate corporately and to share information arising from public consultation. It was recognised that, in the past, consultation exercises had not always been very successful:

> My view is that different bits of the Council do not share information around as they should, sometimes even within departments. We had a bad experience consulting over a new parking scheme in the town centre. We're very keen to consult, but sometimes we make a mess of it, but then I think at least we're trying.

While supporting the idea of involving a wide range of community interests in the planning process, the Councilor felt that in the end tough political decisions are needed:

> To be totally hard nosed about it, I think when it comes to the crunch there are going to be some issues where there's no consensus so judgements will

> need to be made based on professional advice. There are still going to be some difficult decisions on key issues and I guess some of these will be political to the extent that there will be different views. On the other hand, officers are trying very hard to justify everything in the plan on the basis of the consultation.

The Officer's View

A leading officer in the Planning Policy Team was interviewed about the consultation process and the likely outcomes. His view was that the process had been successful in that a wide range of people had been involved. Whilst dramatic policy changes were unlikely to arise, significant suggestions had been made which would improve the quality of the plan and make it more sensitive to local needs. The results of the focus groups and community visioning exercise would also add weight to the Council's case at the Local Plan Public Inquiry:

> I can't honestly say there is a whole raft of policies which has come from the visioning, but I can say they [the public] have informed a large number of policies by the addition or deletion of what we were originally thinking. A good example is care homes, since we realised through community visioning that there is a growing demand for care homes from members of the gay and lesbian communities. They often feel extremely isolated going into "mainstream" homes. The policy will specifically address this issue.

The consultation exercise also produced evidence of how different groups identify often conflicting priorities:

> The other thing which gives the vision strength is the different priorities which different groups attach to things and the contrast between the random sample and the priorities of other groups. With employment space and things like that, you get big differences between what was said by the unemployed, black and ethnic minority groups about the need for jobs and the views of the random sample and those returning leaflets with their comments.
>
> This will strengthen our hand at the Inquiry. We can say in qualitative terms that we got this feedback and we went to this group and they said this was a real priority for them.

On the consultation process, while every attempt was made to include those not normally involved, it appears that with the second meeting on 14 December 1999, only the keener participants attended:

> Most of the community visioning people have felt minded to come to the two sessions but they are going to be among the keener ones. If there'd been more time, we would have liked to explore why some didn't come to the second meeting, on 14 December. We also need to find out why we've not been successful in getting enough people on low incomes and from black and ethnic minorities. With some groups it's been a real struggle to find people – some just got lost on the way. There was an open invitation to the meeting on 14 December but the turnout was low. It may be that many were satisfied with the letter and report on the first meeting.

The Participants' Views

The views of the participants of the community visioning and focus groups have been assessed through interviews and a review of the feedback submitted after the final visioning meeting on 14 December. The outcomes, from the participants' perspectives, can be summarised as follows:

The Process

Overall, the majority of participants welcomed the process and felt "privileged" to be invited:

> Thank you for offering me the opportunity to put forward my ideas.

While not always fully aware of the purpose of a Local Plan, they welcomed the opportunity to express opinions and felt they were genuinely being consulted:

> I do really believe in what's going on and I do think it's really important that the Local Plan is coming from the local community. Before it was very much imposed upon the people, whereas this is coming up from the people and I'm prepared to invest time and energy in that.

Others were realistic. There was a degree of scepticism about who was listening and how long it would take to see results:

> Of course the challenge is putting it into practice and how long it will take.

There was also the view that people felt that they had never been consulted in the past, although, of course, statutory procedures had been carried out. This would suggest that the more traditional approaches to consultation, such as newspaper advertisements or displays, do not engage people's attention. Others, however, felt extremely cross that they had not been invited to these particular events and had only found out about them through personal contacts. They felt intentionally excluded which led to conflict and damaged the trust that had previously been established.

There were many positive comments made about the process at the feedback event on the 14 December. Many welcomed the opportunity to exchange views and debate issues with other residents, even where they disagreed. They enjoyed focusing on particular issues of local concern and felt they had learnt a lot from doing so. The facilitation was highly praised and the overwhelming feeling was that they had been listened to.

Who was Involved?

Considerable effort was put into identifying participants from different community and stakeholder groups, especially targeting social groupings identified as not having the opportunity to participate in the local planning process; namely, older people; people with disabilities; people on low incomes; lesbian and gay people; women; black and minority ethnic groups. Nevertheless, a number of participants pointed to under-representation and non-attendance in some areas. When asked who should be reached out to in the future, black and ethnic minorities, young people (under 25) and those on low incomes were identified.

Questions were also asked about whether it was better to separate out or mix different groups by age, gender, ethnic minority status, income and place of residence. A minority favoured mixing interest groups; the majority appeared to favour the approach adopted in this case of running separate visioning workshops.

There were also predictable problems of group dynamics in some workshops. One participant commented:

> I thought it was rather good – informal but progressive, but it was a small group of seven and some didn't turn up. Two had mental disabilities and seemed to be out of it. Although they were asked, it was not clear whether they couldn't or didn't want to participate. They felt a bit excluded. A couple of others were too vocal and some just agreed and didn't express an opinion. Perhaps they need better briefing or someone to speak on their behalf.

Another respondent said:

> Most of the community visioning people who came are among the keen ones because they have already attended two events. We must find out why we

> were not successful in getting people on low incomes and from ethnic minorities. The process is inevitably self-selecting.

Others suggested more people would be involved if meetings were held at more convenient times or were based in different locations to reduce the need to travel.

Who was Listening?

A number of comments were made suggesting that the officers seemed sincere in taking on board conflicting views. One participant said:

> At this stage this means Council Officers - who always do listen - the problem is the councilors.

Table 9.3 presents the results of a survey of the 12 participants at the feedback event on 14 December which shows that the majority felt that the Council listened "quite a lot" to individual views and the results of the workshops, and that these comments and views were listened to more than the more traditional consultation methods.

Table 9.3: Comments from participants at the feedback event

To what extent do you think the Council has listened to?

	Not at all	A little	Quite a lot	Not sure
Individual views	-	2	10	-
Results of the workshops	-	-	10	1
Results of all Local Plan Consultation	-	2	7	1

Source: Projects in Partnership, "Report on Feedback Event" (1999).

Was Anyone Encouraged to Get More Involved in Community Activities?

The participants who were interviewed appeared to be strongly committed to the principle of public participation and involvement and welcomed this series of initiatives as a genuine improvement on the past:

> The move from consultation (simply being asked for views) to **participation** (enabling people to be part of the solution) was welcomed by

> participants. There was also a willingness to work closely with the council in the future (Brighton and Hove Council, 1999b: 4).

Again most found that officers in particular were listening to them, but some remained sceptical about whether the Council as a whole would act on their views and how long this would take. There was strong support for the way the workshops were organised and facilitated, although some of the workshops were not always fully representative because of difficulties in identifying participants and non-attendance on the day.

An integral part of the community visioning exercise was to identify personal responsibility and next steps for the participants. At the time, they were keen to be an "active part of the solution" (*ibid*: 9). Actions pledged by participants included: finding out more, closer links with the council, action with others, changes in lifestyle and spreading the word (*ibid*). Yet when asked some twelve months later whether their involvement had generated a greater interest in community affairs or a desire to get more involved, almost all said it had not. This may be because most of the participants came from a particular "constituency" – the environment, gay/lesbian rights, disability groups, community fora or, in one case, was practicing as an architect and wanted to know more about local planning policies. Most claimed to vote regularly. One had recently become a councilor, but she had accepted the invitation to stand because she had been a member of the Labour Party for twenty years and had been active in her community long before the consultation process began.

Lessons Learned and Good Practice

This section returns to try to answer some of the questions raised by the White Paper and the current emphasis upon public participation which we set out at the beginning of this chapter. We examine the narrow and wider aspects of democratic renewal and, drawing on theory, we attempt to extract some pointers for good practice.

Are There Ways of Targeting and Encouraging Non-Joiners to Participate?

Evidence from the Brighton and Hove case study certainly indicates that the focus groups and visioning sessions were successful in bringing new people into the local plan consultation process. From the point of view of the officers and members, there was a degree of satisfaction that certain stakeholder groupings were identified, invited and engaged and that this legitimised the process and would add weight to the Council's case at the Local Plan Public Inquiry. Nonetheless, the difficulty of attracting new publics must not be underestimated. Although the Council sought specifically to involve people other than the "usual suspects", it was clear that, in some instances, it was again the experienced, vocal citizens, the "leading players" who attended the sessions. Interviewees commented upon the ways that they became involved - reading an advertisement in the local gay newspaper; opening a letter addressed to someone else who had left the organisation two years previously; being phoned by a friend who worked at the Council. Some highlighted the low turnout, whilst another queried the Council's (poor) organisation in attempting to recruit participants. An officer and a community member also questioned the "representativeness" of those invited. Despite the good intentions, is there evidence that this is still a self-selecting process? Does all this weaken the outcomes?

By targeting older people; people with disabilities; people on low incomes; lesbian and gay people; women; black and minority ethnic groups, the Council was able to ensure that a form of "representation" (in the statistical sense) was given to these groups. In his discussion of the power of professionals, Sanderson has questioned whether the images of particular groups (such as those targeted here) "reinforces assumptions of helplessness and dependency, of a lack of capacity to give an opinion, which constitutes a major barrier to participation" (1999: 332). In this case, the use of high quality facilitators with no stake in the process was clearly an important aspect in redressing the power balance and there was considerable praise for the running of the workshops. The Final Report noted:

> That it is possible to involve disenfranchised groups in strategic policy making, but there MUST be consideration of their needs (e.g. translation services for minority ethnic groups, timing of meetings) (Brighton and Hove Council, 1999c: 19).

In exploring people's experiences, perceptions and aspirations with regard to the places where they live, work and play over the next ten years, the Council was compiling qualitative data for use mainly by the local plan team, but which could also be fed into other relevant organisations and departments. Mechanisms for ensuring that information, regarding both process and content, need to be developed if data and experiential learning are to be shared between departments - both to optimise the effectiveness and efficiency of the Council's service delivery and broader consultation strategy, but also to avoid consultation fatigue. Whilst the Council is not claiming that the participants had any particular mandate to speak on behalf of their social grouping, it was clear from the interviews that officers will use evidence from the consultation process in support of its policies at the Local Plan Inquiry. Sanderson (1999) has argued that much is made by professionals of the "unrepresentativeness" of such qualitative methods, but he points out that this provides yet another institutional barrier to participation. Ironically, it can lead to the systematic use of established networks which may well not speak for the local community and simply reinforces the *status quo*. In this example, Brighton and Hove Council specifically targeted typical non-joiners in order to redress the balance. The objective was to seek out new and different viewpoints.

At the Local Level, is it Just a Small Proportion of Individuals who are Actively Involved in Voluntary and Community Organisations – the "Usual Suspects"?

The interviews showed that, on the whole, the people involved were already actively engaged in community networks, such as pensioner or residents' associations or school governors' boards. Most had had previous experience of local government, either as members or officers; some had worked in planning. There was evidence of the event being fed back into the organisation/group by those with experience of being part of a formal network (the "usual suspects"?), but none that those who had participated on an individual basis "shared" their experience. Thus, for example, although two members of the women's session are parents of children at the same school, they have never once discussed the visioning event. For them it was more of a personal learning experience. This does not, however, devalue the genuine interest in giving views on the future of Brighton and Hove as a community.

A particularly interesting insight related to the fact that interviewees said they valued the face-to-face nature of the exercise and opportunity to share their views and debate issues - "*Voting isn't the same as having your say*". There was evidence of norms and shared values; indeed, one interviewee expressed considerable surprise at how quickly and easily a consensus of important issues was generated. But clearly for some the exercise was a one-off, contributing to self-development, whilst for the more committed community member it was part of a wider social *engagement*. This highlights the individual and collective aspects of citizen involvement. Furthermore, it emphasises the challenge of ensuring that the other quarter of a million local residents each has access to appropriate mechanisms to become involved and to put their views and ideas forward.

Are we right to be wary of the "usual suspects"? Quoting Campbell (1990), Sanderson (1999) suggests that those with power should provide the structures and resources to build representative mechanisms, rather than simply to dismiss self-advocacy. "Leading players" are often the bedrock of a community's social capital. Yet, the visioning workshops explicitly sought to invite participants who do not usually engage in such processes. This rationale was criticised by one particular individual who made the point that it is often the disenfranchised groups who are the very people who do not have the capacity to maximise the potential of such processes. Might there not be a role for the "citizen expert", for example, from the gay/lesbian group, who has, perhaps, built up an individual capacity to articulate the needs of a specific group? There was certainly evidence from the interviews that the Council will need to invest much more in building its own capacity as an institution and that of local people to participate if they want a broader base and a more active citizenship:

> A move towards more effective community engagement will require new skills and competencies in local government - for both members and officers - and in the community at large. The local capacity for community involvement - from the point of view of the consulting, the consulted and the facilitators of that consultation should be developed earlier rather than later (Brighton and Hove Council, 1999c: 19).

Whilst local government can tap into existing networks, if councils are to cultivate a more politically active community, then they need to nurture the social capital and grow networks to replace the apparently weakened civil society. They will also need to invest heavily in the disenfranchised groups.

Is it Possible to Encourage and to Sustain Involvement?

Of the 69 participants in the community visioning, only 12 attended the feedback event. Interviewees remarked on the low turnout at the second session. Possible explanations given were that individuals considered that their views were well encapsulated in the written report and/or that the timing was difficult, being just before Christmas. Nevertheless:

> To ensure continued willingness to be involved in local decision making, consultation of the public must start to encourage participation and

involvement. Community Planning will further require this process (Brighton and Hove Council, 1999c: 19).

Moreover, it might be argued that the mix of individuals at the sessions, that is members and non-members of groups, blurred the "role" played by participants. It is not clear whether they were invited as delegates or representatives or simply as residents of Brighton and Hove with no requirement to share the experience with others. Whilst the primary objective of the exercise was to seek the views of local people, it is interesting to reflect on how the Council might wish to "encourage participation and involvement", perhaps, by promoting the event more widely. Integrating results into the decision making process and keeping participants involved are clearly the minimum administrative requirements if a commitment to the process is to be maintained by those directly engaged. There was certainly considerable favourable comment about the officers and their commitment to listen. There was not a similar belief that members would do the same. How officers and members respond to people's problems and priorities is crucial. It was evident that the concerns of the lesbian and gay community needed to be "interpreted" in the planning context. The Planning Department converted problems articulated by the group into land-use solutions which the group itself was unable to define. Local planning authorities therefore need to be able to offer creative solutions.

Are Members of Local Organisations Likely to Widen Their Interest in Community Activities and Local Democracy on the Basis of Positive Experiences of Participation?

There was no evidence of people intending to take a more active role in community life as a result of the experience; indeed, one "leading player" commented that, at a recent local meeting, the need for consultation was inferred and there was a cry of "Oh no!" Consultation fatigue is a threat. Nevertheless, there was considerable evidence of pride and pleasure at "being asked", as well as anger at "not being asked". The more experienced participants were complimentary of the process and the service and courtesy of officers. The representatives of the disenfranchised groups expressed surprise at being consulted and declared a willingness to be involved again - with the proviso that it should not take up too much time. People are willing to contribute, but this willingness needs to be respected.

Is There Any Evidence That the Newer Methods of Community Participation are Likely to Lead to Greater Community Involvement and Democratic Renewal?

We sought to investigate whether the use of innovative consultation in the preparation of Brighton and Hove Council's development plan strengthened community involvement and/or contributed to democratic renewal. We wanted to examine the extent to which the current ideas of social capital relate to the functions of local government at the micro level, particularly with regards to the local plan. It was evident from the interviews that local people are interested and deeply concerned about aspects of their environment - "hot local issues" such as pedestrian-traffic conflict, waste or town centre decline - and that they have a view. They recognise stakeholder conflict and the need to develop consensus. Nevertheless, as the councilor pointed out, ultimately decisions will have to be made, based upon professional opinion and political argument. There was also evidence that people feel more strongly about local politics "*...because they have more of an effect on my life.... and because of the children*". For those individuals unaware of the existence of a local plan, it was clear that they felt that they had never previously been consulted. Confront people with a draft consultation document of tens of pages and they may well not engage in the process. The community visioning exercise to bring new people together to define a context for, build a "vision" for, the local plan and to help set the overall objectives, illustrates that non-professionals can determine issues at a "strategic level":

> It is simply not true that the public don't want to become involved in difficult issues and strategic thinking. Strategic input comes from bringing together people to discuss issues and take responsibility for change (Brighton and Hove Council, 1999c: 19).

At the community visioning event a number of actions were pledged by participants, for example:

> "take more interest in council activities"
> "contacting other organisations"
> "share findings with colleagues"
> "encourage more people to participate in this process"
> (Brighton and Hove Council, 1999b).

There was commitment on the day. Despite this, there was, however, no evidence of the process directly influencing an intention to vote or inspiring more active engagement in community life. We cannot therefore speak of a single "active community"; rather, it would appear that there is a web of looser networks of active organisations and individuals with an interest. As Wilson has suggested:

> Using a participatory methodology does not by itself ensure the formation of productive social capital (Wilson, 1997: 747).

Public participation is thus only one strand of the Government's attempts to get "in touch with the people". Alongside this, councils need to address issues of leadership, councilor contact with local people and more accountable and open decision making procedures.

Conclusions

Despite the intention to engage non-joiners, this consultation process largely involved those with previous experience of pressure groups, community organisations and party politics, people who are already engaging, directly or indirectly, with the local council. These are probably the people more likely to vote than the general population and certainly those interviewed all claimed to exercise this right. Their involvement in the participation process appears to have *reinforced* previous interests, commitments and degree of involvement in local politics, not changed them. Their involvement is also largely instrumental, in that they wish to promote changes in policy, based on previously acquired values and interests, rather than new values acquired through a deliberative process.

Brighton and Hove Council considers that this form of consultation added value and enhanced the evolution of the local plan. Clearly, the interviewees believed in the operation of democracy and have an interest in local politics. If by "democratic deficit" we mean that people are not voting, then electoral turnout figures show that is true, but our research indicates that there is a real and genuine interest in local politics. The challenge for "modern local government" is to learn how to be in touch with this valuable resource and how to nurture it. If the Council wishes to strengthen local social capital, then it needs to nurture new networks and build the capacity of officers and members to engage and support local people, giving residents mechanisms to voice their opinions and ideas. Important research questions for the future are: What motivates people to get involved and to stay involved? Are those who are generally involved those with established networks in the community? How can we build and sustain new networks of active citizens?

The informal and deliberative participatory methods described here appear to be a useful way to discuss public issues and to make people feel valued. Such processes of social learning can strengthen individuals and groups for their own sake and thus they become more than an instrument to fulfill a council's statutory obligations. This is one way of building the capacity of the community, but such face-to-face contact is costly and resource intensive. Meaningful participation can strengthen relations and has the potential to stimulate new networks. It can encourage and support individuals to become part of an informed and active community, but we need to find ways of sustaining such potential social capital, perhaps, by building on common interests and commitment:

> A move towards more effective community engagement will require new skills and competencies in local government – for both members and officers – and in the community at large (Brighton and Hove, 1999b).

In its narrowest form, participation risks being of purely instrumental value, legitimising the process or strengthening the council's hand at appeal or inquiry, but it needs to engender popular support and commitment if it is to meet the wider agenda of reviving and renewing local democracy.

References

Arnstein, S. R. (1969), 'A ladder of citizen participation', *Journal of the American Institute of Planners*, vol. 35, no. 4, pp. 216-224.

Brighton and Hove Council (1998), *Brighton and Hove Local Plan Consultation Strategy*, by the University of Westminster, Brighton and Hove Council, Brighton and Hove.

Brighton and Hove Council (1999a), *Brighton and Hove Local Plan Consultation Reports: 1. Focus Groups*, Brighton and Hove Council, Brighton and Hove.

Brighton and Hove Council (1999b), *Brighton and Hove Local Plan Consultation Reports: 2.Community Visioning: Final Report & Summary*, Brighton and Hove Council, Brighton and Hove.

Brighton and Hove Council (1999c), *Brighton and Hove Local Plan Consultation Reports: 2a. Community Visioning Workshop Reports*, Brighton and Hove Council, Brighton and Hove.

Commission on Social Justice (1994), *Social Justice: Strategies for National Renewal*, Vintage, London.

DETR (1998a), *Modern Local Government: In Touch with the People*, Cm.4014, HMSO, London.

DETR (1998b), *Guidance on Enhancing Public Participation in Local Government; a research report*, DETR, London.

DETR (1998c), *Enhancing public participation in local government*, De Montfort University University of Strathclyde, HMSO, London.
DETR (1999), *Planning Policy Guidance Note 12: Development Plans*, DETR, London.
DoE (1995), *Involving Communities in Urban and Rural Regeneration: A Guide for Practitioners*, DoE, London.
Forester, J. (1999), *The Deliberative Practitioner: Encouraging Participatory Planning Processes*, The MIT Press, London.
Healey, P. (1997), *Collaborative planning. Shaping places in fragmented societies*, Planning, Environment, Cities Series, Macmillan, Basingstoke.
Leach, S. and Wingfield M. (1999), 'Public Participation and the Democratic Renewal Agenda: Prioritisation or Marginalisation?', *Local Government Studies*, vol. 25, no.4, pp. 46-59.
LRN/PLCRC (London Regeneration Network Pan-London Community Regeneration Network) (1999), *Capacity building: The way forward*, LRN/PLCRC, London.
Projects in Partnership (1999), *Report on Feedback Event* (not published).
Sanderson, I. (1999), 'Participation and democratic renewal: From 'instrumental' to 'communicative rationality'?', *Policy and Politics*, vol. 27, no. 3, pp. 325-341.
Stewart, J. (1996), 'Innovation in democratic practice in local government', *Policy and Politics*, vol. 21, no. 1, pp. 29-41.
Thake, S. (1999), 'Capacity building: Making Year Zero work, Beyond the deficit model', LEPU Seminar, October 13, University of South Bank, London.
Thomas, H. (1996), 'Public Participation in Planning', in Tewdwr-Jones, M. (ed.), *British Planning Policy in Transition: Planning in the 1990s*, UCL Press, London.
Thornley, A. (1977), 'Theoretical perspectives on planning participation', *Progress in Planning*, vol. 7, no. 1, Pergamon Press, Oxford.
Wates, N. (2000), *The Community Planning Handbook*, Earthscan Publications, London.
Wilson, P.A. (1997), 'Building social capital: A learning agenda for the Twenty-first century', *Urban Studies*, vol. 34, no. 5/6, pp. 745-760.

10 Changing Patterns of Social Exclusion in Dundee

KEITH FERNIE

Introduction

Changing political and policy agendas have shaped models of social inclusion designed to alleviate social exclusion. Latterly, for example, the Scottish Executive has focussed on social inclusion reflecting the changes in political and policy direction of previous and incumbent governments. It has put in place specific programmes to address the geographical and thematic aspects of social exclusion. Dundee has profited from being at the centre of the development of the advanced features of overarching inclusion strategy and co-ordinated action intended to ensure these programmes are effectively delivered.

This chapter documents firstly, the most recent forms of regeneration and secondly, changing geographical patterns of social exclusion in Dundee. It demonstrates how social and economic problems have manifested across a transforming social and physical landscape of change.

New Political Directions

It has been argued that following New Labour's election to Government Office, in May 1997, the seeds of a new political direction were at this time planted. These have since then taken root and are now as dominant a characteristic of new forms of governance as the key features of *social inclusion* programmes in the UK today. As such, the emerging political ethos has given birth to, in a more ideological sense, a "third way" in political and economic life, which its proponents claim, will provide the route to the inclusive society (Lister, 2000). The new Labour government has then fostered an interest in devolution and social inclusion as ways of meeting both its ideological objectives and in practical terms some of its related manifesto promises.

Prior to the election, the Labour Party (then in opposition) pledged itself to more open government, greater decentralisation of power, greater accountability in governance and constitutional reform. This reflected a practical concern that the UK constitution is relatively highly centralised when compared to other advanced industrial economies (Hutton, 1994). Critical attention had begun to focus on the consequences of such centralisation on the performance of the economy and, as a consequence, an economic case for constitutional change was developed and initiated (Hutton, 1994). Subsequently, the new Labour Government set out a programme of constitutional reconfiguration that presented itself in the shape of devolution of power. As such, instruments were proposed to reform political systems to achieve greater democratic accountability in the processes of government. This included amongst others a Scottish Parliament and a Welsh Assembly and a strategic planning authority for London.

Giddens (1998) has articulated other principal political arguments associated with the "Third Way". This is presented as the basis of a political programme that transcends the approaches of classical social democracy (1945-1979) and neo-liberalism (1979-1996). In addressing contemporary challenges associated with globalisation, individualism, political polarity, agency and ecology, the "third way" seeks to renew democracy. That is emphasis is placed on a new mixed economy, involving partnership, trans-national systems of governance and importantly, social inclusion (Giddens, 1998). Ruth Lister (2000) does, however, make the point that the "third way" offers a conditional form of inclusion. She maintains it is, "conditional on the willingness of citizens to exercise responsibility to include themselves through paid work and educational training and to grasp the opportunities being opened to them".

Poverty and Exclusion

The debate about poverty and inequality in society is changing quite dramatically. The language employed by politicians, policy makers and researchers is also changing. In short, there has been a relatively seamless shift from the concept of poverty to that of social exclusion and more recently, in Scotland, to the concept of exclusion and *social justice*. Simply stated, the idea of poverty has moved through the evolving process of language and understanding of the subject matter, (e.g. poverty, social deprivation, social and economic disadvantage, social exclusion, inclusion and social justice).

Most significantly, the concept of poverty, particularly income disadvantage, remains at the core of these ideas, although exclusion is perceived as a broader concept (Vidler and Curtis, 1999). On this latter point, Parkinson (1998) puts forward the notion that exclusion is more dynamic and emphasises the way people are closed out of the social, economic and political mainstream of society. Its forms include unemployment and insecure employment, high levels of debt and arrears, homelessness, limited mobility and access to services, restricted rights, low educational attainment and poor health. It is this context that poverty is identified as being one dimension out of four dimensions of exclusion with the other three being distinguished by exclusion from labour markets, services and social relations (Gordon, Adelman, Ashworth et al, 2000). In other words "low income (economic poverty) is a key aspect of social exclusion...lack of money makes participation in society more difficult" (Scottish Office, 1999).

Social exclusion is therefore historical, however, what is new are explicitly defined social inclusion programmes, designed, in part, to alleviate social exclusion. Presently social exclusion is a growing predicament in many European cities, as observed by the Government's creation of the Social exclusion Unit and the European Commission's introduction of URBAN and INTEGRA initiatives (Parkinson 1998). It has been defined by the Prime Minister (Tony Blair, 1999) as a "shorthand label for what can happen when individuals or areas suffer from a combination of linked problems such as unemployment, poor skills, low incomes, poor housing, high crime environments, bad health and family breakdown". It is quite clearly understood as being characterised by poverty, duration of disadvantage, concentration and lack of work and low pay.

Social Inclusion

Acknowledging this, and reflected in the approach now being adopted, socio-economic poverty and the various forms of exclusion are being tackled by policies that prevent social exclusion and promote social inclusion. The Government has said it will promote social inclusion, by:

- helping provide individuals with opportunities to participate in work and learning;
- removing barriers to participation in society;
- helping children and young people develop the skills and attributes which will secure their inclusion as adults; and,
- build stronger communities.

Note the inter-connectedness and mutually reinforcing features. In recognition of this the government claim to understand the complex and inter-related nature of social exclusion, which in turn requires methods that guarantee the different strands of action join together coherently.

Cities and Partnerships

It is asserted that "Scottish circumstances differ from England in that those suffering exclusion in Scotland are disproportionately concentrated in specific communities, and there has been more experience of effective urban regeneration policies originally pioneered in Scotland and maintained in later years by local government and others" (Dewar, 1998: 1). Its basis is that urban centres are characterised by poverty and exclusion, which in turn is centred on mainly dormitory and peripheral housing estates.

Consequently, regeneration instruments in Scotland have been geared to tackling urban poverty, which have been couched in a framework that recognises cities as socially, politically and economically functional structures. There is also a recognition that social and economic exclusion, particularly in the context of area specific dereliction reflect conditions too complex and severe to be tackled by a single agency (e.g., local government) and /or a community body in isolation (Carley, Chapman, Hastings et al., 2000). As such city partnership and local partnership structures have remained central to the designation process; together with an achievable long-term strategy and the involvement of local communities. In this respect the Scottish Executive expects,

> partnerships seeking support to demonstrate that they have convincing strategies, will work together, and can do the job their communities deserve (Dewar, 1999).

The establishment of a partnership reflects a multi-dimensional approach to regeneration. However, the nature of the partnership, particularly the scope, participation and involvement stakeholders will ultimately determine its effectiveness and achievements. Strong partnerships are rooted to the meaningful involvement of a broad range of key regional and local groups organisations, each playing a role in influencing the quality of life in a community (Carley, Chapman, Hastings et al., 2000).

Urban Regeneration in Scotland

Urban regeneration in Scotland has evolved into a particular approach that rests on the geographical targeting of aid, the principles of partnership and empowerment and the implementation of initiatives within a strategic framework. The changing scope and nature of urban policy in Scotland has involved the Scottish Office (now the Scottish Executive) setting the context for active participation by local authorities and other stakeholders, with the private and voluntary sectors working together through partnership and community involvement.

Social Inclusion Partnerships (SIPs) form the geographical focus for addressing social exclusion in Scotland. The SIPs build on and apply the principles of the Scottish Social Inclusion Strategy, which stated that "Scotland today is an unequal society" (Dewar, 1998: 2), where:

- too many people are deprived of the life chances most of us take for granted;
- too many families live in poverty; and
- neglect, decay and crime blight too many communities.

It is clear that SIPs represent the next stage in the development of policy for urban regeneration, while the measure is drawn directly from the Government's approach to social exclusion. SIPs will build on fixed arrangements and experience but with a particular emphasis on seeking to prevent young people, in particular, from being excluded in participation in the economic and social mainstream. The main characteristics of Social Inclusion Partnerships are:

- a focus on the most needy members of society;
- the co-ordination and filling in of gaps between existing programmes in order to promote inclusion; and
- an attempt to prevent individuals becoming socially excluded.

The SIPs initiative involves an innovative feature as it introduces a shift away from an exclusive reliance on geographically focused policy. New approaches to be adopted in the SIPs framework include sectoral measures that can cut across defined areas of disadvantage and address processes of social exclusion in a city-wide context.

Overall the SIPs initiative appears to be an advance on previous policy initiatives for urban regeneration since it attempts to address the causes of urban decline rather than simply ameliorating the effects of such decline.

Governance in Dundee

In the 1920s and 1930s, Dundee experienced relative economic collapse as a consequence of the global recession. The demand for its products fell, its technology became relatively outdated, its relative costs were high and emerging economic activities tended not to locate in the city.

The post war experience of Dundee was no different from other industrial areas in the long boom of the post war period. Subsequently, however, its economy reflected its over-dependence on certain manufacturing activities. There began a process of long term endemic decline in its traditional industrial and corporate sectors from the 1960s to the present. The collapse was not matched or compensated for by significant inward investment or indeed local economic diversification, thereby leading to a steady contraction of the local economy. The economic and corporate restructuring spilled over into the social and community structures in the city, it eroded its skill base, reduced the city income level and undermined the provision of private services and local authority community facilities (Docherty, 1991).

The institutional capacity of the city to respond to the evolving pressures of multiple forms of deprivation and re-development is reflected in the formation of the Dundee Partnership.

The Dundee Partnership was instigated in 1991 and was conducted under a specific political regime that was generally suspicious of development agency intervention and the creation of what was perceived as a dependency culture by disadvantaged communities and localities. In effect, this was balanced by a need to capture the political benefits of being associated with managed recovery of local and regional economies. The Dundee Partnership represents:

- a means of securing the informal co-ordination of the priorities and activities of its constituent participants; and
- an informal institutional arrangement between key players in the regeneration of the city's economy, and it provides a link to other organisations such as the private sector and the trade unions (Dundee Partnership, 1994).

The Dundee Partnership acts as an important conduit of priorities for policy formulation and implementation towards a common objective - the regeneration of Dundee. In 1996, however, there were changes to the composition of the Dundee Partnership. These were an outcome of the Local Government (Scotland) Act 1994. Subsequent local government reorganisation replaced the established two-tier structure with Dundee City Council, a single tier general-purpose body.

As if to celebrate its new persona and composition, the Dundee Partnership laid down a "Vision" statement to guide its operations and activities. The Vision set out the contextual framework to the work of the Dundee Partnership and established a common agenda to the work of the individual partners. The Partnership energised the involvement of other bodies such as the Health Board and Scottish Homes, thereby rounding out the strategic foundations of the organisation. The Vision expresses this in terms of key aims and tasks, identifies gaps and sets out priorities for actions required by individual partners. These were:

- to re-create the city as a thriving regional shopping and service centre (e.g. by improving prime and secondary shopping);
- a major employment centre (e.g., by meeting all land and property requirements);
- a place for the realisation of potential (e.g., by raising workforce skills and qualifications);
- a city to be proud to live and work (e.g., by ensuring a wide range of housing opportunity); and
- a city worth visiting (e.g., by developing the city as a base for the wider area).

Clearly, these elements are inter-linked; for instance the issue of image is a central theme which runs through several objectives of the Dundee Partnership.

Social Exclusion in Dundee

Dundee has the tightest boundaries of any local authority in Scotland. The availability of surplus housing has meant that it is possible for those with social problems to move more easily from one area to another than might be the case in the other cities in Scotland (ref: 1981/91 Census Population). Area regeneration initiatives in Dundee such as the Whitfield Partnership have benefited the area, but have also had the effect of displacing some families with social problems into other areas.

The nature of deprivation in Dundee is that vulnerable families, significantly a high proportion are single parent, are dispersed across the city. They face long-term exclusion from mainstream society. The proportion of Dundee's population whose lives are affected by poverty and who are excluded from mainstream society is undoubtedly significant and is accentuated even further when put in the context of the "eastern Scotland area". The following Table 10.1, which highlights indicators of "poverty and exclusion", illustrates this:

Table 10.1: Comparison between Scottish (East Coast) cities

	% of unemployed (1998) (a)	% or Enumeration District which comes into the worst 10% (b)	% of all households awarded housing benefit (c)	% of population receiving Council Tax benefit (c)	Migration rates 1991-1997 economically active (d)
Dundee	**8.2**	**21.48**	**37.3**	**38.8**	**-4.2%**
Aberdeen	2.1	2.32	19.7	19.0	+0.01%
Edinburgh	3.6	10.00	23.7	24.1	+2.4%
Scottish Average	5.6	10.00	28.3	30.0	+0.2%
	Dundee has the highest rate of unemployment of all Scottish cities	Dundee has the highest in Eastern Scotland area	Dundee is the highest Eastern Scotland area	Dundee highest in Eastern Scotland area	Dundee has the highest out migration rate in Scotland

Sources: (a) NOMIS (1998).
(b) Scottish Office (1994).
(c) Scottish Executive (1998).
(d) GRO (1998).

Further, indicators of "Poverty and Exclusion", some of which are illustrated above show:

Out migration / people leaving the city Dundee has the highest out-migration rate of all Scottish cities – it is predominantly the economically active 18-30 year olds who are leaving the city.

Dependency The levels of benefit dependency, covering most mainstream benefits such as a) Income Support, b) Housing Benefit, c) Council Tax and d) Educational Support benefits linked to dependency and low income, are the highest in the Eastern Scotland Area.

Unemployment Dundee has the highest rate of unemployment of all Scotland's cities and is the highest of the major urban centres in the Eastern Scotland Area.

Extent of Deprivation Dundee is the highest (in the Eastern Scotland area) in terms of the extent of deprivation as a share of the authority's total population (see Table 10.2 below). This is described as:

> striking in its distribution... (with) Just under a third of Dundee's population are located in the worst areas (Dept. of Urban Studies, Glasgow University, 1998).

Table 10.2: The extent of deprivation

Authority	Population	Population in worst 10% of Post Code Sectors	%
Dundee	150,250	46,977	31
Edinburgh	443,600	36,401	08
Perth & Kinross	132,570	6,860	0.05
Fife	349,300	6,456	0.02
Angus	111,020	3,799	0.03

Source: Scottish Office Central Research Unit (1998).

The pattern that emerges is a city with widespread poverty where low wages and part-time employment are as much a concern as unemployment itself. The concentrations of indicators of disadvantage impact on opportunities open to young people. Rates of absenteeism and school exclusion combined with the low levels of achievement of young people at school, point to the long-term nature and high cost to society of not tackling the underlying trends.

This picture contrasts with the many opportunities available within the city and serves to highlight the needs of those who, for a variety of economic and social reasons, are excluded.

Social Inclusion Partnerships in Dundee

In May 1998, the Secretary of State for Scotland announced the introduction of Social Inclusion Partnerships. In practice, it was intended that existing PPAs evolve into SIPs along with the designation of new geographic and thematically focused SIPs. The partnership approach remains central to the designation process, together with an achievable long-term strategy and the involvement of local communities.

These initiatives are intended to extend the principles embedded in the PPA measure to directly confront the dynamics of social inclusion and social exclusion and will focus more closely on promoting inclusion in our communities and preventing social exclusion from developing.

The Dundee Partnership has selected as its priority areas:

- *SIP 1*: a geographical archipelago of neighbourhoods comprising the areas of the Ardler, Kirkton, Mid Craigie & Linlathen and Hilltown.
- *SIP 2*: Census enumeration districts spread across Dundee and designated in the worst 10% and excluded form the SIP1 area.

The key indicators that characterise these communities as disadvantaged, include youth and long term unemployment, low income households, uptake of benefits and education support grants, levels of educational attainment, crime and fear of crime, mortality rates and other health indicators. All of which are significantly worse than the position for Scotland and the city as a whole.

Broadly speaking the SIP strategy addresses the issues of population decline, surplus housing, lack of facilities and environmental degradation, by combining work on physical regeneration with measures which address social and economic concerns, focusing on the development of community capacity and empowerment. More specifically, the Dundee Partnership has established a strategic vision for community regeneration in the city; the creation of stable, sustainable and empowered communities throughout Dundee, in which people wish to live and can prosper.

This strategic vision for community regeneration is translated into practical effect through four themes: stability, sustainability, empowerment and prosperity.

Changing Geographies of Social Exclusion

This section of the paper sets out to present an outline of the trends and patterns of the changing geography of social exclusion in Dundee. The nature and extent of this morphology is linked to orientation of targeted intervention across Dundee, specifically in relation to communities designated under consecutive and rolling programmes designed to alleviate poverty and tackle social exclusion in Dundee. The main purpose of this examination is to attempt to,

- firstly, trace the morphology of poverty and exclusion by analysing a set of indicators over a period of time;
- secondly, show the extent different communities exhibit social exclusion as evolving organisms in this time period; and

- thirdly, couch this in the context of spatially targeted programmes and strategic development and intervention.

For the purpose of examining the nature and extent of change across a physically different landscape, I have chosen two sets of community areas/neighbourhoods. This highlights the different experiences of these communities in terms of both their social and economic infrastructure and the level of Partnership involvement in the areas. The communities are equivalent to the following types:

- Peripheral Housing Areas – Designated UP/SIP status (1990-00).
- Inner City Housing Areas - (Part) designated under SIP (1996-).

The statistical information, which provides the basis for this analysis, allows us to record and examine trends and patterns which include social and economic factors that are commonly flagged as having properties sufficient to measure poverty and exclusion on this scale. The adopted indicators permit an examination of:

- *Crime*: equivalent to recorded crime and a form of social exclusion.
- *Unemployment*: equivalent to income disadvantage and exclusion from the labour market.
- *IS Dependency*: equivalent to low household income and child poverty.
- *Demography*: equivalent to population change and migration patterns across the city.

Incidence of Crime

There are many symptoms of social exclusion, including social fragmentation, youth alienation, drugs abuse and last but not least problems with crime and policing (Parkinson, 1998). The extent and nature of crime can be indicative of other simmering problems associated with inadequate education, housing and support networks.

The crime figures used in this section are grouped together deliberately to show crimes against property (inc. housebreaking & theft), neighbourhood infrastructure and the person. Housebreaking and theft in particular are thought indicative of certain social conditions that correlate with aspects of poverty, itself reflected in material want and, in some cases, dependency.

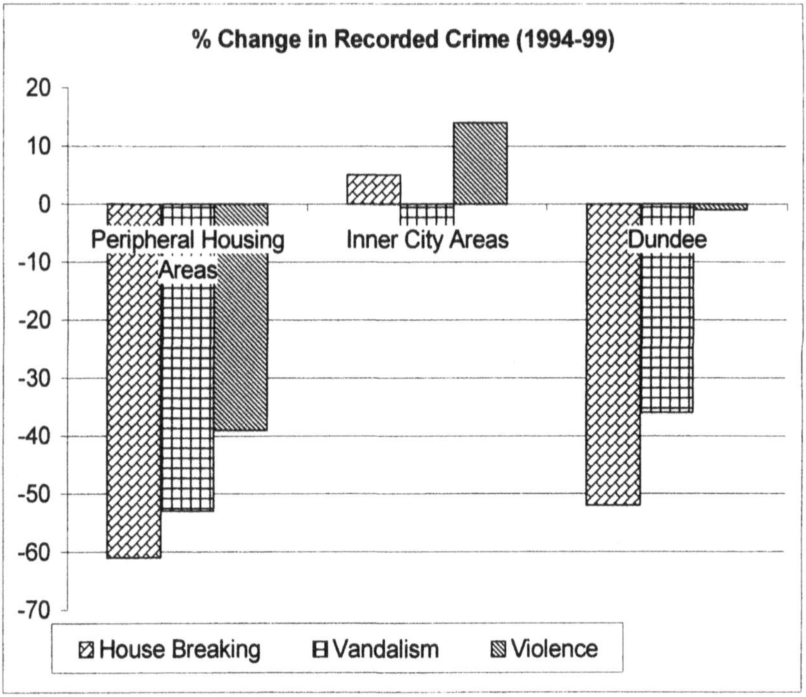

Source: Tayside Police (1994 – 99).

Figure 10.1: Recorded crime data

Crime figures reveal a general reduction in recorded crime at the city-wide level, although the extent of change varies across the city and between areas. This is exhibited by separating the data in terms of geography:

- Significant decreases for each of the categories of crime in the *peripheral housing areas*, which is in excess of the diminishing levels of crime recorded at the city-wide level.
- The reduction in levels of recorded crime significantly below the city average and increases in two of the crime categories (housebreaking & against the person) for the *inner-city housing areas*.
- The increase in levels of crimes of violence and against property in the *inner-city housing* areas is significant and not just because it contrasts with the trend city-wide. It is rapidly becoming a prominent feature of these areas, normally associated with the clustering of deprivation in the peripheral housing areas more recently.

Unemployment Levels

Social exclusion is partly the result of macro economic change, transformation that affects economic environments at both the national and local level. International and corporate restructuring has for most regions, cities and communities led to high levels of structural unemployment, low-paid work and a growing gap between the rich and the poor (Parkinson, 1998).

Taken at an even lower level a major factor of poverty is the level of unemployment and subsequent dependency that results from being unable to obtain a reasonable income. In most cases it leaves individuals and their families economically disadvantaged, which in turn prevents them from achieving an adequate standard of living. Socially, its impact may contribute towards and is perceived in educational under-achievement, poor health and loss of self-respect.

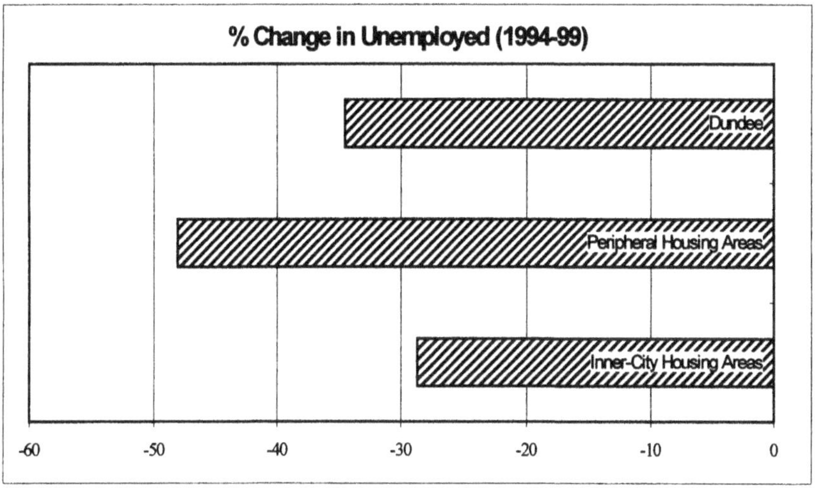

Source: Dept of Employment (1994) and NOMIS (1999).

Figure 10.2: Unemployment

Unemployment levels in the city of Dundee have reduced dramatically in the recorded time span, by more than a third over a period of five years. The forces responsible for this reduction are firstly legislative (e.g., impact of the "New Deal"), secondly a result of macro economic change nationally and thirdly, the forces of global development have impacted on labour markets across all the industrial countries in the developed world:

- Significant decreases for each of the areas, although more significant falls in the *peripheral housing areas*, which is in excess of the diminishing levels of crime recorded at the city-wide level.
- A fall in the levels of total unemployed significantly below the level of change recorded for the city in the *inner-city housing areas*.

Income Support

Income Support data serves to illustrate the extent of dependency for families living in the targeted area, which is in turn a reflection of low income and/or unemployment and therefore exclusion from the labour market. More specifically it shows disadvantage and exclusion amongst "lone parents" and the number of children under five affected, highlighting the extent of child poverty in this age category. *The information relates to Income Support data based on individual claimants at electoral ward level, and is a 100% scan of all cases in Dundee.*

The Income Support claimants' data includes the number of children affected. However, the data is presented as only the number of cases, which does not give any indication of dependency. It covers the categories of pensioners, lone parents, unemployed and disabled.

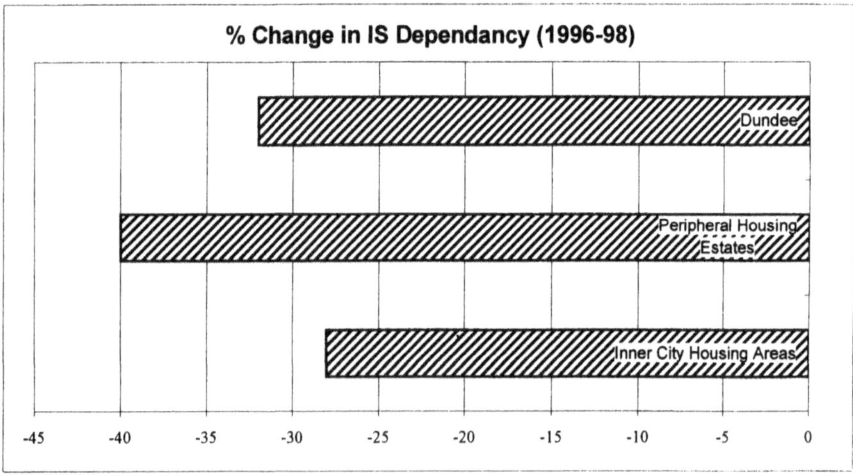

Source: Benefits Agency (1996 – 1999).

Figure 10.3: Income Support

Income Support data reveals a significant reduction at the city-wide level, although the extent of change varies across the city and between communities. This can be clearly illustrated by comparing the areas adopted in this paper as having different properties in terms of their location:

- Significant decreases in the *peripheral housing areas*, which exceeds the levels of crime recorded at the city-wide level.
- The change in Income Support dependency in the *inner-city housing areas* below the diminishing levels recorded city-wide, with the difference if we compare it to the peripheral housing areas even greater.

Population

It has been well documented that the population of Dundee has been consistently falling over a long period of time (i.e. over the last twenty years). However the social dynamic behind the decrease in Dundee's overall population may be less important, in the context of this paper, than the change across the city.

The movement of population between communities and in particular across different types of neighbourhoods is of greater significance to our investigation. The social dynamic(s) responsible for demographic change are to an extent physically driven, although this must be explained in the context of mobile communities and within these household and individual life experiences.

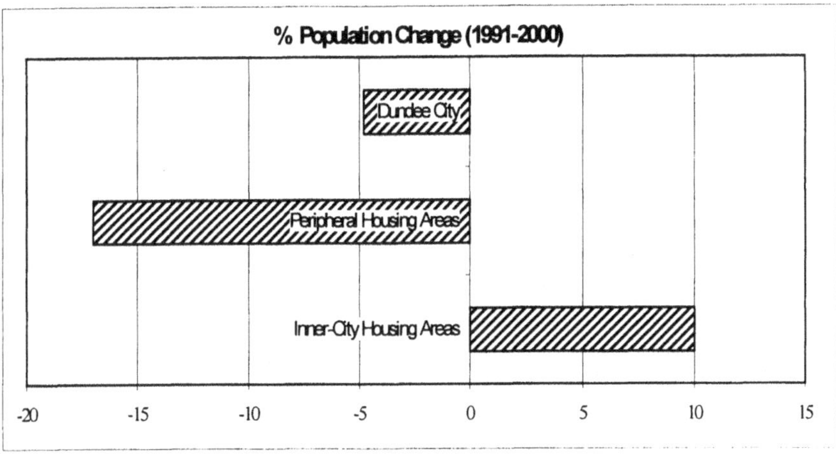

Source: Census Population DCC Electoral Roll (1991).

Figure 10.4: Adult (elector) population change

The Figure 10.4 above clearly shows:

- A dramatic decline in the (elector) population levels in the *peripheral housing areas*, which has exceeded the proportional decrease in the cities population over the same time period by a significant amount.
- A very different picture for the *inner-city housing areas*, in the sense that they have seen an increase in the elector population, which is in contrast to both the peripheral housing areas and the experience of the city as a whole.

Analysis of Change

The following analysis presents both a summary and a description of the changing patterns or morphing of social exclusion in Dundee. This is based on the analysis of specific trends within specific geographical locations, (as described earlier in this paper), and over roughly the same period of time. In terms of the data analysed, the following assertions are made:

- Shifting forms of poverty, disadvantage and exclusion across the city generally and a morphing of a particular set of experiences, not exclusive to dormitory housing estates located in the peripheral areas of the city.
- Patterns of social exclusion spreading from pockets of concentrated deprivation (witnessed in the early 90's) to a greater number of areas more widespread across the city (1990-1996) and latterly form these widespread peripheral housing estates to the inner-city areas of Dundee (1995-2000).
- Significant change in economic conditions a very short time span for inner-city areas, areas which are more heterogeneous and are not traditionally associated with multiple forms of social exclusion.
- Movement of population toward the inner-city possibly as a result of initial physical redevelopment in the peripheral housing areas. A breakdown of this population is unfortunately not understood at present.

Although the scope of the paper is descriptive and doesn't effectively answer questions that draw attention to why this is happening, it does nonetheless seek to show tangible change and elude to forces which have effected this. The implications for policy makers in the context of both shifting patterns and transient forms of social exclusion are reasonably clear cut:

- Continue to target resources in the designated areas that don't cover the majority of the excluded population in Dundee.
- Direct resources into areas, which have transient populations, particularly when housing re-development plays an important part in the regeneration process of the area.
- Move to a more spatial approach covering the city-region and taking advantage of economies of scale in terms of resources and a cost-effective and more realistic strategy.

Conclusions

The new political direction of new labour has engendered a process of ideological change as it has altered the political landscape, particularly in Scotland. The "third way" has set new parameters for the reconfiguration of governance and programmes designed to build different social and economic framework. The former has benefited Scotland, in the shape of a decentralised parliament, where the latter has encouraged more debates on the responsibilities of the state and its citizens in the context of rights, opportunity and inclusion.

The experience of regeneration in Scotland and particularly its cities has contributed to the evolution of governance and the development of strategic instruments for tackling poverty and exclusion in its cities. The advent of the Scottish Parliament and the implementation of social inclusion programmes, which reflect this new political dynamic, have also refocused debate to take account of cities and their role in delivering appropriate structures which are nonetheless not divorced from their regions.

The development of cities is crucial to this debate. If we take Dundee as an example then it has been evidenced that major physical redevelopment in areas has resulted in the displacement of its populations, particularly the young unemployed. It is claimed that Dundee's population decline is centred on this dynamic. Although I could not possibly attempt to refute this assertion in any meaningful way, I do nonetheless believe that the displacement of population within the city and across neighbourhoods is as relevant to this notion.

The evidence also indicates a consistent shift and change of experience for different communities, and although partly explained by population moving out of undesirable areas, it is nonetheless a significant shift. The figures also highlight the different scale of problems experienced by these neighbourhoods over a period of time, which for the purposes of this examination are differentiated by their location. The morphing of poverty and exclusion in Dundee, based on partial data, does show peripheral housing areas to be transient and the inner-city areas affected by and potentially suffering from this displacement. This is exhibited in terms of a relative improvement of social and economic conditions in areas currently targeted through social inclusion and a much slower process or indeed a worsening of experiences in the inner-city areas (see figure below).

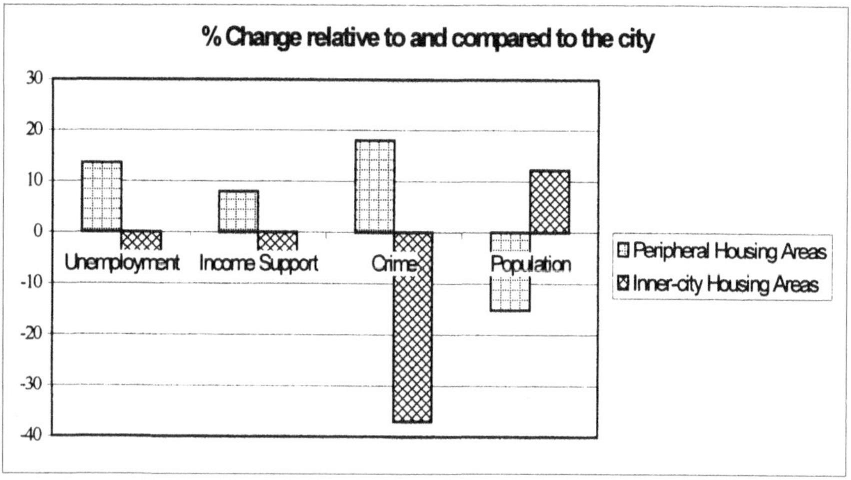

Figure 10.5: Relative change in comparison to the city

The patterns of social and economic change in Dundee do reflect the development of the city in the context of its region, but also the nature of its targeted regeneration and social inclusion programmes. Change limits their effectiveness and prevents analysis of broader forces of change. By addressing the problems of poverty and exclusion in this way the institutions, through the Partnership structures, are instead contributing to the spread of the problem and aiding a process that moves the problem elsewhere in the city.

The magnitude of change and morphing is apparent. The question must be; is this being addressed by the Scottish Executives area-based approach in Dundee or indeed matched by the Partnership, the morphing of this body and the inherent institutional capacity of the city? The rise of the Partnership as a way of harnessing resources, organising people and organisations in the complex business of regeneration and promoting social inclusion and as a way of developing and improving local governance has been positive (Carley, Chapman, Hastings et al., 2000). That said and despite some major success in this area, it is also the case that area specific regeneration has in many respects quite simply shifted exclusion and disadvantaged households elsewhere in the city.

It is in this sense that it might be argued the disadvantage of these area-based programmes is their inherent exclusivity.

References

Carley, M.; Chapman, M.; Hastings, A.; Kirk, K. and Young, R. (2000), *Urban Regeneration through Partnership*, Joseph Rowantree Foundation, Area Regeneration Series, Policy Press, Bristol.

Dewar, D. (1998), *Social Exclusion in Scotland*, Scottish Executive Research Paper, HMSO, Edinburgh.

Dewar, D. (1999), *Social Justice... A Scotland where everyone matters*, Scottish Executive Policy Document, HMSO, Edinburgh.

Doherty, J. (1991), 'Dundee: a post-industrial city', in C A Whatley (ed.), *The Remaking of Juteopolis*, Proceedings of the Abertay Historical Society's Octocentenary Conference, Dundee Abertay Historical Society, pp. 24-39.

Dundee Partnership (1994), *Regenerating Communities in Dundee: towards a long-term strategy*, Dundee Partnership Policy Document, Dundee.

Gibb, K. et al. (1997), *Revising the Scottish Area Deprivation Index*, Scottish Office (Central Research Unit) Report.

Giddens, A. (1998), *The Third Way: The Renewal of Democracy*, Polity Press, Oxford.

Gordon, D.; Adelman, L.; Ashworth, K. et al. (2000), *Poverty and Social Exclusion in Britain*, Joseph Rowantree Foundation, York, pp. 54-71.

Gulliver, S. (1997), 'The SDA: Area Projects', *Town Planning Review*, vol. 37, no. 4, pp. 211-227.

Hutton, W. (1994), *The State we're in*, Jonathon Cape Ltd, London.

Lister, R. (1999), 'Who is included in the Third Way' (article in the Perspectives magazine News and Views of the Democratic Left 'Perspectives' magazine), unpublished.

Parkinson, M. (1998), *Combating Social Exclusion*, Polity Press, Bristol.

Vidler, G. and Curtis, S. (1999), *Measuring Social Exclusion*, The Scottish Parliament: The Information Centre Research Paper 99/11.

11 Neighbourhood Regeneration: Delivering Holistic Area-Based Strategies

ANGELA HULL

Introduction

There is now an understanding that the regeneration of areas of multiple deprivation will require substantial intervention over time if the neighbourhoods are to sustain the investment which has been made. The scale of the problems faced on many disadvantaged estates may take up to 25-30 years to tackle. During this time period, the lifecycle of regeneration will go through several government-funded initiatives. It is important that local stakeholders therefore, develop a delivery plan for neighbourhood regeneration, which recognises both the importance of the time required to implement, and the timing of, different elements of the strategy.

Over the last decade, the British Government has turned to consider both the process and the substantive issues of sustainable area regeneration. This chapter offers a contribution to the policy debate on effective and sustainable neighbourhood regeneration drawing on the rich seam of policy evaluation research funded by charities, research foundations and central government.

The chapter is divided into four main sections. The first section introduces the policy context at the start of the new millennium in summarising the Labour government's policies on neighbourhood regeneration and relating these to the Government's implementation of improvements in public service delivery. Public services are being restructured using managerial strategies of efficiency and control. The second section summarises the "wicked" issues in the regeneration of areas characterised by multiple deprivation. This is then followed by a discussion of what holistic regeneration might entail. The final section builds on this discussion and identifies the building blocks or the infrastructure of successful area regeneration.

Regeneration Policy at the Start of the New Millennium

The Government, since election in 1997, has undertaken a wide-ranging review of regeneration policy to identify new approaches, which are cost effective, and the good practice lessons, which can be disseminated. They have monitored the outcomes and life-chances of residents who live in the most disadvantaged areas and have found that despite substantial, additional public resources to the disadvantaged areas in large towns and cities since the 1960s these areas still lag far behind the national average for the indices on long-term unemployment, crime rates, health chances and educational qualifications (DETR, 1998a).

The continuing geographic concentration of exclusion and deprivation, measured by these indices, is seen both as a failure of the main welfare state programmes and the short-term, single-issue nature of most regeneration initiatives. The Audit Commission's (1989) metaphor of a "patchwork quilt" aptly conveys the lack of programme coherence.

During the 1990s, there have also been changes within local government, with the emergence of unitary authorities and an increasing role for the community in local governance. The Housing Green Paper *Quality and Choice. A Decent Home for All* (DETR and DSS, 2000) introduces the idea of a single source of funds for local authorities, and the transfer of significant numbers of council housing stock by 2012. These changes in housing management will increase the number and type of landlords involved in the most deprived neighbourhoods and will herald future changes in the way that area regeneration is carried out. With the splintering of service delivery the Government aims to integrate and support the actions of public, private and voluntary sector agencies more effectively at neighbourhood level through:

- the targeting of areas and communities where there is a need for priority action;
- support for new, cross-cutting approaches which support the objectives of more than one Department;
- the promotion of effective local partnerships with stronger local involvement; and
- the encouragement of greater flexibility and responsiveness in the operation of public spending programmes (DETR, 1999a).

At national level, the Government has responded to criticism of insufficient co-ordination of the initiatives emerging from separate Government departments with a three-pronged strategy. Firstly, a Social Exclusion Unit (SEU) was established in the Cabinet Office in 1997 to improve Government understanding of the complicated links between unemployment, poor skills, low incomes, poor housing, high crime environments, bad health, poverty and family breakdown (SEU, 1998). Eighteen Policy Action Teams, composed of secondees from Government Departments and "volunteers" from the research and practice communities, have reported on good practice in a number of issue-fields including jobs, skills, neighbourhood management, anti-social behaviour, community self-help and information technology. The Government has drawn extensively on new approaches that seem to be making a difference to produce its long term national strategy on neighbourhood renewal, which aims to provide "joined up solutions to joined up problems" (SEU, 2001).

Secondly, Government regeneration funding is focussed more tightly on the most deprived neighbourhoods and with a greater emphasis on the intensive regeneration of smaller areas than in the past. To tie in with this, rounds 5 and 6 of the Single Regeneration Budget have shifted to concentrate on areas of severe need and to encourage community participation in devising local priorities and solutions. The most important policy development with a spatial focus is the *New Deal for Communities* which is being piloted within 17 pathfinder authorities identified through the Index of Local Deprivation (DETR, 1998e). The initiative, announced in September 1998, encourages a more holistic approach to problem solving, a "mixed market" in the delivery of services and the involvement at all stages of the widest constituency of local interests, including the business community, local community organisations and residents themselves.

The third element of the strategy is a series of short-term programmes to pilot new approaches to client-centred practices, which cut across service departments and provide the missing "glue" where mainstream programmes are working in parallel but with little collaboration. Specific services - health, education and employment - have been chosen to monitor and evaluate innovative approaches to service delivery in Government sponsored Action Zones. Particular client groups (pre-school children, older children, 18-24 year olds, and 16-18 year olds not in full time work, education or training) and the particular crosscutting issues of truancy and school exclusions, and teenage pregnancies are being addressed initially. The development of the Single Regeneration Budget, and the recent launch of several specific area-based programmes, including New Deal for Communities, Health Action Zones, Employment Zones, and Education Action Zones, demonstrate the Government's commitment to area-based programmes where partnership between diverse agencies and the community is essential.

The SEU (1998) is spearheading the promotion of integration and co-ordination between overlapping service deliverers at local authority and neighbourhood level. The Government is implementing a "Best Value" framework to monitor improvements in public service delivery using benchmarks of service quality and effectiveness (Blair, 1997a; Prime Minister and the Minister for the Cabinet Office, 1999). Social care and health care services have been reformed so that hospital, community and family doctor services at the local level are co-ordinated in a more coherent way (Meredith, 1995). Local authorities have been given overall responsibility for the co-ordination of service delivery within their administrative boundaries with new powers to promote the economic, social and environmental "well-being" of their residents.

Working in partnership has been an important component of the publicly funded area regeneration schemes since the mid-1980s. The new strategy for neighbourhood renewal sets local authorities the specific target of improving the life chances of residents in the most disadvantaged neighbourhoods to the average for their area. The Prime Minister, on election to government, called for local authorities and other statutory agencies to work in partnership with local communities (Blair, 1997). He went on to say that regeneration should be a locally appropriate response to locally variable problems, which draws on lessons learned from previous initiatives, and takes a long term and comprehensive view. Local authorities will also need to work closely with the new regional agencies recently established (Hull, 2000). In particular the Regional Development Agencies, appointed by the Government, will decide how best to allocate regeneration resources to implement their Regional Economic Strategies. For local authorities, this introduces another tier of bureaucracy with specific responsibilities and values that they will need to influence and mobilise for the benefit of their disadvantaged communities.

The "Wicked Issues" in Regeneration

Tony Blair has called for new approaches to regeneration, which draw on the lessons learned from previous initiatives. A rich seam of research in social policy analysis has examined the transferable lessons from area-based regeneration schemes. There are two main categories of research relevant to this question. This includes, on the one hand, qualitative research to understand the important components of a "good neighbourhood" (Gilroy 1996; Madanipour and Bevan, 1999; Andersen et al., 1999; Cattel and Evans, 1999; Silburn et al., 1999; Williams and Windebank, 1999; Wood and Vamplew, 1999).

On the other hand, there are both independent and government evaluations of specific government financed regeneration initiatives, including the Priority Estates, Urban Development Corporations, City Challenge, New Life for Urban Scotland and the Housing Action Trust programmes (Power, 1986; Nevin, 1998; Robson et al., 1994; DETR, 1998a; GFA Consulting et al., 2000; KPMG Consulting, 1999; Stewart et al., 1999; Cambridge Policy Consultants, 1999; Hull et al., 2000). This latter seam of research was commissioned to identify the quantitative programme outputs and to evaluate the achievement of programme aims and objectives. In addition to this they provide accounts by regeneration officers of what appeared to work and what did not. It is these experiential accounts from practitioners that provide qualitative evidence based learning to aid the design of subsequent programmes.

Few academic researchers concentrate on the wider structural factors leading to the spatial polarisation of deprivation (Walker and Walker, 1997). This would involve examining the nature of power and authority in society and the distributional effects of the allocation of societal resources on each interest group, for example on gender, race and age. The Government has emphasised material deprivation as the main contributing factor to the life experiences in disadvantaged neighbourhoods (SEU, 2000). Poverty in the United Kingdom, estimated from the number of households with below 50% of the average income (after housing costs), rose from 5 million people in 1979 to just under 14 million in 1993/94 (from 9% of the population to 25%) (Walker and Walker, 1997). The Index of Local Deprivation (DETR, 1998e) monitors the spatial concentration of deprivation down to Census ward level. The wards with the most disadvantaged groups suffering from unemployment, homelessness, poor housing conditions, the effects of disability, sickness and higher than average mortality rates often exist side by side with the most affluent wards (Lee and Murie, 1997).

Table 11.1: The characteristics of disadvantaged estates

- Housing estates, which are physically intrusive and dominant in scale.
- Dwellings with structural and maintenance problems creating poor housing and environmental living conditions.
- Low status reputation evidenced by perceptions of high crime levels, actual vandalism and graffiti, drug dealing and negative media coverage.
- Often high levels of population mobility on the estates leading to tension between longstanding and new households.
- Estate imbalance of household types with either a high proportion of lone or very young parents, and/or ethnic minority households, living adjacent to elderly households.
- Clustering of households with social disadvantages as defined by poverty, benefit dependency, low levels of educational attainment, sickness and disability, high levels of unemployment, etc.
- Insufficient and uncoordinated public service attention to the estates from the police, education, training, health, community care services leading to poorly funded or targeted public resources.
- Insufficient housing management on the estates as judged by rent losses, high levels of voids, low demand relative to other estates, environmental maintenance problems.
- Problems of working across organisational and political boundaries and developing open and effective communication with residents.

Source: Adapted from Figure 1.3, Hull et al. (2000: 13).

Government sees the resolution of the spatial concentration of disadvantage as a task for the localities to address. The main focus in the past has been directed to investment in either the structural and maintenance problems of the estates and/or creating employment opportunities. Table 11.1 shows that the issues are much more complex and interconnected and include the social dynamics of life on the estates, how local services perform for residents and how the estates are perceived by outsiders. Many of these issues point to the need for more effective management of resource targeting, neighbourhood resources and facilities, and partnership working. There are continuing debates about how to target disadvantage in future regeneration programmes, the delivery mechanisms to use, and the causality of actions and interventions. To "break the vicious circle of deprivation" in areas of multiple deprivation (DoE, 1997), requires integrated and problem-focused and responsive government at local, regional and national levels. The next section turns to explore what integrated and holistic regeneration could mean.

Holistic Regeneration

Despite the recognition that a holistic and integrated approach to regeneration is the most appropriate way in which to tackle the problems of areas with individuals suffering from multiple deprivation and significant social exclusion, it has not been easy to achieve. There is now recognition that social and economic goals must be combined to tackle the interaction between the poor physical and environmental conditions and the social and economic problems on the most unpopular estates.

Attention is now focusing on the process issues in regeneration or the mechanisms by which local actors can achieve sustainable regeneration. We now realise that relying on the metaphor of a "trickle down" effect of externally applied investment is unlikely to have other than short-term positive outcomes. There is the recognition in Government policy documents that the community holds the key to sustainability. Community here is defined as the active participants who have a stake in the area either through their domicile or their voluntary and business interests. The Government could go further and widen out the engagement to include all residents within these areas and to address residents' concerns and aspirations for their families. Creating the supportive environment so that residents can play their part in neighbourhood regeneration and to use the experience to enhance their skill levels and confidence requires a fairly radical reassessment of the way that services are presently delivered (Taylor, 1995):

> This sense of powerlessness in the face of authority, of an inability to affect decisions that had major impact on people's lives, was strong. It was not a powerlessness that resulted from a lack of determination, or knowledge of what needed to be done, or lack of ability to do it. On the contrary, "they tell us we need professionals to come and sort us out. No we don't. We want the money so that we can do it ourselves". Poverty is mainly about lack of money, but it is also, crucially, about lack of power. Coping with and confronting this situation then is also about how this power is experienced, and the type of relationships it produces. It is also about establishing

> alternative relationships, and other sources of power (Andersen et al., 1999: 20).

To counteract this sense of powerlessness and apathy in the face of neighbourhood decline, there needs to be a focus on individual problems and support to enable effective engagement with employment and local social networks. Monitoring systems also need to be integrated across services so as to track an individual's progress through their lifetime. A pathways approach to enhance an individual's employability, for example, would trace an appropriate path from unemployment or school, ultimately into a job. The starting point would be to build on an individual's interests and existing skills, possibly through an informal "taster" opportunity in a local environment which residents are comfortable with and which they use regularly. Assistance in the form of steps or modules can them be tailored to the needs of the individual in a progression from motivational support, needs assessment, through pre-vocational training and vocational training, after which they might have gained employment. Complementary to skills training are intervening in the local labour market through social contracting and promoting intermediate labour market initiatives. The same approach can be used for addressing the health needs of an individual, with the improvements made in responding to an individual's mental health problems reinforcing the individual's ability to take up training.

A dedicated regeneration agency with the right focus would have the authority to negotiate the reorientation of mainstream policies so that they really work for the poorest neighbourhoods. The New Deal for Communities programme has an emphasis on linking the regeneration effort of planning, economic development, housing, environment, social services and chief executives' departments. The links to the health service, the police service and the education departments as yet are still rather weak. Area-based regeneration must both interlink with mainstream services and other urban initiatives as well as link into surrounding neighbourhoods. Sustainable regeneration therefore has several interconnecting elements:

- The regeneration solutions must come from within the community. Residents' priorities must be woven into spending decisions and performance assessments.
- The local community needs the support of a dedicated regeneration team who have the legal, financial and management expertise.

- A range of solutions and the expertise and resources of diverse partners are required to deliver services which make the links to residents' needs and tap their creativity in new and innovative ways.
- Regeneration must impact on some of the most important things in residents' lives. They have to have a key role in the process of implementing solutions.
- Regeneration in areas of high unemployment may take at least a decade to have an impact.

Successful Area Regeneration*

In this last section the strategic elements of the lifecycle of a regeneration project are introduced in the form of "building blocks" or infrastructure for regeneration which should be planned into the strategic framework for delivery. A fuller explanation of these concepts is given in Hull et al. (2000). There are four overarching tasks an area-based regeneration programme needs to address: strategic management; neighbourhood management; community development; and economic development. These are introduced sequentially below as building blocks, although in practice they will overlap and intertwine.

Although these tasks are considered separately, it is important to make the links between these activities, highlighting the overlaps and identifying the process issues. This is particularly important for all partners, irrespective of whether they link into the regeneration programme at the strategic management, or the neighbourhood service delivery level. Moreover, agencies involved in programme delivery need to be particularly aware of the overlapping interests with other agencies if opportunities for partnership are to be maximised.

Strategic Management

The long-term management of deprived neighbourhoods is the key to regeneration. This is a task for Local Strategic Partnerships operating at Local Authority level. Across a local authority area there may be several neighbourhoods or ward clusters which the local authority has identified as requiring public intervention to improve the physical, social and the economic environment. Strategic management of regeneration involves developing the connections between the existing public, private and voluntary agencies to mobilise resources in new ways (Table 11.2). These resources can include entrepreneurial problem-solvers, staff with specific expertise, data about the area, funding streams, or buildings and equipment. Additional resources and government requirements can provide a trigger for partnership development. The need to manage and co-ordinate the process highlights the importance of a dedicated manager and team within a regeneration programme. Once the main projects start to be delivered, a more project-orientated and less vertical departmental structure is a more appropriate and effective model to deliver "joined-up" results.

Central to the strategic management of regeneration is the involvement of residents' associations and voluntary and community-based organisations in deciding regeneration priorities, programme organisation, and succession arrangements. These organisations will need to be supported in different ways to ensure they can effectively contribute to district-wide and neighbourhood regeneration strategies. Their "expert" knowledge of what works at neighbourhood level will help inform the programme linkages which will need to be made to progress from the physical area improvements to sustainable employment and personal achievements. In selecting performance indicators, it is important to ensure that they are meaningful to residents, and that they can be effectively maintained and monitored. What is crucial is that the process of appraisal, monitoring and evaluation is agreed with local partners.

Area regeneration is a long-term process, which continues beyond the main government-financed initiatives. The funds and the nature of the succession organisations must be planned for from the beginning. The precise details for sustaining the investment will be influenced by local opportunities that arise and the requirements of external national level monitoring organisations. However, the interests of local partners and the residents must be reflected in the management arrangements. In particular, a residents' association which has been supported and whose members have been trained for the role will be well-placed to play their part in the management of the succession bodies, either as individuals on the board or as the successor body. The regeneration agency needs to build up the asset base and ensure the democratic accountability of any successor organisation proposed.

Table 11.2: Strategic management tasks in area-based regeneration

- **Connect key organisations at local authority level:** Secure the commitment of local service organisations through local strategic partnerships.
- **Use knowledge and expertise:** Ensure regeneration agencies have the knowledge, expertise, and sufficient powers to facilitate the development of networks and the effective implementation of projects.
- **Encourage all partners to work with residents to extend the local networks, community and voluntary groups:** The development of resident involvement requires active co-ordination and commitment by partner organisations.
- **Local outcome monitoring:** Link service standards and performance indicators to the outcomes identified in consultation with local people rather than those defined by funding bodies.
- **Prioritise the succession strategy:** Develop a succession strategy at the outset, which has the flexibility to respond to the community and builds in transitional arrangements from the start.

Neighbourhood Management

The task of management at the neighbourhood level is to more effectively co-ordinate the delivery of services to residents through negotiating new ways of delivering these services with service providers and residents. A key component of this role is to break down both the negative perceptions and ignorance of outside service deliverers about the area, and the institutional barriers, which face residents in attempting to address their needs.

The three key groups, of the regeneration organisation, service deliverers, and residents, must together produce a delivery plan based on the assessment of community needs and the subsequent strategy. Residents' household survey is a good way of collecting the baseline data and residents' priorities for regeneration. The extent of baseline research and pre-planning beforehand should, as far as practical, provide realistic and realisable proposals and targets. The strategy that is agreed must clearly specify who should do what, where, and when, and should identify the monitoring arrangements to put in place from the start of the programme.

Table 11.3: Neighbourhood management tasks in area-based regeneration

- **Democratise governance structures:** Devise an action plan for neighbourhood management, which brings together the key players' assessment of community needs.
- **Improve housing management:** Effective housing management is an essential component of regeneration and provides a focal point for wider community development.
- **Clear service agreements:** Set clear service agreements with providers linked to the neighbourhood plan objectives.
- **Prioritise the succession strategy:** Develop a succession strategy at the outset, which has the flexibility to respond to the community and builds in transitional arrangements from the start.
- **Delegate authority and resources:** Ensure the programme can respond flexibly to resident's needs and opportunities by delegating authority and resources to well-trained, front-line staff.
- **Linked service delivery:** Encourage joint service planning and delivery between local agencies to secure improvements for customers.
- **Support residents' interests:** Give financial support and encouragement for residents' social, cultural and leisure interests. This can be a catalyst for strengthening existing networks and lead to participation in community, training and employment initiatives.
- **Track programme impacts:** Develop individual and cohort tracking systems as a measure of long-term programme impact and to identify future programme needs.

It is important that the strategy makes the links between proposals for physical regeneration and the opportunities, which arise, from this for creating jobs, training and skill development for residents. The plan must also consider how service improvements to the neighbourhood can be sustained over the longer-term. The forward strategy may include service deliverers taking on additional responsibilities and the establishment of community structures managed by residents such as community-based housing associations or community trusts. The decision-making structures agreed for residents must see succession bodies as accessible, democratic and accountable. The regeneration agency may provide initial support to these bodies in the form of interim funding until they can generate their own income to reinvest in the local community.

To engage with the residents in a regeneration area there needs to be a permanent organisational presence which can respond promptly to resident needs on a variety of issues, such as housing and neighbourhood management, health and crime, education and training, and jobs. It is essential to ensure that the residents' agenda is at the centre of an agreed regeneration strategy and to build in sufficient time to establish a good relationship. In neighbourhoods where social housing predominates the housing management function can provide the focal point for interaction with residents. Alternatively, the regeneration organisation can be located in an under-utilised building, which is accessible to residents. Front-line staff requires support and training to be able to effectively work with residents, and ultimately to enable residents to provide and manage service delivery to their neighbourhood. Regeneration is about providing effective services in new ways. There is a critical role here for staff in organisations who can understand and develop links between economic regeneration, community development and effective service delivery to residents. These staff, who have the ability to work across organisations with often hierarchical service structures, and who work well with residents, should be given delegated powers and resources to help improve local conditions. Area regeneration depends on the strengthening of networks between local service managers, for example, the head teacher, the divisional police commander, and the health service manager who have the authority to respond to neighbourhood needs through jointly devised initiatives which span organisations. The potential for mutual benefits, in terms of improved service delivery or more effective problem solving, can provide an effective stimulus to joint working.

Community Development

The task of the community development programme is to find as many different ways of effectively communicating with residents as possible. The importance of two-way communication cannot be over-emphasised. This involves, on the one hand, strengthening the existing channels for communication with residents so that the regeneration agency can respond to residents' concerns and be held accountable for their actions. Communication and trust can be built up through using the estate or neighbourhood office as a focal point for contact, and through the residents or tenants' associations. It is important to give as full information as possible to residents about the resources and options available, and to work at a pace that they set whilst developing the programme momentum. Failure to do this can lead to false expectations and the build up of mistrust.

The regeneration agency can make connections with residents through introducing a system of revenue grants to expand and support residents' interests and networks. From the outset of the programme, community grants should be earmarked for the "less traditional" activities (for example, support for football tournaments and dance clubs, religious and cultural festivals, women's and art events). These are low-cost initiatives, which can effectively contribute to raising the commitment of the residents to the regeneration programme in a positive way. Moreover, they can result in increasing the number of active residents. It is important to provide a continuum of opportunities for involvement, recognising that people may want to be involved to varying extents and in different ways.

It is important to work with the tenants' and residents' representatives in order to help them to broaden their democratic base through annual elections and active resident participation. Significant funds will have to be devoted to provide the residents' representative body with an office, equipment, paid support staff, and training for office holders to develop their capacity as decision makers. The involvement of residents on all working parties, as well as involving them in contractor interviews and in programme evaluation, helps develop their skills and experience. Where the residents' representative body already exists at the beginning of the regeneration programme, it is important to agree how the regeneration agency and residents' body will work together in a written document to develop mutual trust and respect. Sufficient funds need to be earmarked to help resident organisations to expand, to sustain themselves, and to be in the position to play an important management role in any succession bodies established.

Community involvement works most smoothly where residents are seen to directly benefit from the regeneration programme through improvements to the services they have asked for. The key to maximising the benefits to residents is in designing integrated programmes, which provide opportunities to address resident's problems in a broad way. Where residents are involved in the design of programmes this leads to better ownership of policies and enhanced local capacity. Some of the most successful outcomes are the result of low-cost revenue initiatives that have evolved from within the community. For example community businesses, such as the Credit Union, child-care provision, and gardening businesses, provide a neighbourhood service as well as training, capacity building and jobs for residents. Similarly, capital investment in housing and community buildings should benefit residents who are unemployed through creating intermediate labour markets with contractors.

Table 11.4: Community development tasks in area-based regeneration

• **Strengthen residents' associations:** Equip and ensure that the residents' association can effectively represent the diversity of households in their neighbourhood through financial, and other forms of support.
• **Local labour schemes:** It is important to recognise that local residents, paid or unpaid, can deliver essential services.
• **Improve housing management:** Secure effective housing management. It is an essential component of regeneration and provides a focal point for wider community development.
• **Multi-objective physical regeneration:** Stimulate change through investment in physical regeneration, including environmental programmes and community facilities. It is important to maximise opportunities to integrate with social and economic regeneration.
• **Support residents' interests:** Give financial support and encouragement for residents' social, cultural and leisure interests. This can be a catalyst for strengthening existing networks and lead to participation in community and training initiatives.
• **Individual support:** Provide personal support for the most excluded, and alienated, residents. This is essential to success in raising their motivation, on training courses, and starting jobs.
• **Local outcome monitoring:** Link service standards and performance indicators to outcomes identified in consultation with local people, rather than those defined by funding bodies.

Economic Development

The success of economic development programmes relies on securing employment that meets residents' needs, and is integrated with other support programmes. It is important to focus on the demand-side barriers as well as the supply-side measures in economic regeneration. This involves looking outwards and connecting to the labour market, making the links with local business and mainstream job creation programmes. Looking inwards, it is important to gain people's interest so that they want to participate and learn.

Economic development strategies must mesh with other initiatives that support residents' interests (for example sport, local history, cultural celebrations, and computer games) and then make the links to training through a package of incentives, such as child care provision, training allowances, free travel to training. The appropriate framework, including customised training and the provision of mentoring, can contribute to the success of such approaches.

Regeneration

Residents in deprived neighbourhoods often face barriers to participation in training and employment. It is important to change the perceptions held by employers of the employability of residents, and to use a range of incentives and encouragement to private employers to recruit from the neighbourhood. This approach is long-term and will depend on a tailored set of packages, which mesh business human resource needs, and the quality skills of local people in a very competitive labour market. This should include work experience and work trial opportunities as stepping-stones to finding jobs. The regeneration agency can use its own resources to secure or provide jobs for the long-term unemployed residents in delivering neighbourhood services, such as clerical work, cleaning, care-taking, repairs and maintenance, gardening, construction, and security. Other effective measures include incorporating local labour clauses in contracts, negotiating guaranteed job interviews for residents who have undergone training, local procurement policies, and providing a wage subsidy to residents who gain jobs to enhance the wage rate.

Table 11.5: Economic development tasks in area-based regeneration

- **Local labour schemes:** Create jobs and training opportunities for residents in the delivery of neighbourhood services, such as cleaning, caretaking, repairs and maintenance, grounds maintenance, gardening, construction, and security. Self-help training networks can be developed within local communities. It is important to recognise that local residents, paid or unpaid, can deliver essential services.
- **Individual support:** Provide personal support for the most excluded, and alienated, residents. This is essential to success in raising their motivation, on training courses, and starting jobs.
- **Multi-objective physical regeneration:** Stimulate change through investment in physical regeneration, including environmental programmes and community facilities. It is important to maximise opportunities to integrate with social and economic regeneration.
- **Support residents' interests:** Give financial support and encouragement for residents' social, cultural and leisure interests. This can be a catalyst for strengthening existing networks and lead to participation in community and training initiatives.
- **Track programme impacts:** Develop individual and cohort tracking systems as a measure of long-term programme impact and to identify future programme needs.

Training programmes, whether social or vocational, should be designed to meet the individual needs of residents. Job-clubs should be established in the neighbourhood so that they are accessible to residents. It is important to target those least likely to learn early in the initiative through a wide range of interesting projects, since they are the most likely to benefit. Training can be linked to sports management, media and computing courses, arts and health projects, to gain the interest of disaffected youths and the long-term unemployed. The importance of raising training and employment issues with younger people in schools is vital to the culture of lifelong learning.

Personal training plans must be pitched to meet the needs of participants, including those who have learning difficulties or mental health problems. Customised training packages can then be developed with the help of contractors and Further Education colleges. Intensive support is required to help the unemployed to assess their skills and strengths, to examine the opportunities for work or training, and to prepare for interviews. Similar levels of support are required to help residents set up their own businesses, for example, providing training and advice, providing small community investment grants, and supporting applications for business loans. Debt counselling, benefits advice, drug awareness and core skills should be integrated into core training and employment programmes.

Conclusion

This chapter has explored a holistic approach to area regeneration and has argued for making the connections between the various agencies and the different activities, to facilitate cross fertilisation and ensure effective use of resources and maximisation of benefits. Four overarching regeneration tasks have been emphasised and detailed: strategic management, neighbourhood management, community development and economic development.

Disadvantaged areas that have experienced exclusion over a number of years require intensive, long-term management to put them on the upward spiral of regeneration. Housing conditions, crime and security, health, education and training are key issues to address. Residents' and outsiders' perceptions of the area must be tackled from day one. A dedicated team, with staff who are enabled to adopt a "hands on" approach, is essential to bringing together the diverse partners who are all necessary to delivering change. Residents are the key partners in successful regeneration and resources need to be available to maximise their involvement. They must have a big enough input so that they feel they are getting something out of the programme. The significance of an agreed plan with residents and partners which describes the projects and their delivery, but which also indicates the role of each partner in decision-making, is widely acknowledged. The team should set project and programme targets for addressing the problems they find. Whilst targets may not always be met, the key is to understand what has worked and what has not, and this can only be done through monitoring and programme evaluation. This should be part of the democratic process and should involve the residents and local partners.

Real improvements in quality of life depend on acting on the connections between health care, criminal justice and community safety, and between housing, education and employment. Local circumstances will decide the essential balance of the strategy between physical, social and economic regeneration. A succession strategy should be built in from the start. One test of sustainability is the infrastructure put in place to increase the employability of residents. The regeneration programme should link training to "apprenticeships", to local job needs and business investment plans. These links must be made to get residents into new jobs created through regeneration funding, using resident quotas in all contracts.

It is likely that in the lifecycle of area regeneration there will be several government fixed-life, funded initiatives to improve aspects of life in the area. It is paramount therefore, that in each initiative a strategy is developed at an early stage to continue the activities, or protect the investment, after the funding ceases. One critical element of the strategy involves ensuring that the regeneration funds influence the long-term mainline programmes of key departments and external agencies.

Another critical element of sustainability is the residents' ability to continue to have influence, control and leverage over service delivery and the management of community facilities. If the neighbourhood action plan has been effective it will have empowered sufficient numbers of residents through skill training and employment to have the confidence to make their voices heard. Through real purchasing power they will have re-defined themselves as productive members of their community each with a unique contribution to make.

Stemming from this concluding discussion, eight good practice principles can be identified that span across the four overarching components of an area-based regeneration programme:

- Support residents' interests.
- Secure effective housing management.
- Secure multi-objective physical regeneration.
- Provide individual support for the most excluded groups.
- Adopt local labour schemes.
- Implement local outcome monitoring.
- Prioritise the succession strategy.
- Track programme impacts.

These principles are particularly relevant to the co-ordination of the initiatives that will be carried out by different agencies in the New Deal for Community programmes. This chapter has emphasised the importance of building on the networks that exist in an area in order to achieve multiple outcomes through multi-agency working. Central government also needs to consider the efficacy of the specific accounting demands made by government departments on regeneration managers and the legal barriers in place, which stifle innovation. Just providing effective service response to the high levels of crime, low levels of educational attainment, poor health and extensive physical dereliction demands high practitioner and organisational commitment. But to "break the vicious circle of deprivation" in areas of multiple deprivation, (DoE, 1997) requires integrated and problem-focused and responsive government at local, regional and national levels.

Note

* A later version of this section appears in Hull et al. (2000). My thanks go to the rest of the team involved in the work for this DETR project, but particularly to Rachel Kirk, Mike Wilkinson and Keith Kirby who gave helpful comments on the later version.

References

Andersen, H., Munck, R. et al. (1999), *Neighbourhood images in Liverpool. It's all down to the people*, Joseph Rowntree Foundation, York.

Audit Commission (1989), *Urban regeneration and economic development: the local government dimension*, HMSO, London.

Audit Commission (1992), *Developing local housing strategies*, Audit Commission, London.

Audit Commission (1999), *A life's work: local authorities, economic development, and economic regeneration*, Audit Commission, London.

Blair, T. (1997a), Speech delivered at the Stockwell Park School, Lambeth, 8 December 1997, Cabinet Office, London.

Blair, T. (1997b), Speech delivered at the Aylesbury estate, Southwark, 2 June 1997, Cabinet Office, London.

Cambridge Policy Consultants (1999), *An evaluation of the New Life for Urban Scotland initiative*, The Stationery Office, London.

Cattel, V. and Evans, M. (1999), *Neighbourhood images in East London. Social capital and social networks on two East London estates*, Joseph Rowntree Foundation, York.

Clapham, D. and Evans, A. (1998), *From exclusion to inclusion*, Joseph Rowntree Foundation, York.

Department of Education and Employment (2000), *Report of Policy Action Team 2: Skills*, Social Exclusion Unit, London.

Department of the Environment (1984), *Local housing management. A Priority Estates project survey*, Report by Anne Power, HMSO, London.

Department of the Environment (1997), *Regeneration programmes - the way forward*, HMSO, London.
Department of the Environment, Transport and the Regions (1997), *Regeneration programmes - the way forward*, Discussion Paper, HMSO, London.
Department of the Environment, Transport and the Regions (1998a), *Planning for the communities of the future*, Cm 3885, HMSO, London.
Department of the Environment, Transport and the Regions (1998b), *What works - learning the lessons: final evaluation of City Challenge*, HMSO, London.
Department of the Environment, Transport and the Regions (1998d), *Where does public expenditure go? Pilot study to analyse the flows of public expenditure into local areas*, Regeneration research summary no 20, DETR, London.
Department of the Environment, Transport and the Regions (1998e), *Index of local deprivation*, DETR, London.
Department of the Environment, Transport and the Regions (1999), *Area-based and other regeneration-related initiatives-summaries*, http://www.regeneration.detr.gov.uk/policies/area/summaries
Department of the Environment, Transport and the Regions (1999a), *Best Value in housing framework*, Consultation Paper, The Stationery Office, London.
Department of the Environment, Transport and the Regions (1999b), *Interim evaluation of English Partnerships: final report*, report by PA Consulting Group, HMSO, London.
Department of the Environment, Transport and the Regions, and the Department of Social Security (2000), *Quality and Choice. A Decent Home for All. The Housing Green Paper*, DETR, London.
Department of Health (1993), *Implementing community care: population needs assessment. Good practice guidance*, Report by Price Waterhouse, HMSO, London.
GFA Consulting and the European Institute of Urban Affairs (2000), *Regeneration that Lasts. A Guide to Good Practice on Social Housing Estates*, DETR, London.
Gilroy, R. (1996), 'Building routes to power: Lessons from Cruddas Park', *Local Economy*, vol. 10, no. 4, pp. 248-258.
Hastings, A. and Dean, J., (2000), *Challenging Images: housing estates, stigma and regeneration*, Joseph Rowntree Foundation, York.
Holmans, B. (1999), *Kids at the door revisited*, Russell House Publishing, Lyme Regis.
Home Office (1999), *Community self-help, report of Policy Action Team 9*, Social Exclusion Unit, London.
Hull, A.D.; Kirk, R.; Samples, E. and Shaw, K. (2000), *Transferable Lessons in Regeneration from the Housing Action Trusts*, DETR, London.
Jacobs, J. (2000), *The Death and Life of Great American Cities*, Pimlico, London.
KMPG Consulting (1999), *City Challenge - final national evaluation*, DETR, London.
Madanipour, A. and Bevan, M. (1999), *Social exclusion in European urban neighbourhoods, Walker, Newcastle upon Tyne: a neighbourhood in transition*, Occasional Paper Series, no. 2, University of Newcastle: Centre for Research in European Urban Environments, Newcastle upon Tyne.
Meredith, B. (1995), *The community care handbook, the reformed system explained*, Age Concern England, London.
National Audit Office (1997), *Waltham Forest Housing Action Trust: Progress in regenerating housing estates*, Report of the Comptroller and Auditor General, The Stationery Office, London.

Nevin, B. (1998), 'Renewing the Black Country. An assessment of the employment impact of the Black Country UDC', *Local Economy*, vol. 13, no. 3, pp. 239-254.

Page, D. (1993), *Building for communities*, Joseph Rowntree Foundation, York.

Parkinson, M. and Newburn, T. (1998), *A framework for the evaluation of the New Commitment Pathfinder Areas*, Joseph Rowntree Foundation, York.

Power, A. (1997), *Estates on the edge. The social consequences of mass housing in Northern Europe*, 1999 reprint, Macmillan Press, London.

Power, A. with PEP Associates (1991), *Housing. A guide to quality and creativity*, Longman, Essex.

Silburn, R., Lucas, D., Page, R. and Hanna, L. (1999), *Neighbourhood images in Nottingham. Social cohesion and neighbourhood change*, Joseph Rowntree Foundation, York.

Smale, G. (1998), *Managing change through innovation*, The Stationery Office, London.

Social Exclusion Unit (1998), *Bringing Britain together: A national strategy for neighbourhood renewal*, Cm 4045, The Stationery Office, London.

Social Exclusion Unit (2001), *A New Commitment to Neighbourhood renewal*, Cabinet Office, London.

Speak, S. and Graham, S. (2000), *Service not included: social implications of private sector service restructuring in marginalised neighbourhoods*, The Policy Press, London.

Stewart, M.; Goss, S.; Gillanders, G.; Clarke, R.; Rowe, J. and Shaftoe, H. (1999), *Cross-cutting issues affecting local government*, report to the DETR by University of West of England, and The Office for Public Management, The Stationery Office, London.

Taylor, M. (1995), *Unleashing the potential. Bringing residents to the centre of regeneration*, Joseph Rowntree Foundation, York.

Walker, A. and Walker, C. (1997), *Britain divided: The growth of social exclusion in the 1980s and 1990s*, Child Poverty Action Group, London.

Williams, C. and Windebank, J. (1999), *A helping hand: harnessing self-help to combat social exclusion*, Publishing Services Ltd, York.

Wood, M. and Vamplew, C. (1999), *Neighbourhood images in Teesside. Regeneration or decline?*, Joseph Rowntree Foundation, York.

12 Love Thy Neighbour: Good Neighbour Agreements

BARBARA M. ILLSLEY

Introduction

Issues of environmental equity and justice have attracted increasing attention in recent years, stimulated by the emergence of the sustainable development agenda. Research has focused on a number of different dimensions of environmental justice: intergenerational justice, which is linked to the Brundtland definition of sustainable development (Almond, 1995; Haughton, 1999); interspecies, or what Low and Gleeson (1998) describe as ecological justice; distributive justice, concerned with the geographical basis of environmental equity and the spatial relationship of environmental problems and indicators of social exclusion (Cutter, 1995; Bullard, 1999); and procedural justice, where interest is directed at ensuring that decision making processes are designed and operated in an open and fair manner (Lake, 1996; Haughton, 1999).

The environmental justice debate within the field of town and country planning has been concerned, generally, with the last of these dimensions, that is the pursuit of a procedurally just process for reaching decisions about the development of the built and natural environment. Public participation is seen as a vital component of such a process and there has been a long tradition of involving local people in the statutory planning system in the UK, dating back to the 1960s. In more recent times, theoretical debates within planning have stressed the need for new institutional frameworks and arenas within which communities can participate, and innovative approaches to the processes of communication and negotiation (Innes, 1995; Healey, 1997).

Despite the use of a variety of new techniques for involving local people in planning projects, environmental decisions remain fiercely contested. Communities are no longer prepared to sit back and accept without protest the development of a chemical factory, a landfill site or a major road which will have a detrimental impact on their lives. Residents demand to have their views heard. However, as communities become more active in the planning process they discover weaknesses in the system. Local people, who object to a proposal on relevant planning grounds, can be left feeling frustrated and disillusioned when the application is approved contrary to the adopted development plan. Such reality suggests that there is a need to look for new approaches that will aid the creation of an environmentally just decision making process.

In their fight for greater environmental justice, a number of community groups in the USA have sought and negotiated Good Neighbour Agreements with major industrial firms, in order to gain greater influence over their local environment (Lewis and Henkel, 1998). Good Neighbour Agreements are compacts, drawn up by major companies in co-operation with local residents, to raise environmental standards and enhance local accountability. They have been promoted by a number of community groups in the USA as part of a campaign to achieve greater environmental justice. This chapter reviews the Good Neighbour Agreement concept as it has developed in the USA, and considers its application in Dundee, Scotland, where a Good Neighbour Charter has recently been negotiated between Dundee Energy Recycling Limited, operators of a new waste-to-energy plant, and the neighbouring community of Douglas, a predominately local authority council estate situated immediately south of the new plant.

Environmental Justice

Concern for the environment has a long history. During the first half of the twentieth century, environmental groups in the UK sought to conserve the natural heritage and promote the aesthetic values of nature, while in the USA campaigners fought for the preservation of the wilderness (Doyle and McEachren, 1998). These early movements were primarily apolitical, based on acceptance of existing social, economic and political structures. Over time, such assumptions have been challenged, however. Communities have become more aware of the potentially hazardous consequences of locating industrial developments in their neighbourhood, such as increased levels of air pollution or the risks associated with toxic waste dumps. The concept of environmental justice has developed in response to such concerns.

What emerges from a review of the literature on environmental justice is a picture of diversity in the use of the term. However, it is possible to distinguish four separate, although interrelated, interpretations of the concept. The first is intergenerational justice, namely working to meet the needs of today's society without compromising those of future generations (Almond, 1995; Haughton, 1999). This form of environmental justice is fundamental to the achievement of a more sustainable future as set out in the Brundtland Commission Report and Agenda 21. Almond argues that intergenerational justice requires the application of three key principles: conservation of options for future generations; conservation of environmental quality; and conservation of access. The second way in which environmental justice is understood is as interspecies justice or ecological justice (Hayward, 1994; Low and Gleeson, 1998). Here, justice is concerned with the relationship between humans and other species, and it raises questions about the extent of responsibility that people owe to their environment as well as the intrinsic rights of nature.

The third view of environmental justice is a concern with the distribution of environmental quality (Cutter, 1995; Lake, 1996; Bullard, 1999). The emphasis is on identifying and understanding the spatial distribution of environmental problems. Bullard (1999), for example, has found that "the environmental protection apparatus in the USA does not provide equal protection for all communities" (Bullard, 1999: 17) and that environmental and health risks are higher for people of colour and low-income than other sections of society. The fourth and final type of environmental justice is procedural justice. This involves the creation of decision-making arenas that operate in an open and fair way for all. For some, such as Torres (1994), the primary goal of such a participatory process is to ensure that environmental problems are distributed in a more equitable manner than at present. In contrast, Lake (1996) argues that such an interpretation is limited in scope and that procedural justice "entails full democratic participation not only in decisions affecting distributive outcomes but also, and more importantly, in the gamut of prior decisions affecting the production of costs and benefits to be distributed" (Lake, 1996: 165).

Procedural Justice and the Planning System

Town and country planning in the UK has had a long-term interest in issues of procedural justice and public participation. The Skeffington Committee, reporting in 1969, called for planning authorities "to act openly" in engaging the public in the planning process so that planning authorities and the general public would "grasp the opportunities for a new partnership" provided by the then new planning act (Skeffington, 1969: 47). Thirty years later, governments at Westminster and Holyrood have restated their commitment to the underlying principle of community involvement as part of the agenda for modernising government in the UK (DETR, 1998; Scottish Executive, 1999).

Recent theoretical discourse within the planning profession has focused on a collaborative model of planning, drawing on the Habermasian concept of communicative action (Healey, 1997; Hillier, 1999). An active process of participation, where stakeholders are not only consulted but also actively engage in debate and negotiation, is central to this approach. Openness and transparency are vital components in reaching decisions. Healey (1996) identifies three important features in achieving an effective collaborative process: the creation of arenas for communication which are open to all; an approach which ensures that power resides as close as possible to those who will live with the outcomes of decisions; and an open and inclusive style of debate. It becomes important for planners, therefore, to put in place appropriate institutional frameworks which will encourage fair participation, described by Hillier (1999) as comprising "having the ability to express one's opinion and tell one's stories, being listened to with respect, having access to adequate information, being able to question others, having some degree of control over the decision-making procedure and resultant outcome, demonstrating that decisions are made impartially and receiving good feedback" (Hillier, 1999: 236).

Good Neighbour Agreements

The importance of creating and nurturing strong relationships between the numerous stakeholders in a locality, based on openness and mutual trust, is central to the challenge of both sustainable development and environmental justice. This has been recognised in the USA with the increasing use of a new and innovative tool: the Good Neighbour Agreement. A Good Neighbour Agreement is a negotiated agreement or compact, drawn up jointly by a company and its local community, to raise environmental standards and enhance local accountability. The primary purpose of these agreements, according to Lewis and Henkels is "to foster sustainable development in a community by reconciling economic development with the community's welfare, including the health of its environment and individual members" (Lewis and Henkels, 1998: 129).

Lewis and Henkels have reviewed the emergence and use of Good Neighbour Agreements in the USA (Lewis and Henkels, 1998). They suggest that Good Neighbour Agreements developed out of the particular economic and political context that prevailed in the USA in the late 1970s and 1980s. At that time, there was increasing evidence of the negative environmental impacts of continued economic growth and a growing distrust amongst ordinary citizens of large-scale corporations. While government responded to individual environmental crises with changes to legislation, and some businesses took steps to improve their environmental image, many individual communities were still faced with the need to contest the damaging impact of industrial operations on their quality of life. Good Neighbour Agreements provided communities with a vehicle for promoting environmental quality and empowering local people. The experience from the USA indicates that Good Neighbour Agreements have been most commonly pursued where there has been a dispute between a local community and a corporation. A good example of this is the agreement reached between the community of Manchester, Texas and the Rhone Poulenc Plant, following a major pollution incident at the plant in 1992 (Lewis and Henkels, 1998).

While each agreement will vary to reflect local circumstances and concerns, to date most agreements have contained a number of common elements. Firstly, there is a commitment by the company to provide the community with access to information about the operation of the plant or facility. This may include information required by government, as well as internal environmental audits and safety procedures. Typically, the information is provided at a locally accessible site, such as a local library. Secondly, most agreements provide the community with a right of access to inspect the plant on a regular basis, accompanied by an appropriate expert chosen by the community. Thirdly, companies agree to involve the local community in the preparation of accident procedures which may be implemented if a major incident occurs. A fourth element found in many Good Neighbour Agreements is a commitment by the company to reduce pollution from the facility over a given period. Fifthly, companies agree to encourage the growth of the local economy through measures such as recruiting local staff and providing local training. This may be extended to the creation of a community fund which is used to support local projects. All of the proceeding elements emphasise the company's role in the locality. The final element, however, focuses on the community's responsibilities. As part of the agreement, the local community may agree to end local protests against the company, settle outstanding legal action or possibly assist with positive publicity (Lewis and Henkels, 1998).

Good Neighbour Agreements do not replace the legal regulations that have been put in place to protect the environment and safeguard public health and welfare; rather they aim to extend and strengthen them. From the community's view point, an agreement can improve access to information about an industrial operation, limit damage to the local environment and enhance the quality of life. The benefits to the company, although less obvious, may include local goodwill and positive public relations to the wider world. Adopting more environmentally friendly processes can have significant economic advantages as well.

Dundee Waste-to-Energy Plant

The Good Neighbour Agreement tool has been developed and applied in a US context, and it reflects both the regulatory framework in place and the specific environmental concerns of American communities. However, through the initiative of Friends of the Earth Scotland, a Good Neighbour Charter has been negotiated in Dundee, Scotland, between Dundee Energy Recycling Limited (DERL), operators of a new waste-to-energy plant, and the neighbouring community of Douglas, a predominately local authority council estate situated immediately south of the new plant.

Incineration has played an important part in Dundee's strategy for waste disposal. The City of Dundee District Council operated a municipal solid waste incineration plant at Baldovie in the north east of the city, from 1979 until 1996, when the plant was closed due to its inability to comply with new emission standards required by EC Directive 89/429/EEC. When the problem of emission levels was first recognised in 1992, the Council examined a number of possible courses of action, including upgrading of the existing plant, the expansion of alternative methods of waste disposal, and the development of a new waste-to-energy facility. This final option was encouraged by the announcement in 1993 of the Scottish Renewables Obligation (SRO) as a result of which the major electricity suppliers in Scotland were required to purchase a higher proportion of electricity from renewable sources. The waste-to-energy option was preferred by the Council for a number of reasons. Incineration not only reduces the demand for landfill capacity but also renders wastes biologically inactive. In addition, the proposed process recovers energy from wastes which can be used to generate a significant amount of electricity.

In order to progress the project, a joint venture company, DERL, was created, involving the local authority and private sector interests. Planning consent for the new waste-to-energy plant was sought in 1994. Opposition to the proposal was led by the Green Party although, in the event, only four letters of objection were received to the planning application for the plant. The main areas of concern were: the environmental and health impact of the scheme in light of research on the hazards of toxins; investment in large scale incinerators is likely to discourage recycling and waste minimisation initiatives; the consideration that waste-to-energy plants should not be included as renewable energy generators; the proximity of the new plant to Dundee; the lack of consideration of alternative sites and other waste treatment technologies; and criticism of the use of a solidification process prior to disposal of fly ash (Tayside Regional Council, 1994). Although not technically required by the Environmental Assessment (Scotland) Regulations, 1988, DERL commissioned an Environmental Assessment for the proposed development and this concluded that the new plant would have little adverse effect on the local environment (DERL, 1994). It stated that the use of latest technology would reduce emission levels and minimise operational noise and, although traffic flows to and from the plant were expected to be higher than in the past, they were not expected to result in a perceptible increase in noise to local residents. The planning application was approved in December 1994, subject to a planning agreement being put in place. The plant also required Integrated Pollution Control authorisation and this was obtained from the Scottish Environment Protection Agency in 1997.

Work began on the £45 million project following the announcement of the successful SRO2 bid in 1997 and construction work was completed towards the end of 1999. When fully operational, the plant will process approximately 120,000 tonnes of waste each year, consisting of domestic refuse, industrial waste such as oils, paints, packaging and tyres, and clinical and pharmaceutical wastes from hospitals, surgeries and pharmaceutical manufacturers. At full capacity, the plant will output approximately 11 megawatts of electricity to the National Grid.

Good Neighbour Charter

As a result of past worries about the environmental and health impacts of the Baldovie incinerator, the local community in the vicinity of the new plant was understandably wary of the new operation. Given this background, Friends of the Earth Scotland felt that here was an ideal situation to try out the concept of a Good Neighbour Agreement, bringing together DERL and the local community. A series of meetings was held in the autumn of 1999 to gauge whether there was local interest in developing an agreement. Representatives attended from DERL, the Council and a number of community organisations, including residents associations and environmental groups. At this stage there was much discussion about how the "neighbours" of the new plant should be defined and how the wider community could be involved.

While initially facilitated by Friends of the Earth Scotland, the idea was taken up and progressed by one of the local councillors. Following discussion, a draft Good Neighbour Charter was produced between DERL and the community of Douglas (DERL and Douglas Community, 2000). The Charter expresses "commitment to the spirit of good neighbourliness, high standards of practice and open communication" between the parties. It is seen as being a first step in the process of drawing up a more formal agreement. An important aspect of the Charter is the establishment of a

Liaison Group, consisting of two representatives of DERL, three community representatives, two local councillors and two ex officio community advisors, which will be open to public observation and will meet on a quarterly basis.

The Charter reflects the experience of Good Neighbour Agreements in the USA. It contains eight elements:

Community access to information A comprehensive public DERL file will be held at local libraries in the neighbouring communities and at the Central Library. Information will be presented in an accessible and understandable form to lay members of the community.

Right to inspect the facility The Liaison Group has the right to inspect the facility by arrangement and with an agreed period of notice. Other community groups wishing to inspect the facility may apply via the Liaison Group.

Accident preparation DERL will include its procedures for emergency or accident in its public file. These procedures will be subject to consultation with the Liaison Group.

Environmental performance A statement of environmental performance will be publicised in lay terms and agreed by the Liaison Group. Technical reports on emissions will be made available as part of the agenda at Liaison Group meetings. Emission standards and targets higher than the legal minimum will be set and monitored. Procedures adopted when exceeding emission limits will be publicised. Volumes of waste removed for recycling will be published.

Good jobs, local jobs, union jobs Every endeavour will be made by DERL to ensure that local employment is maximised by recruiting locally where appropriate. DERL will abide by its legal requirements in employment practice including relating to trade union recognition.

Local economic needs The Charter includes recognition of the local economy and neighbouring communities and as such DERL and the Liaison Group will discuss matters of mutual economic interest.

Transportation of waste The transportation of waste will be conducted with due consideration to the local communities, with a commitment to design routes and times to take account of local circumstances.

Changing operating conditions Any change in policy on materials will be party to discussion with the Liaison Group prior to test or implementation. The Liaison Group should discuss any change to legislation relating to emissions from the plant (DERL, 2000).

A Significant Step Forward?

The Good Neighbour Charter was signed on 2 May 2000 when the waste-to-energy plant was officially handed over to DERL. At that time, community representatives in Douglas who had been involved in negotiating the Good Neighbour Charter felt very positive about it. Describing the Charter in the local press, Dundee City Council's Environment and Consumer Protection Convener, Neil Glen, said that the agreement represented "a significant step forward in ensuring that the residents and the plant can co-exist in a harmonious fashion" and that it "will allow for open dialogue between residents and the management of the plant" (The Courier and Advertiser, 2000: 5). Friends of the Earth Scotland, who facilitated the initiative, see it as part of their campaign for environmental justice, which has the slogan "No less than a decent environment for all; no more than a fair share of resources" (Scandrett, 1999). However, how useful a tool is a Good Neighbour Agreement? Does it represent a step towards greater procedural equity or is it merely a mechanism for appeasing local objectors?

There is some evidence that local communities up and down the country are losing faith with the process of environmental decision-making, amidst concerns that it is not operating in an open and fair manner. A good example of this is a community campaign against a new hotel and golf development at Kingask, near St Andrews. Despite strong local opposition, the development was approved by Fife Council. Local people were outraged, not just at the decision but also at the way the application had been handled, and as a result they sought a judicial review at the Court of Session.

Unease about the system of decision making is also reflected in recent debates about public access to environmental information and the need for third party rights of appeal. Public access to environmental information is viewed as crucial to both the process of encouraging individual responsibility for the environment and the successful operation of the planning system (Illsley et al, 1997). According to Rowan-Robinson et al., improved public access can aid the promotion of sustainable development, by promoting confidence in Government and industry, informing consumer choice, increasing public scrutiny of industry and hence encouraging environmental practice activities and enabling the public to have a greater role in the decision-making process. The provision of access to environmental information must be seen as a means to an end, whether that end is the execution of the principles of sustainability or enhanced involvement in planning decisions.

The primary arena for debating the appropriateness of a development is within the planning system, whether as part of the process of preparing a local plan or in response to a planning application. Increasingly there have been calls for changes to the current system where local communities have no right of third party appeal (CRPE, 1997). Campaigners for enhanced third party rights refer to the practice in the Republic of Ireland, where for a period of a month after a decision is made on a planning application, third parties as well as applicants can appeal against the decision (Grist, 1996). It has been suggested that in the UK third parties could be given the right to appeal in certain situations, for example where a development is contrary to the development plan, where a development is put forward by the planning authority, or where a person who has commented on an original application and has reason to believe that the development plan has been substantially disregarded (CPRE, 1997).

What the debates about public access to environmental information and third party rights of appeal tell us is that there is a need to promote a more open and fair system of environmental decision making. The Good Neighbour Agreement could provide one possible mechanism for achieving this goal.

The experience of the Good Neighbour Charter in Dundee suggests that the process of its preparation is having a positive impact. It has led to the formation of the Liaison Group, which allows community representatives to meet with company officials on a regular basis and to raise and discuss issues of local concern. Furthermore, the local community has access to information about the operation and impact of the plant as well as the right to visit the facility as required. However, questions remain to be resolved before an assessment can be made on the value of the Charter. Specifically, consideration needs to be given to the way that the community has been defined. Should the agreement only involve those living in one neighbourhood adjacent to the facility or should people from other areas be included? What are the appropriate mechanisms for deciding representation on the Liaison Group? Initially, community representatives have been drawn from established community groups but this may mean that certain sectors of the community are excluded. As presently written, the Charter does not address issues such as the need to enhance environmental standards. It is unclear, therefore, whether the company will be prepared to enter into discussions aimed at improving standards, for example by reducing future emissions to a level that is higher than current European standards, in order to improve local environmental quality. Finally, the Charter is purely a voluntary agreement which is not legally binding on either party. Is this appropriate? It may mean, as found in the USA, that the company can "call the tune" over most issues (Lewis and Henkels, 1998) and, unless the parties move towards a legally binding agreement, implementation of the Charter will be dependent upon the goodwill of the company.

Turning to the bigger picture, it is unclear at present whether the Good Neighbour Agreement is a worthwhile tool in the quest for procedural equity. Lake (1996) argues that full democratic participation is essential to procedural justice and Good Neighbour Agreements can provide an innovative method of engaging with local people. Agreements may assist in developing community capacity, helping individuals to gain environmental knowledge and skills, and they may influence the institutional cultures of industrial companies, stimulating more open and accountable attitudes. However, ultimately the potential of Good Neighbour Agreements will depend upon the way in which they are developed and implemented. To be valuable, the process of negotiation should be inclusive, in the sense that all sectors of the community are engaged. The process should also be transparent, so that the community can feel confidence and trust in the procedures in place, and has real power to influence decision-making.

Conclusion

Good Neighbour Agreements have been used to good effect in the USA for many years in the pursuit of environmental justice. Such agreements offer local communities new opportunities to influence decisions about their environment and they have the potential to act as arenas where a number of elements of fair participation can be met. The Good Neighbour Charter in Dundee is an exciting development since it is one of the firsts of its kind prepared in the UK. Initial evidence suggests that the process of preparing the Charter has been a positive one, giving local people improved information and better access to the plant. However, the long-term value of negotiating such an agreement has yet to be proved.

References

Almond, B. (1995), 'Rights and justice in the environment debate', in Cooper, D. E. and Palmer, J. A. (eds.), *Just Environments*, Routledge, London.

Bullard, R. D. (1999), 'Dismantling Environmental Racism in the USA', *Local Environment*, vol. 41, pp. 5-19.

Council for the Protection of Rural England (1997), *Standards in Public Life: Aspects of Conduct in Local Government*, A contribution by CPRE to the Nolan Committee.

Cutter, S. (1995), 'Race, Class and Environmental Justice', *Progress in Human Geography*, vol. 19, pp. 107-118.

DETR (1998), *Modern Local Government. In Touch with the People*, Command Paper Cm 4014.

Doyle, T. and McEachren, D. (1998), *Environment and Politics*, Routledge, London.

Dundee Energy Recycling Ltd (1994), *Environmental Statement for Proposed Baldovie Waste to Energy Plant*, unpublished.

Dundee Energy Recycling Ltd. and Douglas Community (2000), *Good Neighbour Charte*, Dundee Energy Recycling Ltd., Dundee.

Grist, B. (1996), 'Development Control in the Republic of Ireland - Comparative Aspects', *Oxford Planning Monographs*, vol. 2, pp. 1-8.

Haughton, G. (1999), 'Environmental Justice and the Sustainable City', *Journal of Planning Education and Research*, vol. 18, pp. 233-243.

Hayward, T. (1994), *Ecological Thought: An Introduction*, Polity Press, Cambridge.

Healey, P. (1996), 'Consensus-building across difficult divisions: new approaches to collaborative strategy-making', *Planning Practice and Research*, vol. 11, pp. 207-216.

Healey, P. (1997), *Collaborative Planning*, Macmillan Press, Basingstoke.

Hillier, J. (1999), 'Culture, community and communication in the planning process', in Greed, C. (ed.), *Social Town Planning*, Routledge, London.

Illsley, B. M.; Lloyd, M. G.; Lynch, W. and Burbridge, V. (1997), *Public Access to Planning Information*, Scottish Office, Edinburgh.

Innes, J. (1995), 'Planning theory's emerging paradigm: communicative action and interactive practice', *Journal of Planning Education and Research*, vol. 14, pp. 183-90.

Lake, R. W. (1996), 'Volunteers, NIMBYs, and Environmental Justice: Dilemmas of Democratic Practice', *Antipode*, vol. 28, pp. 160-174.

Lewis, S. and Henkels, D. (1998), 'Good Neighbour Agreements: A Tool for Social and Environmental Justice', in Williams, C. (ed.), *Environmental Victims*, Earthscan, London, pp. 125-141.

Low, N. and Gleeson, B. (1998), 'Situating Justice in the Environment: The Case of BHP at OK Tedi Copper Mine', *Antipode*, vol. 30, pp. 201-226.

Scandrett, E. (1999), 'Good Neighbours for Environmental Justice', *Different Dundee*, no. 10, pp. 8-9.

Scottish Executive (1999), *Report of the Commission on Local Government and the Scottish Parliament: The Scottish Executive's Response*, HMSO, Edinburgh.

Skeffington, A. (1969), *People and Planning*, HMSO, London.

Tayside Regional Council (1994), *Planning Committee Report No 1955/94: Proposed Waste to Energy Plant Baldovie Dundee*, unpublished.

Torres, G. (1994), 'Environmental burdens and democratic justice', *Fordham Urban Law Journal*, vol. 21, pp. 431-460.

13 Public Involvement in Residential Conservation Planning: Values, Attitudes and Future Directions

PETER J. LARKHAM, JOHN PENDLEBURY AND TIM TOWNSHEND

Introduction

Conservation, as an activity, has been around for far longer than most people realise. It certainly predates the usual references to Ruskin and Morris in the mid- and late-nineteenth century (cf. Mielé, 1996); but the early history of conservation, as with antiquarianism and local history (Dellheim, 1982; Schnapp, 1996), was an activity for the élite and wealthy. During the twentieth century the scope of conservation widened: the number and type of protected ancient monuments increased, bomb damage during the Second World War spurred the listing, and then the protection, of buildings of architectural and historic interest, and then from 1967 entire areas could be designated as conservation areas. In the UK we now have over 17,000 monuments, half a million listed buildings and approaching 10,000 conservation areas.

This is a great change from early conceptions. For example, the County Planning Officers' Society (1968) originally felt that there might eventually be "several thousand" conservation areas. And, although there has never been fixed guidance on what a conservation area should be or contain, there was general (and international) agreement that "conservation areas should be coherent, compact and of historic, architectural, artistic, characteristic or picturesque interest" (guideline resolution for the European Architectural Heritage Year campaign, quoted in Reynolds, 1975: 359); and the early, smaller, designations were of key historic town centres and villages. However, relevant to this chapter, in 1973 Lord Sandford stated that conservation "should also have care for the conservation of existing communities and the social fabric, wherever public opinion points clearly towards it" (see DoE, 1973). Delafons (1997) calls this a "temporary effusion" of conservation sentiment, a "billowing out" of the definition of conservation. For a brief moment in this period of active reaction against modernity, conservation broadened from a focus on buildings and fabric to include a social dimension.

In the period since the early 1970s, conservation has become established as a rarely-contested major objective of planning policy (Pendlebury, 2000) and is usually considered to be one of the spheres of planning policy which enjoys most popular public support. There is certainly a massive popular interest in local history and heritage, and, for example, this is an area which has been targeted for funding by the Heritage Lottery Fund. Despite this apparent popularity, underpinning this chapter is a sense that conservation often remains the preserve of a specific set of values, attitudes and procedures belonging to conservation professionals. Others who have a stake in conservation are often assumed to subscribe to these values and attitudes. Research has uncovered differences between the attitudes of professionals and the public in some aspects of planning, particularly related to design (Hubbard, 1994) but, thus far, conservation has been subject to less scrutiny in this respect (although see also Hubbard, 1993; Hobson, 2000).

Public participation has long been a stated concern in planning, and has widely been used to legitimise a wide range of activities and policies, including conservation and design. Indeed, these issues are high in contemporary debates on urban governance and urban quality, involving both academics and policy-makers (see, for example, Healey, 1998; Rydin, 1998). An awareness of the importance of a wider public engagement in conservation has recently been shown by English Heritage (1997, 2000a, 2000b). The RTPI Planning Aid Award for 1999 was given for the work of Planning Aid for London with the London Borough of Hackney and the Hackney Society in involving local residents in the conservation of the borough. This shift towards a more inclusive conservation is part of an international trend. For example, pioneering work in Denmark has involved communities in local surveys of historic architectural values (Ministry of Environment and Energy, 1997). Yet much work on residents and their environmental experiences has tended to rest on the dubious assumption that residents are an homogenous group. However, again drawing from international work, it has been suggested that attitudes towards, and values of, environments will differ if the resident is indigenous to the area or a newcomer (Perähuhta, 2000). Despite these debates and the broadening scope of conservation, in common with much planning activity, most conservation remains in the hands of a select few. This group comprises in particular the educated professional planning officers, including the relatively new breed of "conservation officer", who is usually a member of the newly-professionalised Institute of Historic

Building Conservation. The vociferous, educated and articulate minority who comprise much of the "amenity societies" are held by these officers to represent both the views of the public, and to bring expert knowledge to the conservation planning system. Yet they are also criticised for being unrepresentative of the population as a whole: representing solidly middle-class values and attitudes, self-interest and "NIMBYism" (cf. Eversley, 1974; Lowe, 1977; Rydin, 1998).

Although there is little published comparative data on conservation areas (Pearce *et al.*, 1990, is a now-dated exception), by the mid-1990s the most numerous area designated was the rural village; but the fastest-growing type of designation appeared to be the residential suburb (cf. Larkham, 1999a). Such areas form the "anonymous familiar" for many people, and are, perhaps, worthy of recognition and retention (Meinig, 1979; Tarn, 1975). Their "sustainable renewal" is being discussed (Gwilliam *et al.*, 1999). It is, perhaps, here that many people come into contact with the conservation system on an everyday basis.

Victorian and Edwardian suburbs had been designated from the earliest years of the system, but the designation which brought conserved suburbs to prominence was the extension of conservation to encompass speculatively-built suburbs of the 1930s (which were widely criticised by the proto-conservation movement when built). The designation of School Road, Birmingham, in 1988 led to a feature in *The Times* (Franks, 1988). The resulting debate revolved around the extent to which this, an example of the familiar speculative inter-war semi-detached residential area, could qualify for designation given the legal definition of a conservation area, which requires the area to possess "special architectural or historical" interest. What was special about this average, everyday area? Local planners argued that the area was relatively unaltered since its construction in the late 1930s: few extensions, PVC windows, and so on. While other such areas were changing rapidly, the unchanged, original character of School Road was "special". Protection of this type of area rapidly became popular: even by 1995, for example, the London Borough of Brent had eight "suburban estates by renowned inter-war developers" (London Borough of Brent, 1995).

The conservation of residential suburbs, and the extended relationship with a range of members of the public and organisations that this generates, is an important focus for study. Residential areas bring to the formal processes of conservation planning a wider part of the population than might commonly be involved for commercial areas, where the role of professional groups in representing interests and conflict mediation is likely to be more prevalent. Furthermore, the pressures within residential areas are likely to be different, and probably at a smaller scale, than those met in commercial areas and urban centres (cf. Larkham, 1996, chapters 8 and 9). This is because residential conservation usually involves the conservation of the relatively ordinary, notwithstanding the terms of the statutory definition: in the words of government guidance, the conservation of "the familiar and cherished local scene" (DoE/DNH, 1994, para 1.1). Residential areas will also bring small-scale, but numerous, tensions between the values and aspirations of residents, for example in personalising their homes, and the values embedded in the planning system (Larkham, 1999a: 367-369).

To review these issues affecting residential conservation we now discuss a range of recent empirical research which attempts to explore the nature and extent of public involvement in residential conservation area planning, both from the views of the public – residents in these areas – and of the professionals concerned.

Who Participates, How and Why?

It is surprisingly difficult to discuss the question of "who participates" in residential conservation as so little research has been carried out on the values, attitudes and activities of residents. In broader terms, a number of studies have explored participation in environmental pressure groups and in voluntary amenity societies (Buller, 1981; Lowe, 1977; Larkham, 1985; Pendlebury, 1999a). These are the groups which local planning authorities (LPAs) tend to consult over questions of conservation. Indeed many LPAs see these societies as potential allies (Jennings-Smith, 1977: 245). Informal contact between societies, councillors and planners is often high (Barker, 1976; Buller, 1981), and there is often considerable similarity between the views of local planners and those of amenity societies (Barker, 1976; Larkham, 1985). The influence of amenity societies seems to be greater where they adopt a "responsible style" (Lowe and Goyder, 1983): in other words, they become embedded in the very system which they seek to monitor and review. In their knowledge, skills and engagement with the planning process, many societies can essentially be considered to be "semi-professional", and often attempt to involve retired professionals.

It is a common critique that the members of these societies are the vociferous, educated, middle classes: those "who by reason of education, training, outlook, social milieu and intellectual persuasion, regard quality of environment not only as important in itself but also as a benevolent influence on others" (Cherry, 1975: 3). They are, clearly, a very tiny minority, particularly in the larger towns and cities (Larkham, 1993, Table 1). Their officers certainly possess middle-class occupations and values (Barker, 1976); and many of the suburbs protected were built for, or are now occupied by, the upper and middle classes. Many fewer working-class residential areas are subject to protection and, from the limited empirical work undertaken, the attitudes of their residents appear to be significantly different (Larkham, 2000).

It is clear that self-interest is a strong motivation for participation in residential conservation planning and, indeed, in the planning system as a whole. In fact Samuel (1994) noted that the introduction of conservation areas was the response of government to placate a significant middle-class property interest which wished to secure the pleasant environs of its residence (although less polemic histories give a very different emphasis). This self-interest ranges from objecting to neighbours' planning proposals on the grounds of "detriment to residential amenities", to seeking more, and larger, conservation areas. The latter would seem to stem from the frequent assertion from estate agents (albeit with little hard evidence) that property within designated areas rises in value, perhaps by as much as ten per cent.

This is a long -and widely held view: "too often conservation area status is sought by residents not because they fear the consequence of new development but because they wish to enhance the value of their property" (Lawless and Brown, 1986: 27). Residents' support for conservation is "all about defending their status and their territory and I suppose the bottom line is defending the value of their property" (anonymous planner respondent cited in Hobson, 2000: 106). This would seem to be the major reason why residents of an estate of speculatively-built "executive homes" of the early 1960s in Solihull have successfully lobbied their councillors for designation as a conservation area, despite the strong objection of the professional staff (Wilson, 1999). One resident, writing in support of the proposed designation, stated that "conservation area designation will prevent speculative and inappropriate 'windfall' development, which would destroy the prize-winning character of the estate" (Solihull MBC planning files). Another resident, quoted in *The Times*, stated that "we have already made changes, such as putting in double glazing and prettier entrance porches, so do not have much concern over being able to alter things in the future" (Wilson, 1999). The planning officer's report warned that:

> I believe that the committee should be careful to ensure that it does not devalue the status of the borough's conservation areas by designating, on

> request, new areas that simply are not of sufficient architectural value or historic interest to pass what should be by definition a very exacting test (Solihull MBC Planning Committee minutes, 17/8/1999).

In making the designation, on a vote of 7-6, the committee members felt that "it is perhaps appropriate to consider the area as being a good example of the sort of development from which the modern borough is formed" (Solihull MBC Planning Committee minutes, 17/8/1999).

This single example thus highlights key differences between values of residents, officers and committee members. The activity of the wider population in conservation areas (and the popularity of designations) is often thought to be related to NIMBYism and property values and, although there is some evidence to support this, these reasons are not necessarily dominant.

Also of interest is that, in three suburban conservation areas sampled in the West Midlands, residents have remained far longer than the national average of seven years. This is doubled in the two middle-class areas studied, and tripled in the working-class area. If this pattern is more widely representative, therefore, it would seem that residents are not quickly seeking cash profit from increased property values, and do derive benefits of increased use value from the environmental quality of the area which encourage them to stay.

How Popular are Conservation Areas, and What Understanding of Conservation is Held by Members of the Public?

In the élite origins of conservation, stress was placed on the importance of ancient buildings - often of the upper social echelons - as a continuing part of history, with inherent meaning in their fabric, to be passed from generation to generation in as original condition as possible. To this end, the extent to which ordinary people engaged with conservation, or understood its aims and objectives, was largely irrelevant.

Although attitudes to public involvement in conservation may have changed (see below) and conservation is generally regarded as "popular" (reflected by television coverage, membership of the National Trust etc.), there remain considerable gaps in our knowledge relating to the general public's commitment to historic environments. In particular, little has been written about what people actually perceive to be the aims, objectives and benefits of conserving historic buildings (and environments) and whether these equate in a meaningful way with the priorities and value systems employed by conservation professionals.

The little empirical research which has been carried out in the last two decades has provided a far from clear picture. In 1984, for example, the London Borough of Hammersmith and Fulham undertook a random survey of residents from six of its conservation areas designated over a period from the early 1970s to the 1980s (Wealthy, 1984). The aim of the survey was to establish the levels of knowledge of the conservation areas, whether residents considered living in conservation areas advantageous and what residents considered as suitable improvements for these areas. The results of this research were interesting, if slightly disconcerting. In two areas, which had been designated less than four years previously, for example, fewer than 25 per cent of residents knew that they were living in a conservation area. Furthermore in one area, developed originally as a local authority cottage estate, residents considered picture windows, modern porches and the introduction of metal-framed windows to be "improvements". The survey suggested that residents were generally under-informed about the basic facts of designations and certainly had no comprehension of the nature and extent of additional planning controls (Wealthy, 1984).

One criticism of the Hammersmith and Fulham work, however, is that it attempted to introduce too many variables in a limited study. The differing types of tenure, socio-economic group and physical location involved made sound analysis difficult. Subsequent research, in suburban conservation areas in the West Midlands (Larkham, 1999b), in Chartist villages in Gloucestershire (Larkham and Lodge, 1999) and inner urban areas in Tyne and Wear (Townshend and Pendlebury, 1999a) has attempted to address this issue by undertaking comparative research where factors such as socio-economic status and geographic location have been limited.

In these three studies, the residents' knowledge of, and support for, conservation area status appeared to be high. In the historic Chartist villages, for example, 70 per cent of respondents were aware of the existence of the conservation area and 80 per cent supported the concept (Larkham and Lodge, 1999: 10). In the Tyne and Wear case studies, 92 per cent of respondents supported the existence of the conservation area and 96 per cent thought it important to protect and enhance the historic character of the area (Townshend and Pendlebury, 1999a: 324). Support for conservation was often quite vehement. In one of the north-east studies, respondents rated conserving the historic environment as second only to crime prevention in terms of the most important local issue and, in another, when asked how they would feel if the area lost its conservation status, 87 per cent of the respondents said they would oppose this, using phrases such as "I would fight tooth and nail" (Townshend and Pendlebury, 1999a: 324).

Beyond this general support, however, there was considerable confusion and misunderstanding. When asked in one of the Tyne and Wear case studies to describe what they thought the aims of a conservation area were, of those who attempted a definition, only 43 per cent included some reference to "historic buildings" or "architecture"; 38 per cent included the words "preservation", "restricted development", or "protected against"; and, interestingly, 27 per cent included the words "natural"; "open space" or "landscape". None included words such as "enhancement" or "improvement", and few came anywhere near the legal definition. Such sentiments were reflected in most of the other case studies where, although there was a general appreciation of the aims and objectives of conservation legislation, few respondents had any idea of the precise scope or nature of that legislation.

In the West Midlands case studies, there was a "considerable ignorance of the complexities of conservation area planning controls" (Larkham, 1999b: 8). Here, for example, 50 per cent of respondents thought that permission was needed to change doors and windows in a conservation area. Similarly, in the Chartist villages, 80 per cent of respondents thought, incorrectly, that permission was needed to alter doors and windows (Larkham and Lodge, 1999: 9). In one of the north-east examples, when presented with a list of activities and asked which would require planning permission in a conservation area, only two per cent of respondents correctly identified those which would (Townshend and Pendlebury, 1999a: 324). Most respondents thought that conservation area controls were far more draconian than they actually are and, in general, more closely equated them to those applicable to listed buildings.

Residents also encompassed a far wider range of environmental issues within their conception of conservation than is covered by conservation area legislation. In the north-east research, when asked what aspects of the conservation area were most important to residents, six clear themes emerged. In descending frequency of choice these were architectural appearance (from general styles to details); natural environment (open space, trees etc.); social factors (nice neighbours, the existence of social networks); historical character (referring to distinct historical periods, but also a sense of permanence and continuity); general environmental quality (e.g. peace and quiet); and morphology (size of plots, widths of streets etc.) (Townshend and Pendlebury, 1999a: 325). Many of these issues lie outside the direct control of the local planning authority, yet these factors greatly influence the success, or otherwise, of the conservation area as viewed by the residents (Larkham, 1999b: 20).

Views of Public Involvement Held by the Conservation Professionals

This section is based upon a questionnaire sent to conservation professionals which sought information on the impact of public consultation on decision-making, professionals' views over public comprehension of conservation objectives and more broadly the role of public consultation (for a full account see Pendlebury and Townshend, 1999).

Effects of Public Consultation

Most local authorities (70 per cent) were found on occasion to be modifying the boundaries of new conservation areas in response to consultation; changes to conservation area policy were less frequent (32-33 per cent). Only occasionally had public consultation led to major changes in approach. One respondent reported an area which had been designated as a result of grass-roots pressure. The area had initially been rejected as all the buildings were listed; however, a widely-drawn area has subsequently been designated and this has been important in resisting development. One rural authority commented on a proposal for a new area contained in the district plan to which the parish council objected. The parish council, who appeared as objectors at public inquiry, subsequently had a change of heart and promoted a much larger area which was duly designated. Although the officer saw this shift as beneficial, the suspicion was that "the parish council saw the designation as a means of controlling their [a local landowner's] activities".

Respondents were overwhelmingly positive about the benefits of public consultation: the reasons given for this fell largely into two main themes. First, and most frequent, consultees were seen as a useful resource in providing conservation staff with information, particularly in terms of historical development. Secondly, it was seen as part of a process of engendering support for the designation. Sometimes support from public consultation had been important in securing political support for proposals. In one case, a large-scale public consultation was undertaken on ten new proposed areas due to the scepticism of councillors, resulting in 97 per cent support from the public.

Negative responses to public consultation mainly focused on

pressure for designations without sufficient "special architectural or historic interest", and with councillors responding to these pressures designation became "political". The sense of this is quite nicely conveyed by one respondent who reported that consultation had not in fact changed decisions,

> In my experience the most common approach we receive is people in pleasant, though not necessarily significant, living environments, seeking designation of conservation areas. These proposals are not often justified as they fall short of the qualitative yardstick I apply, but I always judge them carefully.

In one case where the LPA had proposed de-designating part of an area the consequences were considered to be mixed; although this proposal had been fiercely resisted, it had led to a community desire and effort to improve the quality of the area.

Public Knowledge of the Purpose and Practical Implications of Conservation Areas

The questionnaire sought to elucidate from conservation officers their views of public knowledge of the purpose and practical implications of conservation area designation. "The public" is, of course, not an homogenous group and therefore what was sought was a dominant feeling rather than a comprehensive picture.

The overwhelming finding of this part of the survey was the belief of many respondents that the public equate conservation areas with preventing development, and that they tend to over-estimate the restrictions which exist. The degree of control which exists for changes to the exterior of listed buildings was frequently thought to apply in conservation areas, and that the purpose of conservation areas is as a means of preventing all development. This accords closely with the findings from surveys of residents described above. A number of respondents commented that members of the public were often disappointed or annoyed when they discovered the relative lack of controls.

A number of respondents distinguished between public understanding, and support, in different types of conservation area. Typically, there seems to be far more understanding and support for conservation objectives relating to affluent suburbs than for commercial centres. More mixed messages were received for rural conservation areas, with respondents tending to indicate more awareness of conservation areas, but being sharply divided over whether this led to public support or antagonism.

Consultation: Problems and Attitudes

Most respondents stressed the importance of public consultation, although a number of more frankly sceptical views were advanced. However, on closer examination, several quite different views of the role of "consultation" can be discerned. At the most participatory end of the spectrum one respondent stated that "we have found it very beneficial to work with the local community in drafting "character statements" and enhancement proposals rather than simply consulting on our ideas. Local ownership has proved very effective in many ways". In a similar vein, another described their LPA's changing practices,

> It [public consultation] can give inconclusive results. It also often leads to the complaint that Councils have already made their minds up prior to consultation. Agenda 21 principles working *with* communities to produce proposals should result in better understanding and satisfaction and more positive results.

These replies were, however, unusual. A far more common trend was to refer to a process which was less to do with consultation and more to do with raising awareness, and a didactic wish to educate the public in terms of the values of the conservation professionals. To quote a typical response,

> The public consultation process is beneficial as it provides the opportunity to explain exactly what designation means, so people are more informed, and generally more receptive to the idea. It also helps in ensuring the community will respect the policies and controls that will affect them.

A number of respondents explicitly acknowledged a distinction between consultation and information/education, perhaps inevitably returning to issues of "expert"-defined "special architectural or historic interest". For example,

> If an area is of special architectural or historic interest then it should be considered for designation, much in the same way that buildings of special architectural or historic interest are considered for listing, and public debate – as opposed to research – is inappropriate. However, once an area is designated (or a building listed) then education of the public and all other

interested parties is the key to conservation success.

More reflexively, from another respondent,

> It's great if you've got support to start off with; v. hard work if you haven't but the only way forward. I probably shouldn't say this but really it's public "education" here rather than consultation – some may say purveying middle class values?

More sceptical voices made such comments as "costly exercise in terms of local authority resources, often with little or no interest from the public"; "it can be very time consuming with little return in numbers of responses e.g. 10 responses from several thousand residents!", and "value probably over-stated". The issue of low response rates to consultation was frequently mentioned. Other themes which emerged included the domination of consultation processes by a vocal minority; for example, "it [consultation] tends to disproportionately serve articulate interest groups while the general individual or business takes little notice of the process to their possible detriment".

Interestingly, there was no obvious overall pattern to the nature of comments made, either geographically or by the type of LPA responding. So, for example, the most positive and the most sceptical commentaries were both made by extremely rural southern districts. The few discernable patterns were not a surprise, such as the presence of vocal middle-class groups in outer London Boroughs and a consciousness of a political dimension to conservation work from respondents from inner London Boroughs.

Discussion and Conclusion

Conservation areas have, since their inception, encompassed places where people live. But it is clear that a significant change in the emphasis of designation activity has occurred, away from an original emphasis on the historic centres of settlements. The designation of suburban areas, areas which would seem to be increasingly "ordinary", exemplify this. Our work to date suggests that conservation problems in these areas are rarely technical in nature. Rather, they are to do with differences in values and attitudes, and often compounded by difficulties between different groups in the communication of those criteria. So, for example, LPA officers are often principally concerned with architectural form and concepts loosely derived from bodies such as SPAB (the Society for the Protection of

Ancient Buildings), and precise definitions derived from statute and guidance. Even here, although individual officers or authorities may be consistent in their approach, there are quite different values and practices being employed in different localities. The majority of residents questioned often had a looser, but much more holistic, approach to conservation and the qualities of their residential environment, although again this generalisation conceals great diversity.

We must, therefore, return to some basic questions: particularly in these suburban areas, why, how and for whom are we conserving? Attempts to address these issues have, in the past, been unsatisfactory generalisations apparently drawn largely from architectural conservation and town centre experiences. It is hard convincingly to extend arguments predicated on the importance of stewardship of these areas for future generations of inhabitants. The Australian ICOMOS "Burra Charter" (Marquis-Kyle and Walker, 1992) and its concept of "cultural significance" provides some help. However, it has had limited impact on British practice, especially on the sort of relatively modest areas here addressed. A critique of the Danish experience of involving communities in conservation-related surveys and evaluations feels that this process of involvement develops a radically new view of the environment,

> The moment residents enter into a descriptive process concerning their own local identity, they are forced to step outside it... The mapping process is no neutral assessment of the local significance of building culture. It brings disturbance into the observed, everyday world, changing its history through its observing presence (Algreen-Ussing, 2000).

This alerts us to consider whether the very processes of resident involvement are, inevitably, changing the perceptions and values of those residents.

For these suburbs, the issue of what is "special" (in terms of the statutory conservation area definition) seems to focus not on the visual and largely architectural aesthetic, but on the everyday feelings and values of the residents. These are areas that are principally enjoyed by residents, for the quality of amenity they bring, rather by than society as a whole for their enduring contribution to the cultural heritage in general. Such areas are special for those who live there, rather than for the wider community: this is a key difference between these areas and historic town centres and individual monuments and buildings subject to high levels of public use, for purposes of tourism, work and leisure, and which may embody deeper sets of cultural values about society. Yet, evidently, many residents feel that conservation area designation is, *or should be*, equivalent to a preservation order.

It might, therefore, be appropriate to review designation approaches, and develop enhancement and management policies, which distinguish between different types of conservation area. Note that this would be a typological, rather than qualitative, distinction. If the designation of residential areas is essentially for the benefit of the people that live there, then the designation of such areas might, for example, be entirely community-driven. This could be seen to legitimise the approach taken by the Solihull residents described above. However, in terms of resources and strategies for place management, such areas may be considered more in terms applied generally to residential areas rather than key "heritage" areas and the additional resources that this may bring.

The British conservation area system encompasses both major areas of cultural heritage on a world scale and the far more modest environments which are considered here. Thus, at present, an inter-war suburb such as Birmingham's Hall Green is treated under exactly the same legislative provision and official guidance for historic areas as that covering the historic cores of York and Chester. Ultimately it may be desirable to make some formal differentiation between these through, for example, the introduction of a two or more tier system (Pendlebury, 1999b; Townshend and Pendlebury, 1999b). However, it may be premature to consider such a radical shift when there is still much that we need to understand about the nature of people's engagement with the conservation system.

One of the key problems with residential conservation areas seems to be in the differences between public and professional values and approaches, which may not coincide even where they do not actively conflict. At least in part this seems to be an issue of communication. Residents often have a generalised and rather inaccurate perception of the objectives and practical effect of conservation area designation, both in terms of national legislation and their individual local authority. Professionals may be using an approach to conservation which has little relevance to the aspirations of the intended beneficiaries of the designation, i.e. the residents. More communication between residents and professionals and a greater participation of residents in decision-making may help in this respect. So, for example, a revised community-driven designation process in which the residents' values and aspirations for the area are made explicit and agreed could be one stage of this. A second contribution could be to adapt the concept of the Village Design Statement, evolved by the Countryside Commission, under which residents participate actively in drawing up design guidelines and enhancement goals (Owen, 1999).

In Tynemouth, Tyne and Wear, a Village "Character Statement" was launched in 1999 combining elements from both the VDS concept and conservation area character appraisals. Facilitated by the North East Civic Trust, but controlled by a design team (or steering group) of local people, mostly representatives from active local groups, the document has involved widespread local participation through local events such as village character workshops, as well as more traditional forms of consultation such as questionnaire surveys. The document describes the character of the village through the perspectives of local people, and provides guidance for the protection and enhancement of that local character and distinctiveness. The document was adopted as supplementary planning guidance to North Tyneside's UDP early in 2000. These approaches link with broader concepts of urban landscape management (Larkham, 1996) and recent debates on collaborative planning (Healey, 1998).

Research has shown that the amenity groups which are most successful in influencing the planning process are those which work within the framework and adopt the language of the professionals whom they seek to influence (Lowe and Goyder, 1983). Greater public participation in the designation and management of conservation areas may largely reinforce the tastes and values of planning and conservation professionals. However, it may, as with the Solihull example, challenge the basic premises under which the conservation area system has operated. English Heritage (1997) has acknowledged the need to broaden participation in conservation, and this has been further acknowledged in the recent review of the conservation system (DCMS, 2000; English Heritage, 2000a, 2000b). This may produce unexpected outcomes. It has been argued that the extension of participation in countryside conservation "may be bringing about a retreat from the national vision of traditional conservation and a fragmentation of conservation ideas" (Goodwin, 1999: 383).

Finally, however, our research thus far is preliminary and based on relatively small samples. It suggests an agenda for further work which would assist in the development of our understanding of residential conservation areas, the role of residents and professionals, and the refinement of planning procedure. In the recent review of government policy (DCMS, 2000; English Heritage, 2000a, 2000b), a factor given particular emphasis is the community involvement in conservation, specifically in terms of social inclusion and amongst young people and ethnic minorities. It is no longer sufficient to rely upon the judgement of "experts" alone.

A Research Agenda

We consider our conclusions outlined above to be preliminary only because of the need to undertake much more empirical research in this area on what we consider to be an important planning issue. This final section briefly sets out the basis of a research agenda:

- Explore the nature and extent of residential conservation: date its growth, verify the number and types of residential conservation areas.
- Examine in greater depth the motives behind residential area designation.
- Investigate perceptions of, and actual, house price movements inside designated areas.
- Wider examination of residents' understanding and experience of the conservation system, and their aspirations from area designation.
- Explore the nature and function of existing mechanisms of communication between LPA and residents: particularly Conservation Area Advisory Committees, voluntary amenity societies, residents' associations and similar.
- Examine residents' and LPA officers' perceptions of issues and problems arising with residential conservation.

This range of data should enable us to review the working of those conservation areas that are essentially residential in nature. Key issues might include:

- How does residential conservation differ from experience in non-residential conservation contexts?
- How can the legitimate aspirations of residents, in terms of increasing quality of life, improving local environmental quality etc. be met through mechanisms of conservation?
- To what extent is conservation being used to legitimate non-planning-related issues and aspirations such as property price rises?
- If residents do have a more holistic view of conservation than the current legal system, how can this be delivered by the LPA?
- Does the statutory conservation system and its associated guidance (Planning Policy Guidance Notes and advice notes from English Heritage and related bodies) require amendment to better deal with issues of residential areas and their residents' aspirations?

- How can the communication of differing values, attitudes and aspirations between professionals, politicians and residents be facilitated, and how might some consensus be reached?

References

Algreen-Ussing, G. (2000), 'Mapping and Local Identity', paper presented to the 9th International Planning History Conference, Helsinki.

Barker, A. (1976), *The Local Amenity Movement*, Civic Trust, London.

Brent, London Borough of (1995), *Conservation Handbook*, London Borough of Brent Area Planning Group, Wembley.

Buller, H. (1981), *Pressure Groups and the Pluralist Model of Society: the Example of Voluntary Amenity Societies*, Occasional Paper no. 14, Department of Geography, King's College, London.

Cherry, G.E. (1975), 'The Conservation Movement', *The Planner*, vol. 61, no. 1, pp. 3-5.

County Planning Officers' Society (1968), 'Conservation in Urban Areas', survey prepared for the Historic Towns and Cities conference, York.

DCMS [Department of Culture, Media and Sport] (2000), 'Alan Howarth Launches Review of Heritage Policy', press release, DCMS, London.

Delafons, J. (1997), *Politics and Preservation*, Spon, London.

Dellheim, C. (1982), *The Face of the Past: the Preservation of the Medieval Inheritance in Victorian England*, Cambridge University Press, Cambridge.

DoE [Department of the Environment] (1973), *Conservation and Preservation*, Circular 46/73, HMSO, London.

DoE/DNH [Department of the Environment and Department of National Heritage] (1994), *Planning and the Historic Environment*, Planning Policy Guidance Note no. 15, HMSO, London.

English Heritage (1997), *Sustaining the Historic Environment: New Perspectives on the Future*, English Heritage, London.

English Heritage (2000a), *Review of Policies Relating to the Historic Environment*, English Heritage, London.

English Heritage (2000b), *Power of Place: the Future of the Historic Environment*, English Heritage, London.

Eversley, D. (1974), 'Conservation for the Minority?', *Built Environment*, vol. 3, no. 1, pp. 14-15.

Franks, A. (1988), 'The Street They Froze in Time', *The Times*, July 15, pp. 11.

Goodwin, P.P. (1999), 'The End of Consensus? The Impact of Participatory Initiatives on Conceptions of Conservation and the Countryside in the United Kingdom', *Environment and Planning D: Society and Space*, vol. 17, pp. 383-401.

Gwilliam, M., Bourne, C., Swain, C. and Prat, A. (1999), *Sustainable Renewal of Suburban Areas*, Report for the Joseph Rowntree Foundation, York Publishing Services, York.

Healey, P. (1998), 'Collaborative Planning in a Stakeholder Society', *Town Planning Review*, vol. 69, no. 1, pp. 1-21.

Hobson, E. (2000), *Conservation of the Built Environment: an Assessment of Values in Urban Planning*, unpublished PhD thesis, Department of Town and Regional Planning, University of Sheffield, Sheffield.

Hubbard, P.J. (1993), 'The Value of Conservation: a Critical Review of Behavioural Research', *Town Planning Review*, vol. 64, no. 4, pp. 359-73.
Hubbard, P.J. (1994), 'Professional Versus Lay Tastes in Design Control - an Empirical Investigation', *Planning Practice and Research*, vol. 9, no. 3, pp. 271-87.
Jennings-Smith, D. (1977), 'Guide or Rule Book? Rethinking Design Control Application', *Built Environment Quarterly*, vol. 3, no. 3, pp. 245-7.
Larkham, P.J. (1985), *Local Amenity Societies and Conservation Planning*, Working Paper no. 30, Department of Geography, University of Birmingham.
Larkham, P.J. (1993), 'Conservation in Action: Evaluating Policy and Practice in the United Kingdom', *Town Planning Review*, vol. 64, no. 4, pp. 351-7.
Larkham, P.J. (1996), *Conservation and the City*, Routledge, London.
Larkham, P.J. (1999a), 'Tensions in Managing the Suburbs: Conservation Versus Change', *Area*, vol. 31, no. 4, pp. 359-71.
Larkham, P.J. (1999b), *Residents' Perceptions of Conservation: Case Studies in the West Midlands*, Working Paper no. 74, School of Planning, University of Central England.
Larkham, P.J. (2000), 'Residents' Attitudes to Conservation', *Journal of Architectural Conservation*, vol. 6, no. 1, pp. 73-89.
Larkham, P.J. and Lodge, J. (1999), *Do Residents Understand what Conservation Means?* Working Paper no. 71, School of Planning, University of Central England.
Lawless, P. and Brown, F. (1986), *Urban Growth and Change in Britain*, Paul Chapman, London.
Lowe, P.D. (1977), 'Amenity Groups and Equity: a Review of Local Environmental Pressure Groups in Britain', *Environment and Planning A*, vol. 9, pp. 35-58.
Lowe, P. and Goyder, J. (1983), *Environmental Groups in Politics*, Allen and Unwin, London.
Marquis-Kyle, P. and Walker, M. (1992), *The Illustrated Burra Charter*, Australia ICOMOS, Sydney.
Meinig, D.W. (ed.) (1979), *The Interpretation of Ordinary Landscapes*, Oxford University Press, New York.
Mielé, C. (1996), 'The First Conservation Militants: William Morris and the Society for the Protection of Ancient Buildings', in Hunter, M. (ed.), *Preserving the Past: the Rise of Heritage in Modern Britain*, Sutton, Stroud, pp. 17-37.
Ministry of Environment and Energy (1997), *InterSAVE: International Survey of Architectural Values in the Environment*, Ministry of Environment and Energy, Copenhagen.
Owen, S. (1999), 'Village Design Statements: Some Aspects of the Evolution of a Planning Control in the UK', *Town Planning Review*, vol. 70, no. 1, pp. 41-60.
Pearce, G. et al. (1990), *The Conservation Areas of England*, English Heritage, London.
Pendlebury, J. (1999a), 'Civic Societies and Conservation Area Advisory Committees: Children of the 1960s at the End of the Century', paper presented at the XIII AESOP Congress, Bergen, Norway.
Pendlebury, J. (1999b), 'Conservation Areas: an Appraisal', paper presented at the Planning and Environment Training Conference, London.
Pendlebury, J. (2000), 'Conservation, Conservatives and Consensus: the Success of Conservation under the Thatcher and Major Governments, 1979-1997', *Planning Theory and Practice*, vol. 1, no. 1, pp. 31-50.
Pendlebury, J. and Townshend, T. (1999), 'The Conservation of Historic Areas and Public Participation', *Journal of Architectural Conservation*, vol. 5, no. 2, pp. 72-87.

Perähuhta, M. (2000), 'Railway Stations as the Crossing-Points of the National and the Local in Defining Cultural Heritage', paper presented to the 9th International Planning History Conference, Helsinki.

Reynolds, J. (1975), 'Heritage Year in Britain: the Aims and Objectives of Conservation', *Town Planning Review*, vol. 46, no. 4, pp. 355-64.

Rydin, Y. (1998), *Urban and Environmental Planning in the UK*, Macmillan, London.

Samuel, R. (1994), *Theatres of Memory*, Verso, London.

Schnapp, A. (1996), *The Discovery of the Past*, British Museum Press, London.

Tarn, J.N. (1975), 'The Derbyshire Heritage: the Conservation of Ordinariness', *Town Planning Review*, vol. 46, no. 4, pp. 451-65.

Townshend, T. and Pendlebury, J. (1999a), 'Public Participation in the Conservation of Historic Areas: Case-Studies from North-East England', *Journal of Urban Design*, vol. 4, no. 3, pp. 313-21.

Townshend, T. and Pendlebury, J. (1999b), 'Planning the Past in the Future: the Conservation of the Built Environment in the UK', paper presented at the XIII Aesop Congress, Bergen.

Wealthy, P. (1984), *Conservation Areas: Surveys of Residents' Attitudes*, London Borough of Hammersmith and Fulham, London.

Wilson, M. (1999), 'Preserving a Little Bit of the Sixties', *The Times*, October 20, pp. 252.

14 Understanding Sustainability and Planning in England: An Exploration of the Sustainability Content of Planning Policy at the National, Regional and Local Levels

CAROLINE BROWN AND STEFANIE DÜHR

Introduction

There can be little question that the terms sustainability and sustainable development have become firmly embedded in the English planning system over the last ten years. The incorporation of sustainability as the underlying principle in planning policy is of fundamental importance for achieving sustainable development in practice. However, the presence or absence of these terms in policy documents tells us little about the capacity of planning to achieve sustainability because as an abstract concept sustainable development has to be translated – operationalised – into practical actions. Furthermore, the definitions and interpretations of sustainability employed by different actors can (and have) changed over time, and this clearly has its effects on planning policy. In order to understand the definition, interpretation and operationalisation of sustainable development in English planning policy, unpacking the sustainability content of planning policy is therefore a useful first step.

The aim of this paper is thus to examine the way in which sustainability is operationalised in English planning policy at the national, regional and local levels. In doing so the paper identifies the strength and breadth of sustainability coverage in planning policy, and charts the development of this coverage over time. The work is part of a much larger research project – SPECTRA – funded by the European Commission[1] and involving partners in Finland, Ireland, Greece, Italy, Germany, the Netherlands and the UK. The project has been examining the capacity of spatial planning systems to implement sustainability, and trying to uncover the barriers and catalysts to the implementation of sustainability through planning in each of the partner countries.

Approaching Sustainable Development

The term sustainability is one of the most flexibly used words in the English language, interpreted and defined in many ways, and often used by interest groups to support their own points of view (Bruff and Wood, 2000). Generally speaking the concept of sustainable development is understood to mean living within environmental limits and meeting current needs without compromising the ability of future generations to meet their own needs. However, regardless of the merits (or otherwise) of the Brundtland approach, in common with other definitions it fails to offer any concrete guidance about *what to do* to achieve sustainable development. The challenge for policy-makers is thus to translate the concept as a whole into components that can be written into policy and put into practice. This is not a new idea, and one of the characteristics of sustainability has been its association with policy "principles" such as futurity, equity and the precautionary principle.

In the UK, land use planning has been trying to come to terms with sustainable development since the early 1990s (Bruff and Wood, 2000). During that time there has been considerable experimentation and interpretation of sustainability in both planning policy and the development of sustainability frameworks and checklists. Understandings of sustainability have thus evolved over time, changing in response to factors that include: the increased understanding of sustainability issues; the evaluation and monitoring of policy and practice; experimentation, and changes in political priorities. The debates about the priority that should be given to different dimensions of sustainability are well rehearsed and will not be reiterated here. However, in order to uncover the different interpretations and constructions of sustainability used in planning policy, one of the objectives of the SPECTRA project was to develop a free-standing and comprehensive framework of sustainability criteria that could be used as a universal evaluation tool. Using ecosystems theory as a starting point, a series of principles and criteria were built up, based first and foremost on the literature. The resulting framework, illustrated in Table 14.1, has three identifiable dimensions: overarching, environmental, and economic and societal. It is worth noting that the framework is deliberately broad in its interpretation of sustainability in order to illustrate different approaches to the concept.

Table 14.1: The sustainability framework

	PRINCIPLES	CRITERIA
OVERARCHING	Futurity and inter-generational equity	Precautionary principle (no irreversible decisions)
		Include cumulative and long-term impacts in decision-making
	Inter-societal equity	Commitment to equity at local, national and international levels
	Local/regional self-sufficiency	Reducing externality effects (ensure commitment to equity so environmental impacts and the costs of protecting the environment do not unfairly burden any one geographic or socio-economic sector)
		Using close in preference to distant resources
	Risk prevention and reduction	Natural disasters
		Human-made disasters
ENVIRONMENTAL	Maintaining the capacity of natural systems	Absolute protection of critical natural capital
		Defence and improvement of soil quality and stability
		Defence and improvement of key habitats and biodiversity
		Respecting the capacity of natural systems
	Minimisation of resource consumption	Efficient use of renewable resources
		Minimum depletion of non-renewable resources
		Energy efficiency
		Minimisation of waste (through recycling/re-use etc)
	Improving environmental quality	Reduction of pollution emissions (protection of air and water quality, minimisation of noise)
		Protection and enhancement of environmental amenity/aesthetics
		Protection of natural and cultural heritage
ECONOMIC/SOCIETAL	Protect and develop the economic system	Encourage and develop connections between environmental quality and economic vitality
		Satisfy and protect basic needs (shelter, food, clean water etc.)
		Provide entrepreneurial/employment opportunities
	Develop the social system (education, democracy, human rights)	Protect basic human rights
		Ensure health and safety
		Improve local living conditions
		Satisfy the economic and living standards to which people aspire
	Develop the capacity of the political system	Develop an open, inclusive and transparent governance system
		Subsidiarity (allocation of competence at the most appropriate level)

Methodology

In order to uncover the sustainability content of planning policy, individual documents were examined and their contents cross-referenced with the sustainability framework set out in Table 14.1, a content analysis technique similar to that used in other studies (see for example: Bruff and Wood, 2000; Berke and Manta, 1999). The results were recorded using a four-point scale which shows both the strength and breadth of coverage in a visual rather than a numeric way. Where there was some mention of a criterion this received a weak rating (pale shading); where there was ongoing reference to a criterion this received a medium rating (medium shading), and where criteria were covered by many policies this received a strong rating (dark shading). Criteria which were not covered by any of the policies in the document received a zero rating (no shading). In some cases a degree of abstraction was required as it is possible for individual policies to encapsulate a number of criteria. Policy relating to redevelopment of brownfield land in urban areas for example has implicit within it the principles of minimising resource use, enhancing energy efficiency and reducing waste. The analysis thus covers both the explicit and *implicit* sustainability content of policy documents. Clearly this technique is inherently subjective, but when undertaken in a systematic manner it nevertheless provides an interesting perspective on planning policy in the UK; indicating not only the coverage of sustainability criteria by policy, but also the conception and understanding of sustainability on which the coverage is based.

In contrast to other studies (Bruff and Wood, 2000 for example), the evaluation framework used in the SPECTRA project has not been tailored in any way to the scope or coverage of land-use planning systems. It is thus important to note that planning policy is unlikely to address all of the principles set out in the framework. While some principles may be addressed by the processes of planning - public consultation for example - other issues are clearly outwith the scope of the system and must be addressed in other ways.

Planning Policy and Sustainability

Using the method described above, planning policy documents at the national, regional and local levels were evaluated, providing a snapshot of the sustainability content of English planning policy in early 2001. The relatively fast pace of policy evolution at the national level however, means that the analysis dates very quickly. At the national level the analysis includes *all* of the planning policy guidance notes (PPGs) published before February 2001, while at the regional and local levels the analysis is more selective focussing on one region - the South West of England - and one local authority area, Bristol City Council. The reasons for this partial approach to regional and local planning policy are entirely pragmatic (and linked to the scope of the SPECTRA project).

The hierarchical nature of planning policy in the UK means that the national level is extremely important, as policy cascades down to both the regional and local level. The requirement for conformity between different levels of planning policy implies that development plans at the local level will be in line with both regional planning guidance and national planning policy guidance notes. This suggests that the government's ambitions in relation to sustainable development as set out in planning policy should be clearly echoed at the regional and local levels. The reality however, is not quite so straightforward for two main reasons. First, in some cases local and regional planning policy has pushed ahead of national policy on sustainable development – so that policy at lower levels is *constrained* rather than *informed* by the national level. Second, while policy at the national level is constantly evolving, the lengthy policy-making processes in place at regional and local levels mean that current planning documents may reflect a much earlier conception of sustainability. This is particularly noticeable in relation to definitions of sustainability, as there has been a significant shift from the environmental focus of policy in the early 1990s to the broader – and some would argue weaker – interpretation of sustainability made by the Labour government post 1997.

Bearing in mind that planning policy is made up of documents prepared at different times and reflecting different interpretations of sustainability, the following sections discuss the results of the sustainability content analysis for English planning policy at national, regional and local levels.

National Planning Policy

The most important body of government planning policy is found in planning policy guidance notes (PPGs); while policy on sustainable development is set out in the national strategy "A Better Quality of Life" (DETR, 1999a). PPGs - of which there are 25 - cover different aspects of planning, and can best be described as sectoral policy documents, although some deal with procedural rather than substantive issues. One of the strengths of PPGs is that they may be revised relatively quickly at any time. Following the publication of the first UK White Paper on sustainable development "This Common Inheritance" in 1990, and the Rio Earth Summit in 1992, the government has redrafted virtually all of the PPGs to take account of sustainability issues. Many have since been revised again, and more than one quarter of PPG's have been revised since the change of government in 1997. Each revision has provided an opportunity to bring planning policy into line with the latest government thinking on sustainability.

Table 14.2 sets out the results of the sustainability content analysis for all 25 of the PPGs. As the table demonstrates in terms of coverage and operationalisation of sustainability, the vast majority of the PPGs address several sustainability criteria, particularly under the environmental sustainability heading. Only two PPGs fail to address any of the criteria, PPG5 on simplified planning zones and PPG18 on enforcement. The absence of sustainability issues in these documents is linked mainly to their subject matter, and is in stark contrast to the 18 PPGs that tackle at least six of the principles. While the vast majority of PPGs cover a number of the sustainability criteria that is not to say however, that the coverage is either strong or comprehensive. Indeed the results clearly demonstrate that despite a broad covering of sustainability issues, there is little depth or strength to the coverage.

The most distinct characteristic of the way in which PPGs operationalise sustainability is their focus on environmental sustainability issues. The most obvious explanation for this relates to the nature and scope of the English planning system, which emphasises the protection of both natural and cultural heritage, and the quality of the environment. These objectives are long-standing aims of the planning system and help to explain why virtually all PPGs include criteria related to the protection and enhancement of environmental amenity, and the protection of natural and cultural heritage. The protection of the countryside and the pressure for brown-field development as objectives of the planning system are also reflected in the criteria related to the protection of critical natural capital, and minimisation of non-renewable resource consumption (e.g. land). The management of traffic - another key objective of land-use planning in England - is reflected in a number of sustainability criteria, most importantly that related to pollution.

In many ways, it can be argued that the objectives of town and country planning have not changed significantly since 1947. As a result even the oldest PPGs contain some mention of issues relating to the protection and enhancement of the environment. This demonstrates that even before the terms sustainability and sustainable development became commonplace, planning policy was addressing at least some of the principles now underpinning sustainability. What has changed over time however has been the breadth of coverage, and the emphasis given to particular criteria. As Table 14.2 demonstrates, the most recent PPGs have a wider sustainability content than earlier PPGs, regardless of the subject area covered. It is noticeable that only recently has there been any attempt to tackle any overarching sustainability criteria. As the table demonstrates, the commitment to inter-societal equity is the most commonly found overarching idea in national planning policy - although this is usually linked to the provision of affordable housing at the local level, rather than more global issues. Again, the emphasis on this has increased over time, particularly since the government's broadening of both the sustainability and planning agendas to include issues of social exclusion and community development. Surprisingly however, despite the relative longevity of the precautionary principle as a policy idea, it has only regularly been mentioned in English planning policy over the last three or four years. Needless to say, interpretations of what it means in practice are still at an early stage, and there is currently some debate as to the practicalities of the precautionary approach set out in draft PPG25 on flooding.

While few overarching sustainability issues appear to be embodied in planning policy at the national level, there is slightly greater coverage of the economic and societal dimensions of sustainability. Treatment of these is strongly linked to the provision of housing and the improvement of local living conditions, and has deepened over time; a reflection perhaps of the Labour government's broad definition of sustainability which includes economic growth as well as community development (DETR, 1999a). Language linking economic vitality with environmental quality is becoming more common in planning policy, as policy-makers give formal recognition to the economic value of a good environment. These links are best illustrated in PPG6 on town centres and PPG11 on regional planning. PPG6 emphasises the role of a good quality and well-managed environment in maintaining a vital and viable town centre, while PPG11 states clearly that regional planning guidance should address the link between competitiveness and the environment.

As mentioned earlier, the sustainability framework used in this research is a broad one which planning policy alone is unlikely to cover, and there are distinctions to be made between the *substantive* content of planning policy and *procedural* issues. PPGs 11 and 12 relate more to *procedural* issues than *substantive* ones covering the preparation of regional planning policy guidance, and development plans. As a result, both PPGs pay attention to social justice and ideas related to democracy and participation through the requirement to involve *all* sectors of the community in plan-making, or the need to appraise the equity, health and social inclusion impacts of regional planning guidance. These ideas are closely related to overarching sustainability principles concerned with equity, and social sustainability criteria such as the development of open, inclusive and transparent governance systems. PPG12 states clearly that not only should the community be involved in the plan-making process, but that plans should be written in a way that makes them accessible, that is - in non-technical language free from jargon. Other social sustainability criteria also feature in the PPGs, including a requirement to uphold health and safety. PPG12 is especially clear in this respect, setting out the role of the planning system in relation to the Seveso Directive on the prevention of major accidents involving hazardous substances (DETR, 1999e).

As a body of policy, PPGs demonstrate the increasing sustainability content of national planning policy and the changing definition of sustainability adopted by government. As the government interpretation of sustainable development has changed, the sustainability content of planning policy has also changed, moving away from an entirely environmental approach to a broader understanding that includes economic and social objectives. However, the results of this analysis also show that there are serious gaps in the coverage of sustainability by planning policy guidance notes. Overarching sustainability criteria, and issues related to human rights and subsidiarity are all notable in their absence at this point in time, although further evolution of ideas about sustainability and planning may lead to their inclusion in future policy documents. Quite clearly, the incorporation of the International Treaty on Human Rights into English law in October 2000 may have a significant impact in this respect. Debate about the potential impact of the Human Rights Act is at an early stage, and conclusions cannot be drawn until current test cases have been resolved. However, it seems likely that future revisions to PPGs may have to take greater account of human rights issues and issues of equity, inclusion and transparency in decision-making.

Planning Policy at the Regional Level

Since the early 1990s, central government has taken steps towards expanding regional planning guidance in England. The government programme of devolution, and the decentralisation of decision-making to regional government has led to the set-up of Regional Development Agencies (RDAs) - responsible for regional economic development strategies, and eight Regional Planning Bodies (RPBs) since 1997. RPBs (Regional Chambers or Regional Assemblies), are made up of representatives from local government and other regional organisations, and are required to draw up a comprehensive regional strategy, which should provide the framework for statutory development plans. These strategies, or regional planning guidance (RPGs), are the principal source of planning policy at regional level, and must be subject to a sustainability appraisal.

Despite the obvious devolution efforts, however, central government keeps a close eye on developments and planning at regional level (Gilg, 2000; Baker, 2000). The new PPG11 on regional planning (DETR, 2000b) for example, clearly sets out the process for drawing up RPGs in each of the eight regions, and requires RPBs to work in close co-operation with the Government Offices in the regions. Central government is also still responsible for issuing the final RPGs, after these have been subject to public examination and consultation.

This section will focus on the policy content analysis for the RPG for the South West (RPG 10). The current RPG 10 was published in 1994 (GOSW, 1994), although the process of revising of the guidance has been going on since 1998. The final version of the new RPG is expected sometime in 2001, and will cover the period up to 2016. A first draft of the revised RPG was published in 1999 (SWRPC, 1999), and a revised draft emerged in 2000 after the public examination (GOSW, 2000). The lengthy RPG preparation process provides an opportunity to compare the sustainability content of the existing RPG 10 (1994) with the two draft versions of the new RPG *and* the two sustainability appraisals prepared by external consultants (SWRT 1999; Baker Associates 2001).

Table 14.3 sets out the results of the policy content analyses of the three versions of RPG for the South West. The table reveals the gradual broadening and widening of the sustainability content both in comparison to the 1994 RPG, but also between the two draft versions over time. The second draft RPG (2000) demonstrates a significantly broader and stronger sustainability content than the 1999 draft, and certainly the old RPG 10 (1994). The revised RPG (2000) sets out a series of broad policy principles, it comprises stronger and clearer guidelines for lower tiers of government than the first draft, and is less liable to wide-ranging interpretation. The key issues (e.g. prudent land of use, increased use of public transport, urban renaissance, and the conservation and enhancement of environmental assets) reflect the developing agenda for sustainable development in England. Whereas the 1994 RPG and the 1999 draft cover the same sustainability criteria (although the latter in most cases with a stronger coverage), the revised draft planning guidance (2000) shows a far more comprehensive coverage of sustainability criteria. The revised draft places strong emphasis on overarching principles, particularly on cumulative and long-term impacts in decision-making, the commitment to equity, and the reduction of externality effects. In terms of environmental sustainability criteria, both the revised draft and the consultation draft of the RPG place strong emphasis on the protection of non-renewable resources (especially land), and the reduction of traffic congestion and pollution by improving public infrastructure, encouraging walking and cycling, and by promoting mixed use and brownfield development.

Similar to the broadening and deepening of the sustainability content of overarching criteria over time, the revised draft RPG (2000) also features much stronger coverage of economic and societal sustainability factors. The provision of housing and employment opportunities is a central theme in all three versions, reflecting the emerging role of RPG in the English planning system. RPGs are expected to provide for the right climate for business investment and the creation of employment opportunities. However, mirroring the development in national planning policy, the revised draft RPG (2000) places increasing emphasis on the relationship between economic vitality and environmental quality. Soft location factors, such as environmental and cultural assets, are openly acknowledged as important for attracting business activity to the region. The revised draft RPG furthermore places stronger emphasis on partnerships and networking than its predecessors, reflecting the growing demand for transparency and accountability in regional governance. The revised draft also shows stronger consideration of themes to do with social inclusion and economic regeneration in comparison to the two earlier versions.

In comparison with the existing and the consultation draft RPGs, the revised draft is stronger when incorporating sustainability objectives into all of its policies. Where the draft RPG continued to support further growth of the airport and improvements to the highway network, the revised draft at present lacks a regionally specific strategy for the region's ports and airports. In contrast, the overall approach of the revised draft RPG (2000) seeks to improve sustainable access (rail, bus, cycle and pedestrian), in association with demand management where appropriate, in order to secure modal shift. Nevertheless, some of the transport proposals involve the provision of new road space on some of the region's trunk roads, which will clearly have economic benefits but environmental disbenefits, thus demonstrating the ongoing rivalry between economic competitiveness and environmental concerns.

Planning Policy at the Local Level

Planning policy at the local level is found primarily in development plans, which include: structure plans, local plans, unitary development plans, minerals local plans and waste local plans. Planning policy is not only set out in statutory documents however, and most local planning authorities have other strategies and documents that are relevant to planning and sustainability. Bristol City Council is no exception to this, and the policy content analysis discussed here includes the city's nature conservation strategy, local transport plan and city centre strategy, as illustrated in Table 14.4. Examining these documents is useful for two reasons; first, it gives us clues about the development of policy over a reasonable length of time, and second, it provides a useful context within which to position the development plan.

At first glance the results of the sustainability content analysis for the local level reveal a marked difference between the sustainability content of sectoral strategies and the sustainability content of the development plan. Both the structure plan and the local plan are noticeably broad and strong in their coverage of sustainability issues, while the local transport plan, nature conservation and city centre strategies operationalise sustainability in a weaker and much more uneven way. Quite clearly the non-statutory documents included in the analysis tend to operationalise sustainability in a way that links directly to the scope of the document. For example, the local transport plan concentrates on transport-related sustainability issues, with a strong emphasis on reducing traffic-related pollution and making places accessible to all. The city centre strategy on the other hand tends to cover criteria linked to the improvement of local living conditions and economic development, while the nature conservation strategy covers criteria related to environmental protection. Perhaps the most striking feature of the nature conservation strategy however, is the overall weakness of its sustainability content. In many ways this is surprising given the clear links that nature conservation has with environmental sustainability, although it may be that the date of the document, *before* the Rio Earth summit, is enough to explain these results.

In contrast to the narrow operationalisation of sustainability by topic-based policy documents, Bristol's development plan approaches sustainability in a broader and more robust manner. Both the local plan and the structure plan for the Bristol area cover sustainability criteria across all three dimensions of the evaluation framework, with only a slight bias towards environmental sustainability issues. The local plan places particular emphasis on transport-related issues, the protection of natural and cultural heritage as well as housing, shopping and employment objectives and the improvement of local living conditions. The structure plan on the other hand takes the operationalisation further by embedding sustainability indicators into the plan and the policies.

The reasons for the broad and strong operationalisation of sustainability by local planning policy in Bristol are not entirely clear. When compared with the analysis results for national level policy, it is obvious that the operationalisation of sustainability seen in both the Bristol local plan and the structure plan is not derived in any way from the national level. The notion of a policy cascade from the national level to the local level can thus be questioned, at least in relation to the sustainability content of planning policy, and the case of Bristol. The relationship - or lack of relationship - between the sustainability content of national and local policy in this instance thus highlights the contested nature of sustainability and the lack of consensus about the way in which it can be translated into policy and action.

The Understanding of Sustainability in Planning Policy

The results of this investigation into the sustainability content of English planning policy have revealed a variable pattern of coverage at the national, regional and local levels. In addition, the results also reveal something of the understanding of sustainability upon which planning policies are constructed. Generally speaking, planning policy operationalises sustainability as an environmental issue - focussing on protecting habitats and enhancing environmental amenity – and revealing an understanding of sustainability rooted in an environmental paradigm. Such an observation is unsurprising given the narrow scope of the English planning system and its focus on land use issues. However, as illustrated above, the way in which planning operationalises sustainability is neither static nor confined to environmental sustainability issues.

In order to investigate these issues further, a selection of PPGs was arranged in chronological order to illustrate changes in the way that sustainability is operationalised. The results of this exercise are illustrated in Table 14.5. The table demonstrates that there has been a noticeable broadening of the criteria covered by the PPGs since the early 1990s, although the broadening has not necessarily been matched by any strengthening of the coverage. While the earliest PPGs address fewer principles, later PPGs cover more and more of the sustainability framework. The revised PPG3 for example addresses almost all elements of the sustainability framework, covering many more criteria than the relatively robust PPG20 on coastal planning. The explanation for these changes may be linked to the changing definition of sustainability set out by government. While in the early 1990s, the environment white paper "This Common Inheritance" reflected an understanding of sustainability focussed on environmental criteria, the Labour government has made it clear that its understanding of sustainability goes much further. The aims of sustainable development set out in the national strategy for sustainable development "A Better Quality of Life" emphasise social progress; environmental protection; prudent resource use; and, economic growth (DETR, 1999a). The widening view of sustainable development appears to be reflected in the broadening coverage of sustainability in planning policy at the national and regional levels. This suggests that more recent planning policy documents are based on, and reflect, a broader understanding of sustainability than policies written at an earlier time.

Although such observations are entirely plausible, the pattern of sustainability coverage seen at the local level slightly confounds the idea of a gradual evolution in understandings and operationalisations of sustainability. Despite the fact that the Bristol local plan was written in the early 1990s (it was *adopted* in 1997), the way in which it operationalises sustainability is much broader and deeper than the equivalent operationalisation in planning policy at the national level. As a result it seems reasonable to suggest that the way in which planning policy interprets sustainability may be influenced as much by political objectives as it is by practical experience.

Conclusions

The purpose of this paper was to examine the sustainability content of English planning policy in order to better understand the way that planning operationalises the concept of sustainability. The results of the analysis, set out and explored above, have revealed a number of things about English planning policy and approaches to sustainability.

The first conclusion to emerge is that despite more than ten years of policy developments on sustainable development, the translation of sustainability into planning policy is enormously variable. At the national level in particular, coverage of sustainability criteria is very uneven, with noticeable gaps in the coverage of overarching sustainability principles. While some of these variations are explained by the subject matter of individual PPGs - not all topics lend themselves to ideas of futurity or self-sufficiency - other variations are explained by the changing approach to sustainability seen over time.

The second conclusion that can be drawn about planning policy is about the different *strength* of coverage that exists at different levels. Regional policy, as noted above, appears to promote sustainability in a fairly robust manner – more strongly than coverage at the national level.

This pattern, repeated to a degree at the local level, raises a question about the policy hierarchy, which implies a cascade from national to regional and local. Where regional and local coverage of sustainability is stronger (but not necessarily wider) than coverage at the national level, this challenges the notion of a hierarchical relationship between plans. Having examined policy in only one region and one local authority area however, the pattern observed here might simply be an anomaly created by the ambitions of Bristol City Council and the South West regional planning body. Further research is required to see whether similar patterns are evident in other regions.

Alongside the variation in strength of coverage at different policy levels, there is also a noticeable variation in coverage between documents. At the national level, the sustainability content of PPGs varies according to the topic covered by the guidance and the time at which the guidance was prepared. However, bearing in mind that regardless of their date *all* of these PPGs are *current* policy, this raises questions about the performance of such a disjointed body of policy in promoting sustainability. Is it possible for planning to successfully promote sustainable development if there are gaps and variations in the interpretations of sustainability set out in national planning policy?

The results described above demonstrate two further issues in relation to the operationalisation of sustainability in planning policy. First, that there are very few principles which are tackled consistently in all planning documents; and second, that despite the generic nature of the evaluation framework there is only one criterion (protecting human rights) which is not tackled at all by planning policy. The implication is thus that there are no intrinsic limits to the dimensions of sustainability that can be tackled by planning policy. Generally speaking however, planning policy tends to operationalise sustainability in an environmental way, focussing on the protection and enhancement of the natural and built environment and the traditional role of town and country planning.

Finally, and most significantly, examining the sustainability content of planning policy in this way reveals the changes in the approach to sustainability over time. While documents from the early 1990s tend to reflect a narrow understanding of sustainability, later documents are broader in their coverage; a pattern demonstrated most convincingly by the changing sustainability content of planning policy guidance notes. The gradual broadening of the way in which planning policy operationalises sustainability illustrates two things: changing definitions of sustainability and changing approaches of planning to the translation of the concept into practical action. All in all, the sustainability content of planning policy is both dynamic and variable, rooted in a largely environmental view of sustainable development. However, more than a decade after sustainability was first explicitly incorporated into planning policy, consensus about the definition of sustainability, or the way in which it can be operationalised has yet to emerge.

Note

1 Contract no: ENV4 CT970644. Further information about the SPECTRA project is available at: www.uwe.ac.uk/fbe/spectra

References

Baker, M. (2000), 'PPG11 and the future of regional planning guidance', *Town and Country Planning*, vol. 69, no. 12, pp. 338-340.

Baker Associates (2001), *Sustainability Appraisal of Proposed Changes to Draft RPG10. Final Report*, Government Office South West, Bristol.

Berke, P. and Manta, M. (1999), *Planning for Sustainable Development: Measuring Progress in Plans*, Working Paper WP99PB1, Lincoln Institute of Land Policy.

Bristol City Council (1997), *Bristol Local Plan. Written Statement*, Bristol City Council, Bristol.

Bristol City Council (1998a), *Bristol City Centre Strategy. Section 1. Introduction and Summary*, Bristol City Council, Bristol.

Bristol City Council (1998b), *Bristol City Centre Strategy. Section 2. Functions and Themes*, Bristol City Council, Bristol.

Bristol City Council (1998c), *Bristol City Centre Strategy. Section 3. Neighbourhood Statements. Consultation Draft*, Bristol City Council, Bristol.

Bristol City Council (2000), *The Bristol Local Transport Plan*, Bristol City Council, Bristol.

Bruff, G. E. and Wood, A. P. (2000), 'Local Sustainable Development: Land use Planning's Contribution to Modern Local Government', *Journal of Environmental Planning and Management*, vol. 43, no. 4, pp 519-539.

Department of the Environment (1990a), *PPG14 Development on Unstable Land*, HMSO, London.

Department of the Environment, Transport and the Regions (DETR) (1997a), *PPG1 General Policy and Principles*, HMSO, London.

DETR (1997b), *PPG7 The Countryside - Environmental Quality and Social Development*, HMSO, London.

DETR (1998), *Planning for sustainable development: towards better practice*, TSO, London.

DETR (1999a), *A Better Quality of Life: A Strategy for Sustainable Development for the UK*, TSO, London.

DETR (1999b), *Revision of PPG13, Transport. Public consultation draft*, TSO, London.

DETR (1999c), *PPG10 Planning and Waste Management*, TSO, London.

DETR (1999d), *PPG12, Development Plans (revised)*, TSO, London.

DETR (2000a), *PPG11 Regional Planning*, TSO, London.

DETR (2000b), *Guidance on Preparing Regional Sustainable Development Frameworks*, TSO, London.
DETR (2001), *PPG25 Development and Flood Risk. Consultation Paper*, TSO, London.
DoE (1990b), *PPG16 Archaeology and Planning*, HMSO, London.
DoE (1991a), *PPG17 Sport and Recreation*, HMSO, London.
DoE (1991b), *PPG18 Enforcing Planning Control*, HMSO, London.
DoE (1992a), *PPG20 Coastal Planning*, HMSO, London.
DoE (1992b), *PPG21 Tourism*, HMSO, London.
DoE (1992c), *PPG4 Industrial and Commercial Development and Small Firms*, HMSO, London.
DoE (1992d), *PPG5 Simplified Planning Zones*, HMSO, London.
DoE (1993), *PPG22 Renewable Energy*, HMSO, London.
DoE (1994a), *PPG15 Planning and the Historic Environment*, HMSO, London.
DoE (1994b), *PPG23 Planning and Pollution Control*, HMSO, London.
DoE (1994c), *PPG24 Planning and Noise*, HMSO, London.
DoE (1994d), *PPG9 Nature Conservation*, HMSO, London.
DoE (1995), *PPG2 Green Belts*, HMSO, London.
DoE (1996), *PPG6 Town Centres and Retail Development*, HMSO, London.
Gilg, A. (2000), 'A welcome rebirth – but genetic engineering required', *Town and Country Planning*, vol. 69, no. 6, pp. 174-176.
Government Office for the South West (GOSW) (1994), *RPG 10. Regional Planning Guidance for the South West*, HMSO, London.
Government Office for the South West (GOSW) (2000), *Draft Regional Planning Guidance for the South West (RPG 10). Proposed Changes. Public Consultation*, HMSO, London.
Joint Strategic Planning and Transportation Unit (JSPTU) (1998), *Joint Replacement Structure Plan. Deposit Plan*, Joint Strategic Planning and Transportation Unit, Bristol.
Nature Conservancy Council, South West Region (1991), *The Greater Bristol Nature Conservation Strategy*, Nature Conservancy Council, Taunton.
South West of England Regional Development Agency (SWRDA) (1999), *Regional Strategy for the South West of England 2000-2010*, South West of England Regional Development Agency, Exeter.
South West Regional Planning Conference (SWRPC) (1999), 'Draft Regional Planning Guidance for the South West', South West Regional Planning Conference, Taunton.
South West Round Table for Sustainable Development (SWRT) (1999), *Sustainability Appraisal of Draft Regional Planning Guidance for the South West. Second Draft*, South West Round Table for Sustainable Development, Bristol.

Appendix

Table 14.2: Sustainability content of national planning policy

	CRITERIA	PPG1 (97)	PPG2 (95)	PPG3 (00)	PPG4 (92)	PPG5 (92)	PPG6 (96)	PPG7 (97)	PPG8d (01)	PPG9 (94)	PPG10 (99)	PPG11 (00)	PPG12 (99)	PPG13d (99)	PPG14 (90)	PPG15 (94)	PPG16 (90)	PPG17 (91)	PPG18 (91)	PPG19 (92)	PPG20 (92)	PPG21 (92)	PPG22 (93)	PPG23 (94)	PPG24 (94)	PPG25d (01)
OVERARCHING	Precautionary principle																									
	Include cumulative and long-term impacts in decision-making																									
	Commitment to equity at local, national and international levels																									
	Reducing externality effects																									
	Using close in preference to distant resources																									
	Natural disasters																									
	Human-made disasters																									
ENVIRONMENTAL	Protection of critical natural capital																									
	Defence and improvement of soil quality and stability																									
	Defence and improvement of key habitats and biodiversity																									
	Respecting the capacity of natural systems																									
	Efficient use of renewable resources																									
	Minimum depletion of non-renewable resources																									
	Energy efficiency																									

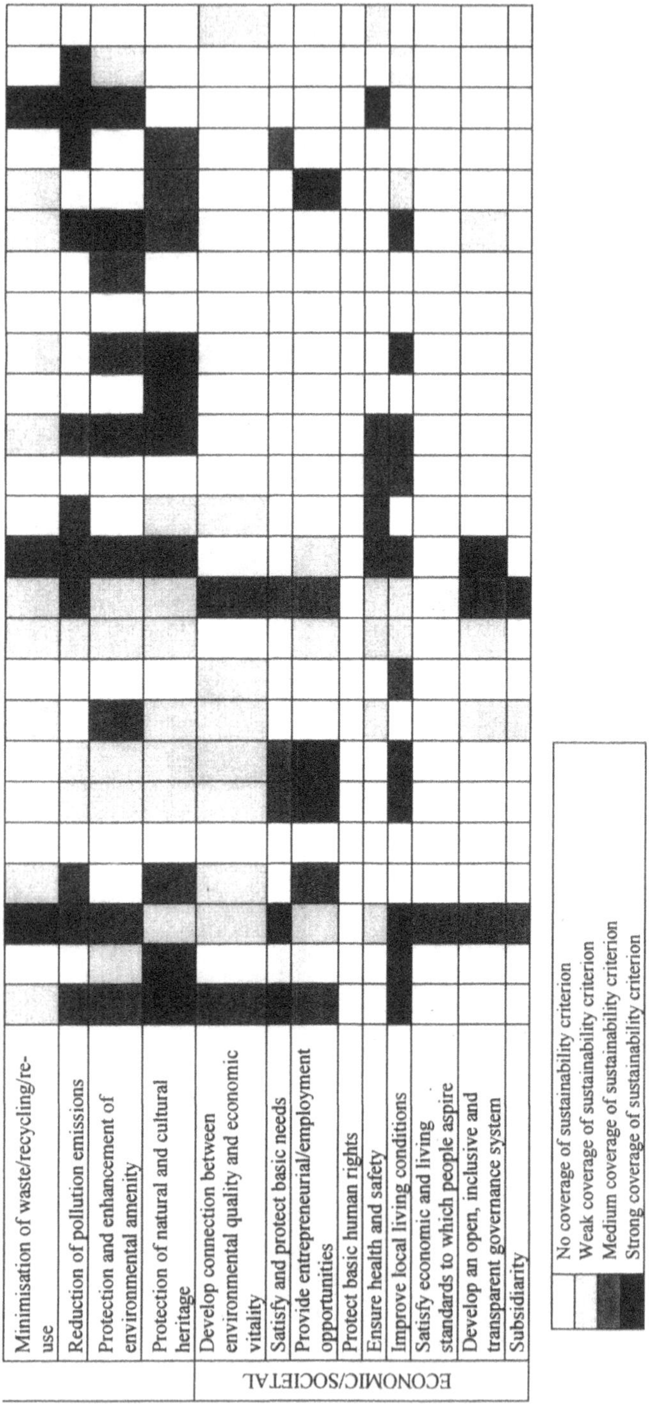
Minimisation of waste/recycling/re-use
Reduction of pollution emissions
Protection and enhancement of environmental amenity
Protection of natural and cultural heritage
ECONOMIC/SOCIETAL
Develop connection between environmental quality and economic vitality
Satisfy and protect basic needs
Provide entrepreneurial/employment opportunities
Protect basic human rights
Ensure health and safety
Improve local living conditions
Satisfy economic and living standards to which people aspire
Develop an open, inclusive and transparent governance system
Subsidiarity
No coverage of sustainability criterion
Weak coverage of sustainability criterion
Medium coverage of sustainability criterion
Strong coverage of sustainability criterion

Table 14.3: Sustainability content of regional planning policy – South West region

	CRITERIA	RPG10 1994	Draft RPG 1999	Draft RPG 2000
OVERARCHING	Precautionary principle			
	Include cumulative and long-term impacts in decision-making			
	Commitment to equity at all levels			
	Reducing externalities			
	Using close in preference to distant resources			
	Natural disasters			
	Human-made disasters			
ENVIRONMENTAL	Absolute protection of critical natural capital			
	Defence/improvement of soil quality and stability			
	Defence/improvement of habitats and biodiversity			
	Respecting the capacity of natural systems			
	Efficient use of renewable resources			
	Minimum depletion of non-renewable resources			
	Energy efficiency			
	Minimisation of waste			
	Reduction of pollution emissions			
	Protection and enhancement of environmental amenity/aesthetics			
	Protection of natural and cultural heritage			
ECONOMIC/SOCIETAL	Encourage and develop connections between environmental quality and economic vitality			
	Satisfy and protect basic needs			
	Provide entrepreneurial/employment opportunities			
	Protect basic human rights			
	Ensure health and safety			
	Improve local living conditions			
	Satisfy the economic and living standards to which people aspire			
	Develop an open, inclusive and transparent governance system			
	Subsidiarity			

Table 14.4: Sustainability content of local planning policy - Bristol

	CRITERIA	Draft Structure Plan 1999	Local Plan 1997	Local Transport Plan 2000	City Centre Strategy, 1998	Nature Cons Strategy, 1991
OVERARCHING	Precautionary principle					
	Include cumulative and long-term impacts in decision-making					
	Commitment to equity at local, national and international levels					
	Reducing externality effects					
	Using close in preference to distant resources					
	Natural disasters					
	Human-made disasters					
ENVIRONMENTAL	Protection of critical natural capital					
	Defence and improvement of soil quality and stability					
	Defence and improvement of key habitats and biodiversity					
	Respecting the capacity of natural systems					
	Efficient use of renewable resources					
	Minimum depletion of non-renewable resources					
	Energy efficiency					
	Minimisation of waste/recycling/re-use					
	Reduction of pollution emissions					
	Protection and enhancement of environmental amenity/aesthetics					
	Protection of natural and cultural heritage					
ECONOMIC/SOCIETAL	Develop connections between environmental quality and economic vitality					
	Satisfy and protect basic needs					
	Provide entrepreneurial/employment opportunities					
	Protect basic human rights					
	Ensure health and safety					
	Improve local living conditions					
	Satisfy the economic and living standards to which people aspire					
	Develop an open, inclusive and transparent governance system					
	Subsidiarity					

Table 14.5: Sustainability content of selected PPGs

	CRITERIA	PPG20 Coastal Planning (92)	PPG22 Renewable Energy (93)	PPG23 Pollution Control (94)	PPG13 Transport (94)	PPG1 General (97)	PPG7 Countryside (97)	PPG11 Regional Planning (00)	PPG13D Transport (99)	PPG3 Housing (00)
OVERACRCHING	Precautionary principle									
	Include cumulative and long-term impacts in decision-making									
	Commitment to equity at local, national and international levels									
	Reducing externality effects									
	Using close in preference to distant resources									
	Natural disasters									
	Human-made disasters									
ENVIRONMENTAL	Absolute protection of critical natural capital									
	Defence and improvement of soil quality and stability									
	Defence and improvement of key habitats and biodiversity									
	Respecting the capacity of natural systems									
	Efficient use of renewable resources									
	Minimum depletion of non-renewable resources									
	Energy efficiency									
	Minimisation of waste/recycling/re-use									
	Reduction of pollution emissions									
	Protection and enhancement of environmental amenity/aesthetics									
	Protection of natural and cultural heritage									

ECONOMIC/SOCEITAL	Encourage and develop connections between environmental quality and economic vitality									
	Satisfy and protect basic needs									
	Provide entrepreneurial/employment opportunities									
	Protect basic human rights									
	Ensure health and safety									
	Improve local living conditions									
	Satisfy the economic and living standards to which people aspire									
	Develop an open, inclusive and transparent governance system									
	Subsidiarity									

15 Mainstreaming Sustainable Development into Local Politics

SUSAN PERCY AND VICTORIA HANDS

Introduction

Local government has been charged with the delivery of many sustainable development policy goals. This responsibility stems from the Rio Earth Summit of 1992 at which Agenda 21 together with the local version – Local Agenda 21 – were launched. Local authorities have been called upon to develop action plans for sustainable development at the local level, which need to be backed up by local political support if sustainable development is to be implemented.

This chapter will focus on one local authority's response to the sustainable development challenge by examining a specific initiative –The Sustainable Southwark Members Initiative (SSMI), introduced to promote sustainable development action at a ward level and requiring the commitment and support of local elected Members. The chapter will explore, in respect to SSMI and sustainable development, issues around levels of understanding, "greening the greens", lay knowledge, capacity building, local politics, and institutional barriers. It will conclude by assessing the extent to which sustainable development is a mainstream policy goal at the local political level.

The Policy Context

Sustainable development is a widely used phrase but is often misunderstood due to the rhetoric that surrounds this concept. It is also a contested notion which can often lead to "a muddled confusion and a swirl of partial thought reflecting the complexity of the sustainable development concept" (Crilly et al., 1999). The World Commission on Environment and Development's publication – the Brundtland Report of 1987, first defined sustainable development as "...meeting the needs of our generation without compromising the ability of future generations to meet their own needs" (WCED, 1987). The term covers inter-related issues as diverse as global climate change, biodiversity, human health and social equity. It is therefore a far-reaching and over-arching term and when translated into local policy it involves supporting the following:

- intra- and inter-generational equity;
- in the environmental, social, cultural, political and economic fields;
- based on capacity building;
- and empowering the local community;
- whilst acknowledging transfrontier responsibility.

The Local Institutional Context

Local government in the UK has been actively trying to deliver sustainable development policy goals in response to the Rio Earth Summit of 1992 at which Agenda 21 (a blueprint for global sustainable development) was launched. Chapter 28 of Agenda 21 calls upon local authorities to develop action plans for sustainable development at the local level through, for example, Local Agenda 21.

Clearly, local authorities have a primary role to play in implementing the principles of sustainable development at a local level, since they are the level of government that is closest to the people, and play a vital role in educating, mobilizing and responding to the public in respect of sustainable development (Chapter 28 of Agenda 21, 1992). Furthermore, "The overall objective is to improve or restructure the decision making processes so that consideration of socio-economic and environmental issues are fully integrated" (Chapter 28 of Agenda 21, 1992). Local government is central in delivering many of the sustainable development policy goals due to their "role as a policy organisation, thereby connecting sustainable development into spatial, economic and service plans" (Selman, 1998).

Mainstreaming Sustainable Development

"Mainstreaming is now a recognised tool / strategy to achieve equality. It has been promoted by the European Commission and has been adopted by the UK government" (Reeves, 1999). Whilst mainstreaming has been used mainly in the context of gender equity, in this chapter mainstreaming is concerned with the integration of a sustainable development dimension in all local authority activities, as a routine consideration. Reeves (1999) points out that mainstreaming will involve changes in approach within local authorities supported by a multi-pronged strategy in which:

> ...the process is likely to be iterative, resistance may be found at all levels of decision – making...The issue of organisational or institutional change really is a key element and it is not just a technical process, but perhaps, more importantly, a political process which will affect the organisational cultures, structures, ways of thinking and working (Reeves, 1999: 6).

New Opportunities at the Local Level

There is widespread support for the view that sustainable development will only be achieved at the local authority level by some form of internal reorganisation and refocusing of policy making (Patterson and Theobald, 1996). The modernisation agenda of local government (see DETR's Modernising Local Government, 1998 and 1999) rests on the goal of improving the quality of peoples' lives by encouraging the local authority to be in touch with its local community, provide high quality services through Best Value and provide vision and leadership. The modernisation agenda is also about introducing new political structures, which aim to provide management structures, which are effective and improve local democracy.

This new agenda is likely to offer opportunities for local authorities to reinvent themselves and incorporate sustainable development into local politics and create new political space for local government. The government believes that "Best value will also help councils to address the cross cutting issues facing their citizens and communities, such as community safety or sustainable development, which are beyond the reach of a single service or service provider" (DETR, 1998). Since local government activities have a political dimension, mainstreaming sustainable development into local politics is vital, if meaningful change is to occur.

Sustainable Southwark Members' Initiative (SSMI)

London Borough of Southwark[1]

A major local sustainable development initiative in the London Borough of Southwark (LBS) is the SSMI, designed to increase participation of Members and community at ward level in Southwark's Local Agenda 21 Programme[2] by supporting projects developed at a local scale illustrating the principles of sustainable development and improved quality of life. It therefore provides a good basis for exploring the extent of local political support for sustainable development in Southwark.

The initiative involved the LBS inviting local voluntary and community groups to submit project proposals to the Environmental Development and Education Unit for consideration by Members for the award of a grant. The proposals were judged against a number of criterion which included:

- the project must contain an element of community consultation involving the relevant ward Members;
- the project must involve a wide representative section of community participation;
- the project must contribute towards sustainable development and an improved quality of life in the area.

A total of 36 projects were submitted and 32 projects were each given a £500 award voted for by 55 of the 60 LBS Members.

Methodology The research was carried out towards the end of 1999 and reviewed the SSMI projects approved at the beginning of that year. The analysis was based upon both quantitative (LBS project documentation) and qualitative information (telephone interviews) helping to reduce problems of construct validity associated with in-depth interviews, i.e. respondents answering with a response most likely to please the interviewer which may not represent the "true" picture. The research methodology aimed to give maximum validity and reliability to the research by bringing together different sources of evidence in a single study.[3]

Successful Projects

The focus of projects approved in the 1999 SSMI (shown in Figure 15.1) followed traditional environmental and/or green projects, providing an insight on to and indication of, the extent of understanding of the term sustainable development by the community and Members.

Community and school gardens
Development of window box schemes
African and Caribbean planting/propagation
Community composting schemes
Establishment of organic vegetable gardens
Tree and shrub planting
Mural on school wall
Accessibility to local facilities survey
Strategy to safeguard local playing fields
Support of Muslim cultural events
Recycling scheme
Cultivation skills for homeless people
Community clock building
Energy efficiency training
Placement of a yurt in schools/shopping centre
Installation of sanitary facilities in an allotment
Developing local construction skills
Prize giving to encourage gardening

Figure 15.1: The focus of successful projects in the SSMI 1999

Understanding the Rhetoric

In order to participate in Local Agenda 21 and sustainable development initiatives, participants need to have an understanding of these terms and how they work in practice. This seems an obvious statement but in fact understanding the vocabulary of sustainable development is a common challenge to the success of its implementation. Irwin (1995) has written on how publics are often excluded from discussions about health, safety and environment because of a lack of understanding of the language of policy discourse. "In the case of the sustainable development debate, the language and concepts employed are notoriously complex and often widely removed from day to day experience" (Evans and Percy, 2000). The research showed that it is not only the general public who find the vocabulary and concepts of sustainability difficult, many of the Members have limited understanding of the concept. "Whilst sustainable development has been widely accepted as a theoretical concept, the translation of this abstract idea into operational practice gives rise to problems" (Counsell, 1999).

Figure 15.2 is a way of representing the range of understanding of the terms sustainable development and Local Agenda 21 amongst Southwark Members. A scale of 0-5 has been used in order to categorise responses into bands and examples of Member responses have been given to illustrate the sorts of levels of understanding.

The analysis of Figure 15.2 suggests that respondents in band 0 who did not identify the term cannot therefore engage with the issues. Respondents in band 1 are able to name the terms but cannot apply anything to them. Respondents in band 2 understand the terms in so far as they can apply them to general environmental issues. This can be attributed, in part, to current environmental education campaigns, which have succeeded in underlining the importance of ecological and environmental issues. This finding is in fact reflected in and supported by the range of project applications, which focussed on traditional environmental issues (see Figure 15.1). Respondents in band 3 understand the terms, apply them to environmental issues and are further able to apply them to personally relevant activities in time - this generation and future generations - and in space – the immediate locality though generally lacking an understanding of global issues. Respondents in band 4 recognise the importance to applying the terms to all sectors of the community and all aspects of social life. This encompasses an understanding of exclusionary issues and the need to include the traditionally excluded groups including e.g. women, ethnic groups, the young and the elderly. Respondents in band 5 take action to address the range of issues through innovative enterprise and recognise the all-embracing holistic nature of sustainable development in both space and time. It is clear that there is a very variable level of understanding, as Table 15.1 further demonstrates.

Table 15.1 illustrates the range of Member responses as broken down into the 5 bands described in Figure 15.2 and compares these with the responses of community groups assessed in the same manner. The research shows that the community respondents had a clearer and more fully developed understanding of the concepts of sustainable development and Local Agenda 21. The majority of Member respondents fell within bands 2 and 3 showing partial but limited understanding of the issues.

0 – no knowledge whatsoever
"What do they mean to me? Nothing at all, both terms are far too vague for me to waste any energy upon." (Member respondent 4)

1 – heard of the terms but not aware of the meanings
"Never been clear to me. Arising from Rio international conference." (Member respondent 12)

2 – able to associate the terms with general environmental issues
"...energy conservation, non-renewable resources, recycling." (Member respondent 11)

3 – also aware of the long-term implications of the terms and local application
"Built upon the community's long term vision..." (Member respondent 3)
"Environmentally friendly actions, lasting for the future, protecting the environment for future generations." (Member respondent 10)

4 – also good concept of the social aspects
"...projects which don't have to be propped up by external bodies, but able to sustain and regenerate themselves in the local community...world education on environmental issues for all walks of life." (Member respondent 7)

5 – also fully aware of the holistic nature and the philosophy behind the terms in a very innovative fashion (global implications)
"Community support through development which forms a long lasting improvement ensuring environmental effects are considered in all aspects of life, not something seen as apart or separate, but all embracing." (Member respondent 7)

Figure 15.2: Member understanding of the terms sustainable development and Local Agenda 21 on a scale of 0-5

Obstacles to Mainstreaming Sustainable Development

The SSMI research highlighted several examples of obstacles to the mainstreaming of sustainable development into local politics. These included: understanding the principles of sustainable development, "greening the greens", lay knowledge, capacity building and empowerment; local politics and institutional barriers.

Table 15.1: Distribution of member and community perceptions of sustainable development and Local Agenda 21

Score	0	1	2	3	4	5
Members	• •	•	• • • •	• • •	•	•
Community		•	• •	•	• • •	• • •
	LESS		KNOWLEDGE			MORE

Understanding the Principles of Sustainable Development

Interpretations of sustainable development when translated to community projects are for the most part focused on traditional green and environmental issues, as the trend for gardening and planting projects demonstrate (see Figure 15.1). Sustainable development in fact encompasses the social, economic, cultural and political as well as traditional environmental issues. Some of the SSMI projects did encompass the holistic nature of sustainable development as Table 15.2 shows: those which provide skills training, integrate minorities into the community and encourage inter-cultural, and inter-generational exchanges.

The traditional projects were of no less value, since in addition to the outcomes of individual projects, the actual process of establishing the project has enormous benefits:

> The visible, tangible result of the project will represent a testimony to local cooperation.... (Community respondent 1) (Outcome).

> The process of creating the project in a fragmented community allowed cross-generational and cross-cultural interaction, which wouldn't otherwise have taken place (Community respondent 8) (Process).

It is clear that sustainable development initiatives, like SSMI, must seek to reduce the use of rhetoric to enable the scope of sustainable development projects to expand beyond traditional green issues. This would facilitate the inclusion of projects which embrace wider social aspects such as: equal opportunities, health, regeneration, future generations, ethnic groups, the elderly, the young, demonstrating the "think globally, act locally" principle by showing that small local projects on a national level have a potentially large impact.

"Greening the Greens"

Any awareness raising as a result of the SSMI, impacting beyond those who were already involved in sustainable development issues, has been relatively limited mainly due to problems in understanding the terms used. Those aware of or familiar with the terms sustainable development and Local Agenda 21 were much more easily targeted and better equipped to be able to participate in such an initiative. In this respect, the rhetoric limited the range of citizens that the SSMI could reach and prevented a ripple effect of knowledge through the community. This relates to an area of common concern in awareness raising, as the methods employed inevitably define the range of citizen's targeted and unavoidably marginalise some sections of the population.

One way of ensuring the maximum possible target audience is to make vocabulary and methods of dissemination as locally appropriate as possible:

> Not sure if its fair (SSMI) or that it attracts everyone, in reality only those who understood it [the terms] could participate... get rid of jargon and make it Cockney English... a simple level of explanation... need to talk bread and butter issues... (Member respondent 10).

In much the same way, the extent to which citizens are open to such awareness raising is governed in large by their previous experience around an issue. The research demonstrated that those who considered themselves to be aware of, or sympathetic to, quality of life issues were more enthusiastic or personally committed to becoming involved in SSMI than those for whom sustainable development was not personally engaging. Personal motivation can be stimulated by emphasising the relevance of sustainable development in everyday life and captured in order to involve more people in the SSMI. Greening the non greens and enabling people to make connections between everyday concerns and worries and sustainable development is essential to its success.

Table 15.2: SSMI project types mapped against sustainable development principles

Sustainable development principles	SSMI project types						
	Traditional green issues	Waste management	Job Opportunities	Education	Improving local surroundings	Local access/transport	Community development
Intra-generational equity	•	•	•	•	•	•	•
Inter-generational equity		•		•			•
Environmental	•	•		•	•	•	
Social		•	•	•		•	•
Cultural			•	•			•
Political		•		•		•	•
Economic		•	•			•	
Capacity building			•	•			•
Community empowerment		•	•	•	•	•	•
Transfrontier responsibility							

Interviews with Members, revealed that personal views on environmental and quality of life issues affected their approach to participation in sustainable development initiatives. Members who already considered themselves to be green admitted to being more willing to participate in the SSMI than those who were not already interested in environmental issues. By the same token, Members who already had a strong relationship with local community groups were personally motivated to enable the local community to participate in the initiative.

The research showed that unfamiliarity with the terms sustainable development and Local Agenda 21 sat alongside a good familiarity with the practical manifestations of the terms, prompting associations with practical examples of sustainable development such as re-housing, regeneration, skills training, open space issues and waste management. The key must lie in bridging this gap in understanding by constructing methods to demonstrate that sustainable development is relevant to every day concerns (of both Members and community). Encouraging the links and connections between everyday personal living and sustainable development is proving to be a major challenge in mainstreaming sustainable development into local politics.

Lay Knowledge

The research highlighted that there were feelings, amongst community and Members, that community views are not listened to (in spite of consultation) and that lay knowledge is not valued by authorities. To close the loop in the consultation process local views need to be incorporated in decision-making. The research indicates a lack of understanding of linking or making connections between, for example, regeneration of the local area in practical terms and sustainable development as a policy goal. The danger here perhaps is one of missed opportunities, an inability to link up everyday concerns with sustainable development, and it takes part on two levels.

From a top-down perspective, local authority decision making processes currently rely on a combination of "expert judgement, democratic pragmatism, and economic rationalism" (Davies, 1999), and are not well-equipped to incorporate lay knowledge which is largely based upon personal values and a sense of place, both of which are non-quantifiable, intangible and not scientifically proven. Consequently, a wealth of local vernacular knowledge is not effectively communicated and is not always utilised in decisions affecting the local area. Equally, from a grass roots level, although much is already being done, especially at ward level, which falls under the auspices of sustainable development, there is the same inability to make connections,

> ...the current wording divorces us from the true meaning and sets up a barrier... (Community respondent 2).

Sustainable development and Local Agenda 21 in particular, could provide a very important vehicle for incorporating lay knowledge in decision-making processes overcoming community cynicism towards participation which is based on many instances, still continuing in the eyes of some community members, of no one listening to local community views:

> They are despondent, angry…they tell the Member what they want but nothing or even the opposite happens and they get a kick in the teeth….there is very little trust now to do things to improve the most impoverished areas… (Member respondent 10).

Sustainable development offers a positive opportunity for local authorities to win back the trust and support of the local community and gain increased credibility. Encouragingly, the SSMI has proved successful in stimulating the exchange of local expertise in at least one instance. One SSMI project encouraged local residents to develop their gardening skills through local competitions and,

> … even on a very small level, people are exchanging seeds and cuttings and garden knowledge… (Member respondent 8).

Promoting the value of lay knowledge, even horizontally, within the community will have a positive effect on top-down interactions. Vernacular knowledge and life experiences, of all sections of the community, especially cross-cultural knowledge (35% of the population of LBS is made up of African, Bengali and Irish communities) provides a valuable resource for sustainable development and for the local authority generally in policy and decision-making processes which are locally appropriate.

Another positive impact is gained from projects, which remain very relevant on a local level with the advantage of local independent management. The SSMI was run on a ward basis and was very locally appropriate as it targeted locally identified areas of importance and made a resource available for local action. More than that, community respondents felt that the SSMI allowed individual projects to remain open to local dynamics, in other words to remain responsive and flexible throughout. This meant that in at least one case, the SSMI project was able to grow, according to the demands of the local community, into something with much more local impact than the original project could have foreseen.

Capacity Building, Community Empowerment and Education

Current drives to increase participation need to address the "skilling up" of the local population, raising awareness of the issues and building confidence within the local community:

> The citizen finds it overwhelming to suddenly have power and is very anxious to ensure that any ideas are democratic, wide-ranging and long-

> term. There is a strong distrust of the Council and of the "usual suspect" interest groups who often serve their own interests and in fact compete with each other instead of working for the good of the whole community... (Community respondent 5).

Members recognised the importance of having effective partnerships between themselves and the community to advance sustainable development, but there was a sense of frustration that certain sections of the community did not engage with Members in this way because they lacked the necessary skills –for example communication skills, confidence and ability to articulate views in a formalised system:

> So far it seems sustainable development has been aimed at people with money and hence other skills which enable participation, but they need to focus on what working class local people need (Community respondent 4).

> People think that it [participation] will be more and harder work than it is because they lack confidence and have low self-esteem (Community respondent 5).

It is true that many sections of society do not have the skills to participate in sustainable development issues in a meaningful way. Some sections of the community do not possess the skills or confidence to put their views across in a formalised system. A focus on capacity building through environmental education and community development will enable local authorities to reach out to all sections of the community on sustainable development issues. In order to build up community skills to empower effective participation in the consultation process, environmental education needs to go hand in hand with community development work emphasising democratic principles. There are of course resource implications for such a programme of sustainable development promotion, which focuses very much on officer time and involvement.

It was felt that the Council and Members should help encourage and support joint responsibility in sustainable development by addressing these exclusionary issues and working to enable and empower all sections of the community to participate. However, this means that Members must be fully aware of the broadest meaning of sustainable development, which will require regular and extensive internal environmental education within in the Council administration to spread awareness of sustainable development to a wider community.

Local Politics

As Lusser and Riglar (1999: 225) point out:

> Recent research by the Local Government Management Board (LGMB, 1998) suggests that the key to minimising effort while maximising returns, lies in harnessing the "corporate drivers" ... A characteristic common to all Councils with a reputation for environmental excellence is that each has progressed because the cause has been championed by one, or more, individuals inside the organisation. It is no surprise that progress is most often made where the "champion" is an elected politician with the status to influence events, assemble resources and galvanise colleagues.

A major concern, however, of over half of the Member respondents was that SSMI seemed to have the effect of politicising sustainable development along party lines despite its holistic nature. In mixed party wards, Members from all parties felt that they were always competing on party lines and that the way that SSMI was presented made it another politically competitive issue. This lead for calls to

> ...get the Members out of the equation...especially in respect of the voting process...(Member respondent 11).

Where there was a higher level of understanding of sustainable development, there were good working relationships, not only in same party wards but in mixed party wards with a feeling of overcoming party boundaries to work for a higher shared goal.

Sustainable development presents a challenge in that party politics do not often sit well with an over-arching societal aim. Political bargaining and the balancing of priorities will always influence politicians' views and as an elected politician, the Members have to be sensitive to local party politics. This issue was recognised by the majority of the Members:

> Sustainable development is greater than the promotion of local politics (Member respondent 9).

In addition, the competitive nature of the SSMI funding meant that involving Members in the voting process opened them up to potential criticisms of favouritism based on their personal motivations. Unquestionably, Members played an important role in putting forward as many local community groups as possible for local funding initiatives and indeed in providing many types of support for the application in general.

However, following on from discussions on personal motivation, it was recognised by community and Members alike that the involvement of Members in the SSMI voting process – of an obviously competitive nature - created potential difficulties, especially for Members with close links to specific organisations:

> It must be like choosing between your children....much too difficult especially for such a small amount of money... (Community respondent 8).

> I have an anxiety about local Members voting on local projects...they are going to disappoint other groups...which doesn't do much to enhance relationships at a local level... (Member respondent 9).

This was a major issue for all Members, who were very politically sensitive to the possibility of being perceived to be supporting one community group in preference to another. Fear of favoritism was quoted as the single reason why 5 of the Members failed to vote.

Institutional Barriers

The current highly departmentalised structure of local government does not support the holistic approach necessary for achieving the goal of sustainable development. Those respondents familiar with the concept of sustainable development felt that the Council perpetuated and supported a major barrier to greater understanding through its highly departmentalised structure:

> The biggest problem for the understanding of sustainable development is Council departmentalism...sustainable development could play a central strategic role in combating departmentalism (Member respondent 8).

There is a need to establish cross-departmental sustainable development initiatives, encouraging effective inter-departmental communications, such as an environmental checklist or a sustainability matrix – similar to an Equal Opportunities checklist - to be incorporated within all Council literature and all decision-making to aid in mainstreaming sustainable development.

The problem of dealing with an all-embracing issue within a highly departmentalised system is not entirely unexpected, as meeting the challenge of joined-up thinking is a difficult task for all local authorities. During the course of interviews respondents expressed the view that despite the commitment of the Environmental Development and Education

Unit, any acknowledgement of the holistic character of sustainable development needed to be translated into visible practical measures spanning the entire Council:

> Environmental considerations must be incorporated within every aspect of Council life (Member respondent 3).

Effective internal Council communication was named as the most pressing issue. It was felt that,

> ...until all the council departments learn to act together as one, sustainable development will be caught up in the internal confusions of Council work (Member respondent 4).

As of December 1 2000, the function of the Environmental Development and Education Unit of London Borough of Southwark, which before influenced all other departments, have now been absorbed into the heart of those departments. This is an encouraging first step to integrating sustainable development into decision-making processes and local politics.

Conclusion

The main barriers to mainstreaming sustainable development identified in the SSMI research include: understanding the rhetoric, community empowerment - feeling comfortable articulating views to Members and decision-makers, personal interest and greening the greens. Linked into these issues are concerns about social exclusion and the need for institutional change. In addition, as noted by Evans and Percy (2000: 3):

> The short-termism of (local) politics is a key hurdle, especially highlighted by funding regimes, and is linked inextricably to the very heart of local government. There may be some encouragement in the modernising local government agenda which seeks to encourage local authorities to change their decision making structures to facilitate greater public participation and for initiatives such as Best Value which aims at departmental cross-cutting in areas such as sustainable development.

SSMI acted as a unifying force and provided an opportunity to bring together diverse sections of the community. It is essential that,

> ... having planted the seeds of community participation in sustainable development, they are nurtured to grow, because without care they will die (Community respondent 2).

There is a need for a commitment to sustainable development targets in local authorities' corporate plans and all policies and strategies. The right skills are also needed and most councils have a corporate training framework in to which sustainable development should be incorporated. The goal of sustainable development should be written into all terms of reference of all committees, panels and groups. Annual reports should be produced outlining how sustainable development is being promoted and best value indicators should relate to sustainable development goals.

Ultimately, sustainable development has to be put on an equal footing with Members' traditional concerns and ideally to become the overriding policy goal out of which all other policy initiatives will flow. As yet, however, most local politicians still associate sustainable development with traditional environmental concerns, rather than with socio-economic issues, but with improved training and understanding of the concept, sustainable development, as the main local government goal may be possible. The research demonstrates that sustainable development is not currently mainstreamed into local politics in the LBS, although innovative funding mechanisms like the SSMI can help to reaffirm the value and strengthen mainstream support of popular and effective activities. Equally such initiatives can help focus new directions for action to be rolled out into mainstream programmes.

Notes

1. LBS is an inner city borough immediately to the south of the River Thames. LBS has the Boroughs of Lambeth and Lewisham on either side with Croydon to the south and the City of London on the opposite side of the Thames. The population (approximately 24,000) is mixed in terms of both its socio and economic profiles.
2. The London Borough of Southwark has responded to the Local Agenda 21 challenge by launching in November 2001 Essence, a Local Agenda 21 Strategy for Southwark. This found its roots in the Local Agenda 21 Action Plan (adopted in September 1997) based on low level actions stemming from London Borough of Southwark's Environmental Strategy which covers twelve key areas including transport, environment, health, regeneration, housing, economic development and leisure. In line with central government policy, all the strategies are based on the principles of sustainable development, democratic renewal and best value. The draft Local Agenda 21 Action Plan underwent a widespread consultation programme using a wide range of community consultation techniques, the results of which were

incorporated into the final Local Agenda 21 Action Plan – Essence. Essence will, in turn, facilitate the setting of priorities, targets and indicators for improvements in social, economic and environmental well being in Southwark.

3 The research was commissioned by the Environmental Development and Education Unit of the LBS. 10 of the 32 community groups (31%) and 12 of the 60 members (20%) spanning the political range (Labour, Liberal Democrat and Conservative), were interviewed. Thanks go to all those individuals who participated in the research.

References

Counsell, D. (1999), 'Sustainable development and Structure Plans in England and Wales: Operationalizing the Themes and Principles', *Journal of Environmental Planning and Management*, vol. 42, no.1, pp. 45-61.

Crilly, M.; Mannis, A. and Morrow, K. (1999), 'Indicators for change: taking a lead', *Local Environment*, vol. 4, no. 2, pp. 151-168.

Davies, A. R. (1999), 'Where Do We Go From Here? Environmental Focus and Planning Policy Formulation', *Local Environment*, vol.4, no. 3, pp. 295-316.

Department of Environment, Transport and the Regions (1998), *Modernising Local Government In Touch with the people*, White Paper, DETR, London.

Department of Environment, Transport and the Regions (1999), *Modernising Local Government: Local Democracy and Community Leadership*, DETR, London.

Evans, B. and Percy, S. (2000), 'Obstacles to local Sustainability: Actors, Agencies and Organisation in the English Local Planning Process', unpublished paper, South Bank University, London.

Irwin, A. (1995), *Citizen science: a study of people, expertise and sustainable development*, Routledge, London.

Local Government Management Board (1998), *Corporate Approaches to Local Agenda 21 through the implementation of EMAS*, LGMB, Luton.

Lusser, H. and Riglar, (1999), 'Corporate Drivers: the key to environmental management success', *Local Environment*, vol. 4, no. 2, pp. 225-230.

Patterson, A. and Theobald, K. (1996), 'Local Agenda 21, Compulsory Competitive Tendering and Local Environmental Practices', *Local Environment*, vol. 1, no. 1, pp. 7-20.

Reeves, D. (1999), *Mainstreaming Gender Equality in Planning*, Paper Presented to Royal Town Planning Institute Full Council 10th November.

Selman, P. (1998), 'Local Agenda 21: substance or spin?', *Journal of Environmental Planning and Management*, vol. 41, no. 5, pp. 533-553.

United Nations Commission on Environment and Development (1992), *United Nations Conference on Environment and Development Agenda 21*, UNCED, Geneva.

World Commission on Environment and Development (WCED) (1987), *Our Common Future (The Brundtland Report)*, Oxford University Press, Oxford.

16 Mitigating and Monitoring Ecological and Visual Impacts of EIA Projects

ELAINE QUINN

Introduction

Environmental impact assessment (EIA) is a procedure for systematically assessing the effects of certain projects on the physical, biological, social, cultural and economic components of the environment. The main purpose of EIA is to provide a systematic analysis of the environmental implications of a proposed action to aid decision-making in the development control process. The EIA process aims to provide the local planning authority, the developer and the public with objective information on the likely effects of a development.

EIA is mandatory for certain projects and is determined on the basis of specified criteria which include the size and type of project and sensitivity of the location and potential impacts. EIA became mandatory in the UK in 1988 under the Town and Country Planning (Assessment of Environmental Effects) Regulations 1988 SI. 1199, in response to Directive 85/337/EEC. This legislation was subsequently replaced in 1997 and 1999 by Directive 97/11/EC and the Town and Country Planning (Environmental Impact Assessment) (England and Wales) Regulations 1999 SI. 293, respectively, with various other regulations for projects not subject to planning permission.

In theory, the EIA process has a role to play in achieving sustainable development through the prediction and mitigation of adverse environmental impacts (Glasson et al, 1994). Assessment of the ecological and visual impacts of a development is a statutory requirement of Directives 85/337/EEC and 97/11/EC, and the UK's EIA regulations. The assessor is required to predict the magnitude and significance of impacts of the proposed development on habitats, plants and animal species and the visual amenity at the site and surrounding landscape (DoE, 1989; DETR, 1999). If significant, adverse impacts are predicted, it is a mandatory requirement for mitigation measures to be proposed to avoid, reduce or remedy the impacts. Mitigation measures are of little value unless they are implemented and prove to be effective. However, at present, under UK EIA legislation there are no statutory requirements for monitoring of the implementation and/or the effectiveness of mitigation measures proposed in environmental statements.

Monitoring is perceived as having numerous benefits for the environment and for the development of the EIA process. Monitoring alerts project managers to potential adverse impacts before they occur, thus facilitating environmental management of a development and has the potential to provide information to feedback into the EIA and decision-making process (Bisset and Tomlinson, 1988). Despite these benefits monitoring is rarely undertaken in practice, particularly for ecological and visual impacts (Wood et al, 2000).

This ESRC (no.R00429834637) funded research examined the practice of mitigating and monitoring ecological and visual impacts of EIA projects through the review of existing research and literature, analysis of 50 environmental statements and the examination of ten case studies. This chapter presents the results of the ESs reviewed and discusses the ecological and visual impact mitigation and monitoring in one of the case studies: a 9.5km bypass.

Background - Ecological and Visual Impact Mitigation Measures

"Mitigation includes any deliberate action taken to alleviate adverse effects" (Treweek, 1999). Measures to mitigate predicted impacts can be broadly separated into four categories: avoidance, reduction, remedial and compensatory measures (DETR, 1997). Mitigation should firstly be aimed at avoiding adverse impacts through the consideration of alternative sites, designs and processes. If adverse impacts cannot be completely avoided measures should be proposed to reduce the significance and magnitude of the impact. In addition, measures to remedy unavoidable temporary effects of development should be proposed where possible. In cases where impacts are unavoidable and remediation is not possible, measures should be proposed to compensate for the ecological and/or visual resource affected.

Avoidance Measures

Measures to avoid impacts, such as the development of alternative strategies, locations, site layouts and processes, have to be incorporated into the initial stages of development design. According to Treweek (1999), "siting and design based on least damage criteria is the most effective way of ensuring that the integrity of natural resources is preserved". However, there is evidence indicating that this method is rarely employed in practice. In ESs reviewed by Treweek and Thompson (1997) the majority made no attempt to avoid areas of ecological importance, particularly at the design stage. Only three out of 192 ESs based the siting of the development on "least damage criteria", with 212 sites of either international, national importance potentially affected by the proposed developments. Of 112 Sites of Specific Scientific Interest (SSSI) to be directly affected, mitigation measures to avoid the areas were proposed at only eight sites. In Byron et al's (2000) review of road ESs prepared between 1993 and 1997 only a small minority made reference to route realignment as a way of avoiding ecological impacts.

Mills' (1992) review of visual impact assessments revealed that 30% of ESs discussed alternative options to the proposal including re-routing and re-positioning. However, the extent to which visual impacts were primary considerations in these options was unclear.

Reduction Measures

In some cases it may be impossible to completely avoid a significant, adverse impact, therefore detailed design and management measures to reduce or remedy the impacts must be proposed. Extensive research has been undertaken in the Netherlands on mitigation measures to reduce ecological impacts of roads including (Roads and Hydraulic Engineering Division, 1995):

- fences to reduce the exposure of the receptor to the impact;
- planting to reduce disturbance to birds from noise and visual intrusion;
- tunnels and bridges to reduce effects of habitat severance and isolation;
- ecologically designed culverts to allow use by mammals;
- wildlife corridors and visual clues to encourage movement and reduce effects of fragmentation and isolation.

The results of previous ESs reviewed revealed that there is a greater reliance on mitigation measures to reduce rather than avoid ecological and visual impacts. In ESs reviewed by Treweek and Thompson (1997) the majority of mitigation measures were associated with reducing impacts through screening and/or landscaping at the site. Similarly, in more recently prepared ESs, reviewed by Byron et al (2000), the most common measures proposed were those aimed at reducing ecological impacts through landscaping, tree planting and the installation of tunnels and fences.

With regard to visual impact mitigation measures, all of the ESs reviewed by Mills (1992) proposed measures to reduce the severity of the visual impacts predicted. Measures included: screening earth bunds, vegetation screening, landscaping of cuttings and embankments; colour of turbines, speed of turbine rotations, absence of permanent access roads and underground positioning of power cables.

Remedial Measures

In cases where it is impossible to avoid or reduce significant, adverse ecological and visual impacts, measures should be proposed to remedy or restore the habitat or visual resource affected. Reinstatement of soil and vegetation to restore a site after temporary damaged has occurred may be proposed in ESs for extraction and pipeline construction. When restoring sites and habitat the aims and objectives should be clear. Restoration of a site for purely ecological purposes will require a different set of objectives compared to the restoration of a site specifically for amenity purposes and in some cases a balance between the two will be desirable (Gilbert and Anderson, 1999).

Compensatory Measures

Compensatory measures, in the form of the creation of new habitat, will be required in situations where a site has been damaged beyond the point of repair. Although habitat creation can be successful in certain instances, particularly to compensate for sites of low nature conservation value, it must not be used to justify adverse impacts on high biodiversity receptors (Byron, 2000). The extent to which newly created habitats or ecosystems compensate for lost resources varies considerably and is heavily dependent on the complexity of the habitat (Treweek, 1999). Generally techniques in the creation of wetland habitats are more advanced, particularly in comparison to woodland habitat creation which remains under-developed (Gilbert and Anderson, 1999). Research has revealed a general agreement that habitat creation does not compensate, and thus is not considered to be acceptable for areas of high ecological value e.g., SSSIs (English Nature, 1994). The general principles for successful habitat creation are that it is important to have clear objectives, sufficient resources available, detailed methodology and implementation, and short and long-term management programmes (Parker, 1995).

In previous ESs reviewed by Treweek and Thompson (1997) descriptions relating to the implementation of the proposed measures were limited and many of the measures were "vague recommendations...with emphasis on more cosmetic measures such as landscape and tree planting".

Background - Monitoring Ecological and Visual Impacts

The monitoring of ecological and visual impact mitigation measures is essential to ensure that measures proposed are actually implemented in practice and to determine the effectiveness of the measures in mitigating the impact. "Evaluating the effectiveness of ecological mitigation is essential to the pursuance of sustainability objectives through the EIA process" (Treweek, 1999). Despite this, impact monitoring is rarely carried out and remains under-developed particularly when compared to "pre-decision activities" (Sadler, 1996). Frost's (1997) study of ESs, prepared between 1988 and 1993, revealed that none of the ESs referred to visual impact monitoring and only 6% made specific reference to ecological impact monitoring. In comparison, 5% of ESs reviewed by Thompson et al (1997) suggested ecological monitoring as a "possibility for the future", but none included a commitment to monitoring. ESs of roads prepared between 1993 and 1997, reviewed by Byron et al (2000), showed a marginal improvement in that 10% recommended ecological monitoring and 5% made a firm commitment to ecological monitoring.

Although past research has revealed a limited number of ESs make reference to monitoring, some monitoring does take place in practice through planning conditions and pollution control legislation. The examination of development projects, either at the construction or operational phase, by Wood et al (2000) revealed that although in practice monitoring was limited, some monitoring was required under planning conditions and pollution control legislation for noise, air and water impacts. However, ecological and visual impact monitoring was particularly limited and only undertaken for a small minority of impacts predicted: 2 and 15% respectively.

Methodology

Review of Environmental Statements

A sample of 50 ESs was selected to represent those submitted between 1988 and 1999 and reflected a cross-section of:

- main project types for which ESs were prepared;

- dates of ESs submitted;
- geographical regions of the developments;
- EIA regulations under which the ESs were prepared.

A checklist was prepared to review the mitigation measures and monitoring proposed for ecological and visual impacts in the 50 environmental statements.

Case Study Analysis - Implementation, Monitoring and Effectiveness of the Mitigation Measures Proposed

For the second phase of the study ten projects were selected from the sample of 50 environmental statements. The selection criteria for the projects was based on the:

- type of development;
- geographical location;
- status of ecological and visual resource affected (i.e., designated sites/landscape, protected species);
- types of mitigation measures proposed;
- monitoring proposed in the ES, required under planning conditions or pollution control legislation.

Site visits and interviews were undertaken to ascertain if the mitigation measures proposed in the environmental statements had been implemented and were proving to be effective. For monitored projects, the ecological and visual impact monitoring data was evaluated to assess its reliability and adequacy for evaluating the effectiveness of the mitigation measures. For projects without specific ecological and visual impact monitoring, site visits and interviews with statutory consultees, NGOs and interest groups were undertaken to determine whether the absence of monitoring was permitting unnecessary adverse environmental effects.

Discussion of the Preliminary Findings - Review of the Environmental Statements

Ecological Impact Mitigation Measures Proposed in the Environmental Statements Reviewed

All of the ESs reviewed made reference to ecological impact mitigation measures. However, there was much variation between the types of measures proposed.

Measures aimed at **avoiding** impacts included: the provision of buffer strips, timing to avoid breeding seasons, water pollution prevention measures, fencing to protect sensitive areas and avoidance of flight paths (see Figure 16.1). A number of ESs (34%) stated that the remaining vegetation would be protected specifically during construction. However, only 10% actually stated that fencing would be provided. Approximately one third of the ESs reviewed proposed measures for preventing the pollution of water courses (30%), these included oil interceptors, settlement lagoons and bunding of construction compounds.

Although none of the ESs stated that the development had been sited on "least damaging criteria", a number did state that areas of higher ecological value would be avoided where possible. Siting to avoid the disturbance of sensitive areas was the third most common type of avoidance measure and was stated in 26% of the 50 ESs reviewed. "Siting" was generally proposed when the exact positioning of the development remained undecided at the time the ESs was being prepared. However, "siting to avoid sensitive areas" was often a general statement with limited delineation of the more ecologically sensitive areas.

Measures to **reduce** the severity of the impacts that could not be avoided included; appropriate construction methods to reduce impacts; warning markers on overhead power lines to reduce incidents of bird strike; tunnels for amphibians and badgers and translocation of habitats/species to areas unaffected by development (see Figure 16.2). Translocation was referred to in approximately one fifth (22%) of the ESs reviewed and ranged from individual species to whole habitats. Translocation is a measure that is increasingly proposed as a way of reducing the ecological impacts of development projects. However, the complexity of translocation should not be underestimated. "Translocation is rarely 100% effective" and problems with the success of translocation occur when those involved lack detailed ecological knowledge of the habitats and species requirements (Treweek, 1999). A number of factors are vital if the translocation is to have a chance of success including (Bullock, 1998):

- Selection, preparation and monitoring of an appropriate receptor site.
- Careful removal and handling of the habitat or species during translocation.
- Appropriate management and monitoring of the site post translocation.

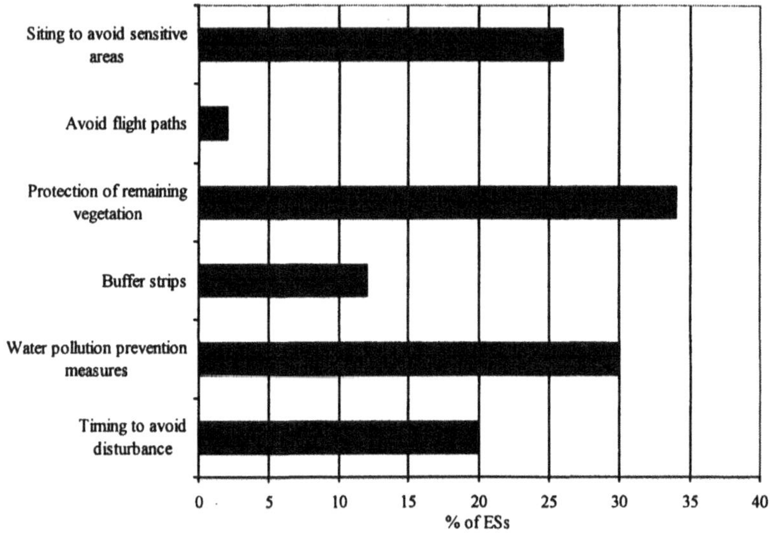

Figure 16.1: Measures proposed to avoid ecological impacts predicted in the ESs reviewed

Unfortunately none of the ESs reviewed in this study, which proposed translocation, considered any of the above factors. In addition none of the ESs predicted the success of the translocation or estimated the effect the translocation would have on the ecology of the receptor site.

Remedial and **compensatory** measures were proposed in a large majority of the ESs reviewed and included: landscaping/amenity planting; compensatory planting with native vegetation; management of remaining vegetation; reinstatement of soils and vegetation; habitat creation and restoration (see Figure 16.3). Habitat creation and restoration were proposed in a large majority of the ESs reviewed (59%) (see Figure 16.3). This percentage is comparable with research by Byron et al (2000) in which 53% of the ESs reviewed proposed habitat creation to mitigate ecological impacts. Both these studies revealed an increase in the percentage of ESs proposing habitat creation/restoration in comparison with Thompson et al's

(1997) review of ESs prepared between 1988 and 1993 in which 34% of ESs proposed habitat creation/restoration.

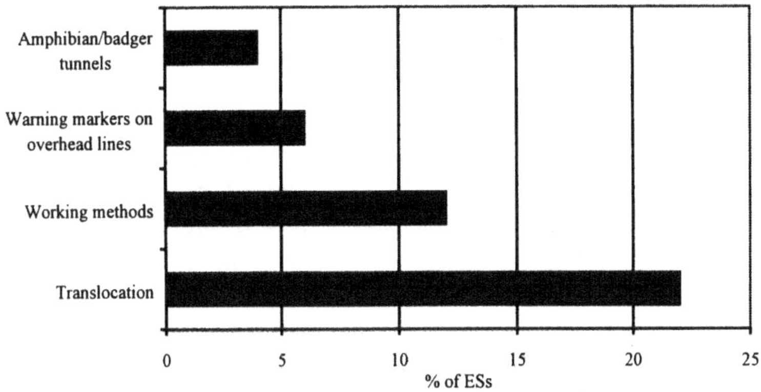

Figure 16.2: Measures proposed to reduce the severity of the ecological impacts predicted in the ESs reviewed

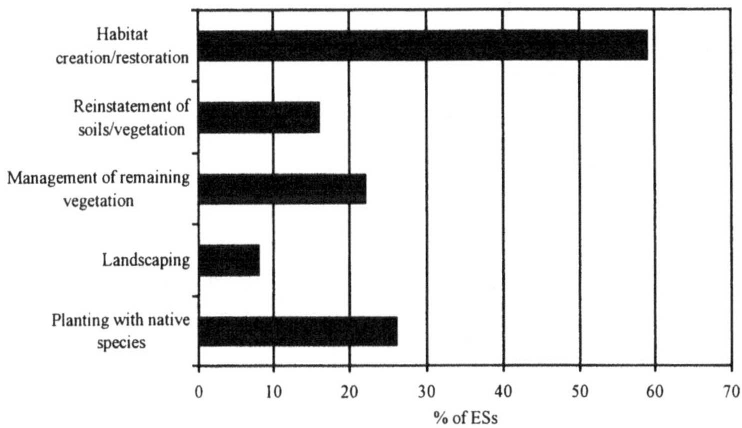

Figure 16.3: Measures proposed to remedy and/or compensate for the ecological impacts in the ESs reviewed

Although habitat creation can be successful in certain instances, particularly to re-create sites of low nature conservation value, it must not be used to justify adverse impacts on high biodiversity receptors (Byron, 2000).

Visual Impact Mitigation Measures Proposed in the Environmental Statements Reviewed

Visual impact mitigation measures were proposed in 92% of the ESs reviewed. This finding is comparable with the ESs reviewed by Mills (1992) in which 97% of the ESs reviewed made reference to visual impact mitigation measures. Of the 8% of ESs which made no reference to mitigation measures, two concluded that no visual impacts were predicted and two failed to undertake a visual impact assessment. The measures proposed to mitigate potential visual impacts were separated into two categories: those related to the design and positioning of the development and those relating to the "cosmetics" of the site, such as planting and landscaping.

Measures included in the former of these two categories were aimed at **avoiding** and **reducing** impacts through design (25%), re-routeing (10%) and positioning the developments to fit in with the existing landscape features such as landform and vegetation (10%) (see Figure 16.4). These measures were either stated in the ESs as having already been considered or recommended for consideration in the final design layout.

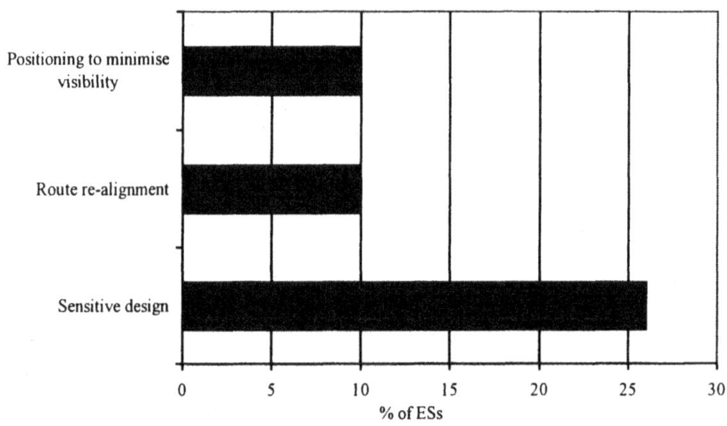

Figure 16.4: Measures proposed to avoid visual impacts in the ESs reviewed

Route re-alignments were proposed in ESs prepared for roads, pipelines and overhead power lines. Positioning, or siting, was proposed in 10% of ESs for developments in which no conclusion had been made on the exact layout of the development e.g., "the [windfarm] turbines will be sited to minimise visibility" (Yorkshire Windpower, 1990).

Approximately one quarter of the ESs reviewed (26%) stated that visual impacts would be minimised through "sensitive design" of the proposed development (see Figure 16.4). According to the Institute of Environmental Assessment and the Landscape Institute (1995) avoidance of an impact through design and consideration of the environmental constraints is the "ideal strategy for each identifiable impact". It is important for visual impact mitigation to be considered from the initial design stages of the development and not be viewed as the "green icing on a newly baked cake" by the developer (Goodey, 1995).

Mitigation measures relating to the cosmetics of the site with the purpose of screening visual receptors and **reducing** visual impacts of the developments included: landscaping (26%), screen mounding (16%) and screen planting (36%) (see Figure 16.5).

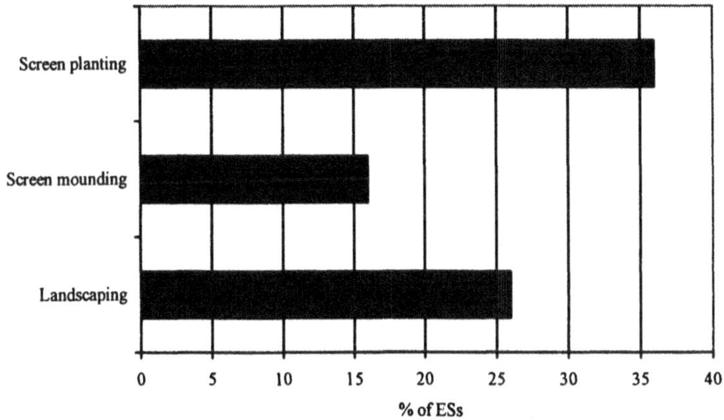

Figure 16.5: Measures proposed to reduce the visual impacts predicted in the ESs reviewed

Important factors to consider when planting is proposed as a measure to screen visual impacts is the time it takes for the planting to become effective and the seasonal differences in the effectiveness of the planting (Institute of Environmental Assessment and the Landscape Institute, 1995). The ESs reviewed in this study generally assumed that planting would become effective immediately. Only one ESs provided photomontages to demonstrate the effectiveness of the proposed planting at the time of commissioning and ten years later. None of the ESs reviewed made reference to the seasonal differences in the effectiveness of the planting proposed.

Ecological and Visual Impact Monitoring Proposed in the Environmental Statements Reviewed

Approximately one fifth (18%) of the ESs reviewed highlighted the need for ecological and/or visual impact monitoring, during the construction, operational and/or de-commissioning phases of the developments (see Figure 16.6). This percentage is higher than the results of previous ESs reviewed by Frost (1997), Thompson et al (1997) and Byron et al (2000). However, in the ESs reviewed in this study there was considerable variation between the type of monitoring proposed and the level of commitment to monitoring.

Monitoring to prevent the pollution of water-courses was the type of monitoring proposed in the most ESs (10%) (see Figure 16.6). This was generally in the form of a statement anticipating what would be required under pollution control legislation. This type of monitoring was recommended in ESs for waste projects i.e., landfill site, sewage treatment plant.

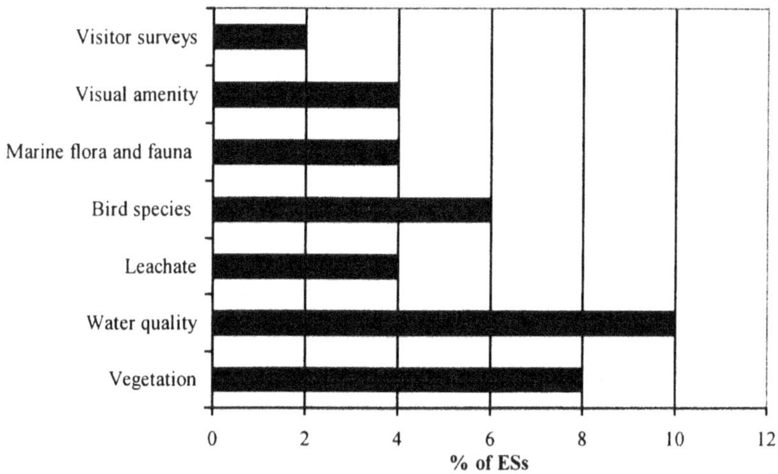

Figure 16.6: Type of ecological and visual impact monitoring referenced in the ESs reviewed

A minority (10%) of the ESs reviewed made reference to ecological impact monitoring. Ecological monitoring included bird, vegetation and marine flora and fauna surveys. Vegetation monitoring was recommended in the highest percentage of ESs (8%). Monitoring of bird populations was recommended in 6% of the ESs reviewed. However, only two of the ESs actually stated the species of bird to be monitored. Monitoring of marine flora and fauna was recommended in the ESs for the sea defence projects (4%).

Visual impact monitoring was recommended in a very small minority of the ESs reviewed (6%). One ES, of an afforestation project, stated that the visual appearance of the vegetation change would be monitored annually using repeat photography. Another ES stated that the success of the screen planting on the site would be monitored annually for a five year period. One ES, for a sea defence project, recommended monitoring the numbers of visitors using the beach before, during and after construction. Although this was not direct monitoring of the visual impacts it could be used as an indication of the effects of construction on visual amenity.

Although more ESs in this study made reference to ecological and visual impact monitoring than in previous ES reviews, the majority were recommendations rather than commitments. They failed to state specific information such as the species or area to be monitored, frequency and duration of the monitoring and methods to be used. Three (6%) of the ESs with references monitoring were more specific in stating the area and variables to be monitored, frequency and duration of the surveys and one also suggested survey methods. However, none of the ESs contained a comprehensive monitoring proposal which included the following points (Spellerberg,1991; Treweek, 1999; Glasson et al, 1994; Arts, 1998):

- aims and objectives,
- time-scale/duration,
- frequency,
- variables and indicators to be monitored,
- survey methods,
- costs,
- responsibilities for data collection,
- reporting procedures,
- remedial actions,
- data disseminated strategy.

Open-ended monitoring programmes can be costly and inconclusive, therefore, "the importance of identifying precise objectives cannot be over-emphasised" (Hellawell, 1991). The aims and objectives of the monitoring programme will subsequently determine the duration/timescale, frequency, variables and indicators to be monitored. Survey methods to be used in monitoring should be appropriate to the variable(s) being monitored.

Ideally standard methods should be used and detailed in the monitoring programme e.g., the British Trust for Ornithology's Common Bird Census Technique. The availability of funding for long-term monitoring programmes should be secured at the outset of the project, e.g., through planning or financial agreements. Responsibilities for data collection and reporting in the long-term should also be decided upon and clearly outlined in the monitoring programme. The reporting and interpretation of the significance of the data is essential for determining future management of the mitigation measures and the need for remedial measures. "In monitoring, one is concerned with setting limits and in deciding what action may be necessary when the monitoring reveals that the current situation is wide of the target" (Hellawell, 1991). Dissemination of information on the effectiveness of mitigation measures is essential to "fine-tune management operations and add to the sum of knowledge on the state of the art" (Gilbert and Anderson, 1999). The publication of results is particularly important if one of the benefits of monitoring - to provide information to strengthen the EIA and decision-making process - is to be realised.

Discussion of the Results - Analysis of the Implementation, Monitoring and Effectiveness of the Mitigation Measures

Case Study History

A 9.5km bypass was selected for case study analysis. Although the route for the bypass was determined in the early 1970s, limited public sector finance prevented the County Council from building the road until 1991 when financial agreements between private developers and the respective Council enabled work on the project to resume. During this 20 year period the route had been protected from development through the local plan process. However, various other planning permissions had been granted along the route boundary thus restricting modifications to avoid ecologically or visually sensitive areas.

Assessment of Ecological and Visual Impacts

Between 1989 and 1991 the ecological impact assessment and landscape proposals were prepared. The ecological impact assessment identified and predicted the effects of the bypass on primary and secondary ecological receptors and proposed measures to mitigate the impacts. The landscape proposals provided information on the landscape and visual receptors to be affected by the route and described the planting and design proposals to mitigate those impacts.

The ES revealed a number of areas of both visual and ecological interest affected by the route, including:

- an Area of Great Landscape Value;
- the green belt;
- public right of ways;
- residential properties;
- ancient and semi-natural broadleaved woodland;
- unimproved rough grassland;
- two major river courses;
- species-rich scrub;
- three pond clusters containing great crested newts;
- two badger setts.

A public inquiry was held in 1991 during which time concerns were raised regarding the presence of amphibians in the vicinity of the site, particularly the communities of great crested newts. Therefore the County Council commissioned consultants to undertake an amphibian survey which was carried out in 1992. The impacts of the bypass on the amphibian population were predicted and an extensive amphibian mitigation package, the largest at the time for a road scheme, was put together.

Further ecological surveys revealed a number of badger setts close to the route, some of which would be directly affected by the bypass. Apart from directly destroying a number of the badger setts, foraging routes would be severed, feeding territory isolated and foraging habitat reduced. These effects combined with the direct road kills were predicted to significantly reduce the badger population in the area.

Monitoring the Implementation and Effectiveness of the Mitigation Measures

Visual The effectiveness of the visual impact mitigation measures was assessed at the locations identified in the ES where it was predicted that properties and valued landscapes would be affected by visual intrusion.

In general, the deep cuttings and planted mounds were effective in screening the road from many of the properties along the route and reducing the visual impacts. However, in areas where the cuttings were shallow the road was visible, particularly in winter when the vegetation planted would not provide an effective screen.

The Area of Great Landscape Value was most affected by the road. The sensitivity of the landscape combined with the magnitude of the development resulted in significant visual intrusion. Trees and shrubs were planted to reinforce the existing vegetation in an attempt to reduce the visual intrusion. However, the combination of hard painted footpaths, linking up the route of the valley severed by the road, and vandalised concrete bridge structure do little to merge the road into the existing environment and cause significant visual intrusion.

Funding for three years landscape maintenance post-construction was included in the budget for the bypass, therefore until March 1999 the vegetation had been managed and regularly maintained by the County Council's Environmental Planning Department. Following this date the funding for vegetation maintenance expired and it is now the responsibility of the Highways Department to finance further landscape management from the Highways Maintenance Fund. Although the Landscape Management Report, prepared by the Environmental Planning Department and presented to the Highways Department in March 1999, recommended a further 25 years landscape management it would seem that due to financial constraints landscape management is low on the list of priorities for the Highways Department.

Ecological Many of the ecological mitigation measures were non-specific and referred to compensatory planting with native species or, where water courses had to be culverted, "early and adequate culverting" with additional measures to prevent run-off. The measures to compensate for the loss of woodland, scrub and unimproved grassland were implemented as part of the landscape proposals and thus consisted of mainly amenity planting. This was checked and managed as part of the landscape management scheme. However, was no monitoring or surveys of the ecological value of the remaining or compensatory vegetation provided.

The mitigation measures proposed for the badgers and great crested newt population were more specific as these species are both afforded legal protection.

The Protection of Badgers Act 1992 makes it a criminal offence to directly injure badgers or disturb an inhabited badger sett. Thus, a number of measures were implemented to reduce impacts of the bypass on the badger population: 3,300m of badger fencing was provided in the areas where there were known setts and two 50m tunnels were provided under the bypass to provide routes to foraging areas on the other side of the bypass. However, no monitoring of either the use of the badger tunnels or badger numbers after construction has taken place and the effectiveness of the mitigation measures could not be assessed.

The great crested newt and its habitat are protected under the Wildlife and Countryside Act 1981 and The Conservation (Natural Habitat and c) Regulations 1994 (SI. 2716). Therefore the County Council were required by law to relocate and provide habitat for the amphibian populations affected by the bypass. The mitigation package included the creation of new ponds, restoration of existing ponds, capture and release of approximately 11,000 amphibians and the construction of amphibian tunnels.

Consultants were commissioned to monitor the population changes of amphibians in the new and restored ponds during the period 1992-1996. In 1996, despite considerable annual variation, great crested newt populations were larger than prior to construction and it was clear that all replacement ponds were being used in 1996. However, the results fluctuated and it was impossible to determine any significant trends. Monitoring of the use of the amphibians tunnels took place in spring 1996 and concluded that although there did not seem to be any fundamental problems with the tunnel design, only small numbers of amphibians were actually passing through them. The low usage was attributed to the low numbers of amphibians on the east side of the road. It was predicted in 1996 that as the amphibian population increased and the loss of terrestrial habitat continued, as a result of further development, more amphibians would be forced to use the tunnels to search for new areas of terrestrial habitat.

With regard to long-term monitoring, the original, restored and new mitigation ponds are located outside of the boundary of the land controlled by the County Council's Engineering Services and therefore they have no responsibility for the maintenance of the ponds. As many of the ponds are in public ownership, on-going management of the ponds was not guaranteed and since 1996 no maintenance of the ponds or surrounding habitat or monitoring of the great crested newt population has taken place. In addition, although subsequent development has taken place further reducing terrestrial habitat around the ponds provided for the great crested newts, no surveys have been undertaken to monitor the usage of the amphibian tunnels.

Conclusions

All of the ESs reviewed acknowledged the need for ecological impact mitigation measures. Although measures to avoid impacts were proposed in a number of ESs, detail relating to the implementation of these measures was limited. Translocation of species and/or habitat was the most common method proposed to reduce the severity of impacts. However, as with avoidance measures, detail of the translocation proposed was limited. The most common compensatory measures were those for habitat creation or restoration. Again, detailed information relating to the implementation, management and effectiveness of the habitat creation was consistently omitted.

Visual impact mitigation measures were proposed in a large majority of the ESs reviewed. The measures most commonly proposed were those aimed at reducing visual impacts through screen planting and landscaping. However, limited consideration was given to the time required for the planting to become effective or seasonal differences in the effectiveness of the planting.

With regard to ecological and visual impact monitoring, only a limited number of ES made reference to the need for ecological and/or visual impact monitoring, of those that did:

- water quality and vegetation surveys were the variables most commonly recommended for monitoring;
- limited consideration was given to the monitoring of visual impacts;
- recommendations were non-specific and none of the ES included a comprehensive monitoring proposal.

Examination of the case study revealed that the visual impact mitigation measures proposed were implemented as part of the landscape proposals for the scheme. The planting provided a visual screen during the summer months but little consideration had been given to seasonal effects. The deep cuttings were effective in reducing the visual intrusiveness of the road, particularly where they were combined with planted buffer zones between the residential properties and the road. However, due to the lack of any long-term commitment to management of the vegetation by the developer, it was difficult to predict the future success and effectiveness of the planting.

With regard to the ecological impact mitigation measures, the planting proposed to compensate for the loss of habitat (woodland, scrub, and unimproved grassland) was provided as part of the landscaping proposals. Thus, the emphasis was on planting for visual amenity purposes and to provide visual screens rather than to provide ecologically valuable habitat. In addition the planting has not been monitored or surveyed for its ecological value.

In contrast, the measures proposed to mitigate the effects on the amphibians were implemented as part of a specific package. A significant amount of time and money was invested in the creation of new ponds, habitats and tunnels and the translocation of the amphibians. Unfortunately the lack of adequate baseline data and the short time-scale of the monitoring limited the usefulness of the data collected. It was impossible to identify long-term trends and thus difficult to determine whether fluctuations were as a result of the project or not. In addition the limited duration of amphibian monitoring failed to demonstrate the success of the mitigation measures in the long-term.

With regard to the badgers, there were no monitoring results for determining the success of the measures to reduce the impacts on the population.

In summary, these results have revealed that the implementation of ecological and visual impact mitigation measures and monitoring of the effectiveness were rarely covered in detail in the ESs reviewed. However, analysis of the case study revealed that, despite the lack of attention to implementation in the ESs, the proposed visual impact mitigation measures were generally well implemented. Similarly, although impacts on the amphibian and badger populations were not assessed in the ES, mitigation measures were specified in detail after the ES had been submitted and implemented accordingly.

The effectiveness of the mitigation measures in the long-term could not be determined due to the lack of monitoring of the badger population and the inadequacy of the amphibian monitoring data collected. This case study has emphasised the importance of having a well-designed monitoring programme with clear aims and objectives, adequate long-term funding and open and regular reporting procedures.

References

Arts, J. (1998), *EIA Follow-up*, Geopress, Groningen.

Bisset, R. and Tomlinson, P. (1988), 'Monitoring and auditing of impacts', in Wathern, P. (ed.), *Environmental Impact Assessment: theory and practice*, Unwin Hyman, London.

Bullock, J.M. (1998), 'Community translocation in Britain: setting objectives and measuring consequences', *Biological Conservation*, vol. 84, pp. 199-214.

Byron, H. (2000), *Biodiversity and Environmental Impact Assessment: A Good Practice Guide for Road Schemes*, RSPB, WWF-UK, English Nature and the Wildlife Trusts, Sandy.

Byron, H.J.; Treweek, J.; Sheate, W.R. and Thompson, S. (2000), 'Road developments in the UK: an analysis of ecological assessment in environmental impact statements produced between 1993 and 1997', *Journal of Environmental Planning and Management*, vol. 43, no.1, pp. 71-97.

Department of the Environment (DoE) (1989), *Environmental Assessment: a guide to the procedures*, HMSO, London.

Department of the Environment, Transport and the Regions (DETR) (1997), *Mitigation Measures in Environmental Statements*, HMSO, London.

Department of the Environment, Transport and the Regions (DETR) (1999), Circular 02/99, DETR, London.

English Nature (1994), *Nature Conservation in Environmental Assessment*, English Nature, Peterborough.

Frost, R. (1997), 'EIA monitoring and audit', in Weston, J. (ed), *Planning and Environmental Impact Assessment in Practice*, Longman, Essex.

Gilbert, O.L. and Anderson, P. (1999), *Habitat Creation and Repair*, Oxford University Press, Oxford.

Glasson, J., Therivel, R. and Chadwick, A. (1994), *Introduction to Environmental Impact Assessment*, UCL Press, London.

Goodey, B. (1995), 'Landscape', in Morris, P. and Therivel, R. (eds.), *Methods of Environmental Impact Assessment*, UCL Press, London.

Hellawell, J. M. (1991), 'Development of a rationale for monitoring', in Goldsmith, F. B. (ed.), *Monitoring for Conservation and Ecology*, Chapman and Hall, London.

Institute of Environmental Assessment and The Landscape Institute (1995), *Guidelines for Landscape and Visual Impact Assessment*, E & F. N. Spon, Chapman and Hall, London.

Mills, J. (1992), *The Adequacy of Visual Impact Assessments in Environmental Statements*, unpublished Msc. dissertation, School of Planning, Oxford Brookes University, Oxford.

Parker, D. M. (1995), *Habitat Creation – a critical guide*, English Nature, Peterborough.

Roads and Hydraulic Engineering Division (1995), *Nature Over Motorways*, Ministry of Transport, Public Works and Water Management, Delft.

Sadler, B. (1996), *Environmental Assessment in a Changing World: Evaluating Practice to Improve Performance*, Final Report, International Study of the Effectiveness of Environmental Assessment, Canadian Environmental Agency, Ottawa.

Spellerberg, I. F. (1991), *Monitoring Ecological Change*, Cambridge University Press, Cambridge.

Thompson, S.; Treweek, J. R. and Thurling, D. J. (1997), 'The Ecological component of environmental impact assessment: a critical review of British environmental statements', *Journal of Environmental Planning and Management*, vol. 40, no. 2, pp. 157-171.

Treweek, J. (1999), *Ecological Impact Assessment*, Blackwell, Oxford.

Treweek, J. and Thompson, S. (1997), 'A review of ecological mitigation measures in UK environmental statements with respect to sustainable development', *International Journal of Sustainable Development and World Ecology*, vol. 4, pp. 40-50.

Wood, C.; Dipper, B. and Jones, C. (2000), 'Auditing the assessment of the environmental impacts of planning projects', *Journal of Environmental Planning and Management*, vol. 43, no. 1, pp. 23-47.

Yorkshire Windpower (1990), *Ovenden Moor Windfarm*, Environmental Statement.

17 Brownfield Land: Owner Characteristics, Attitudes and Networks

DAVID ADAMS, ALAN DISBERRY,
NORMAN HUTCHISON AND THOMAS MUNJOMA

Introduction

Over recent years, the debate on brownfield land has intensified. There has been increasing recognition of the crucial role of land ownership in strategies and programmes designed to achieve urban regeneration. It has become apparent that the implementation of redevelopment projects depends very much on successful land assembly including the acquisition of lesser ownership interests. Both costs and delays in concluding this crucial phase in the development process can harm the feasibility of redevelopment projects and prevent the realisation of redevelopment opportunities.

This chapter examines brownfield ownership interests in selected cities of the United Kingdom within the context of "growth coalition theory" as developed from the United States. The growth coalition approach regards ownership interests as dominant forces in local economic growth. The chapter thus concentrates on the characteristics, attitudes and networks of owners of brownfield land in the UK, exploring their contribution to local property development and economic growth. The rationale for investigating land ownership interests is that it improves understanding of the operations of land and property markets in relation to matters of wider economic importance.

The chapter is based on research on ownership interests in 20 potential redevelopment sites in each of four selected cities in the United Kingdom. The results establish clear patterns of owner characteristics and relationships with other property-related interests. These findings have important repercussions for the application of the concept of growth coalitions in the United Kingdom and have wider significance regarding the strategies and mechanisms appropriate to the redevelopment of land and the regeneration of localities.

Landowners, Growth Coalition and Urban Networks

Landowners do not represent a homogenous category of property interests. The plethora of approaches to land ownership bears testimony to the range of attempts to reveal the significance of such interests in both the property development process and local economic development. In addition to conferring economic and political power, land ownership shapes many aspects of urban development. Kivell (1993), for example, contends that the principles and supposed effects of ownership are hotly debated from time to time but empirical evidence is thin.

Adams and May (1991) attempt a classification of landowners based on their contributions in seeking to promote the development of their land. While "active" landowners take action that leads ultimately to the development of their land, "passive" landowners hold on to their land without development. More recent research has revealed a range of ownership constraints to redevelopment (Adams *et al.*, 1999). Some of the constraints identified are clearly the hallmarks of passive ownership interests. In its vision of an urban renaissance, the Urban Task Force (1999: 250) recognised the harm such ownership impediments can cause to redevelopment potential, and recommended that compulsory purchase legislation be consolidated and streamlined. In the subsequent Urban White Paper for England (Department of the Environment, Transport and the Regions, 2000), the Government accepted this recommendation and promised the necessary legislation as soon as Parliamentary time allowed.

Since land ownership is regarded as pivotal to local economic development and growth, the strategies and actions of land demand investigation. The focus on ownership interests as the engines for both the initiation and promotion of local economic growth is embraced in the concept of "growth coalitions". The growth coalition model has been developed within the American context of economic growth as the essence of virtually any given locality (Molotch, 1976). Despite its distinctive American origin and flavour, the concept shows relevance to situations in the United Kingdom where it has spurred research interest, particularly in urban regeneration and local economic growth (Harding, 1991; Lloyd and Newlands, 1990).

The fundamental proposition of the concept of growth coalition is that the political and economic forces in any one locality co-operate to promote local economic growth (Molotch, 1976). The "growth imperative" is the focal issue upon which local forces of different political flavour and diverse social interest are mobilised and achieve consensus. It is this unity of purpose that knits together various private and public sector interests and strategies.

The growth coalition, therefore, represents an alliance of such interests whose common agenda is the achievement of local economic growth. However, it is precisely the need to achieve growth that constrains the range of feasible options for local social and economic reform.

The growth coalition approach emphasises the importance of property-based interests, in particular land ownership, in the promotion of economic growth. Unlike other forms of capital, land is immovable. Ownership interests are place-bound since they relate to specific sites within certain locations. They derive profits from an intensification of the use of their land and buildings. Growth coalition theory therefore suggests that these interests coalesce and remain at the forefront of strategies designed to achieve economic growth, the benefits of which accrue to coalition members. Other forms of capital are relatively mobile, making transfer by owners possible from region to region, depending of course, on their assessments of the strength of development opportunities.

However, there are other interests, besides ownership, that are related to property directly or indirectly. These include organisations that derive benefit from construction and servicing of property. Additional interests include those concerned with regulations regarding property ownership and development, and a much wider variety of organisations that derive benefits from local economic development. In growth coalition theory, ownership interests in any one locality are deemed to be consciously working to achieve growth even in competition with others in different locations. The place-bound interests attempt to attract mobile forms of capital from both the private and public sectors through the establishment of stable networks and political alliances.

Landowners are often at the core of the alliance of interests that constitute the growth coalition and whose purpose is to promote local economic growth. They can, therefore, be expected to have established an array of formal and informal relationships and contacts with the objective of enhancing their interests through local economic growth. Such contacts would seek to redevelop the properties in question or to enhance their exchange values. The nature and extent of the contacts remain diverse, and in some cases are constituted through fluid relationships that may be hard to prove (Lowndes, *et al.*, 1997). Networks are dynamic due to the changing intensity and nature of individuals' relationships with other members within the network.

The importance of networking by any one ownership interest reflects the weight of property development and management within the organisation's strategy. Organisations whose interests in property are incidental to some primary interest are less inclined to take the initiative in the development of their properties unless there are reasons compelling them to do so. On the other hand, organisations that have property development interests that are central to their business can be expected to be actively seeking to redevelop their properties.

In addition, the pressure to develop a particular piece of land derives from the expected marginal impact of that decision on the organisation's property portfolio. Harding (1991: 298) contends that "The commitment to redevelopment on the part of owners will also reflect the relative importance of particular holdings for the larger interest in question." The more significant property holdings become within an organisation's assets, the more likely are properties to be actively managed so as to realise overall benefit to the portfolio. On this basis, it is important to set discussion of owner networks within the context of owner characteristics and attitudes. After explanation of the research method in the next section, the chapter will follow this approach.

Research Method

The research on which this chapter is based explored the characteristics, attitudes and networks of 120 owners of vacant urban land or obsolete urban buildings in four British cities. Two of these cities, Aberdeen and Nottingham, were selected for investigation because they had witnessed strong development pressure in the previous decade, while the other two, Dundee and Stoke-on-Trent, were picked because they had experienced weaker development pressure. The choice of two Scottish and two English cities was also intended to reflect significant differences in development policies and land law north and south of the border.

The research concentrated on the owners of 20 substantial redevelopment sites[1] in each city which remained undeveloped at 31 December 1995 and which were of at least 2 hectares in area or on which at least 5,000m^2 of gross floorspace was then under active consideration. Although such larger sites can make a disproportionate contribution to urban regeneration, they were favoured in the research design because, despite notable exceptions, the more significant redevelopment is, in terms of area and floorspace, the greater the number of existing owners likely to be affected and the more complex the process of negotiation with each one.

Such potential redevelopment sites were identified through discussions with local authorities, prominent chartered surveyors, Chambers of Commerce and, in Scotland, local enterprise companies. Across the four cities, only 88 sites that fully met the research criteria were discovered by this exercise. Random sampling reduced the number for detailed investigation to 80, or 20 in each city.

The research aimed to contact all non-residential owners of freehold or long leasehold (above 99 years) interests in English sample sites on 31 December 1995, or of corresponding interests in Scottish ones, together with those who had owned such interests during the research period of between 1991 and 1995[2]. Although many potential owners were identified through local authorities, chartered surveyors or informal on-site enquiries, a more comprehensive picture of ownership was obtained only through formal searches (at negotiated fees) to the Land Registry in England and Register of Sasines in Scotland.

Exhaustive probing and cross-checking of these various sources of information revealed an initial 298 separate ownership interests in the 80 research sites. However, more detailed investigation, including direct contact with potential owners, eliminated 86 of these interests as inapplicable because, for example, they either owned the site in question outside the period 1991-95, or owned land beyond the site.

Successful contact was made with 140 of the 212 owners in the final research population, representing a response rate of 66%. As Table 17.1 reveals, a full research questionnaire was completed for 120 of these 140 owners (normally by a member of the research team at interview, although occasionally by owners themselves and returned through the post). More limited notes were obtained from the other 20 owners successfully contacted, usually through telephone conversations but occasionally from meetings with a member of the research team.

Only 13 owners openly refused to participate. A further 59 owners proved impossible to contact, including 30 who had already relinquished ownership by the end of 1995. Several of these were businesses that had previously gone into receivership. Such earlier ownership was particularly significant in accounting for the proportionately lower response rate in Stoke-on-Trent. Of the other 29 owners not contacted, many were small companies in owner-occupation who failed to respond to the persistent efforts of the team. This was particularly so in Dundee.

The results presented in this chapter refer to the 120 owners for whom a full research questionnaire was completed. Although the research was designed to ensure that owners investigated were reasonably representative of those who own interests in large redevelopment sites in British cities, there is likely to be some limited bias in the results against smaller companies and private individuals, owing to the greater difficulties experienced in successfully contacting them. This should be borne in mind in interpreting the results.

Table 17.1: Owner response rates

	Aberdeen		Dundee		Nottingham		Stoke		Total	
	No	%	No	%	No	%	No	%	No	%
Successful Contact										
Full questionnaire	29	83	35	48	30	63	26	46	120	57
Notes from owner	4	11	3	4	7	15	6	11	20	9
Non Response										
No contact	1	3	28	38	9	19	21	38	59	28
Refused	1	3	7	10	2	4	3	5	13	6
Total	35	100	73	100	48	100	56	100	212	100

Owner Characteristics

At a rudimentary level, it was possible to divide the 120 owners in two different ways. As Table 17.2 shows, it is first evident that private-sector owners (72.5% of total) significantly outnumbered public-sector owners (27.5% of total). This immediate discovery runs counter to the popular and political perception that most brownfield land is in public ownership. However, it will be noted that the 87 private-sector owners identified in Table 17.2 contained 17 privatised PLCs, which included such well-known companies as British Coal, British Gas, Powergen and Railtrack. In a sense, then, the apparent dominance of the private sector in the ownership of substantial brownfield sites within Aberdeen, Dundee, Nottingham and Stoke-on-Trent partly reflects the privatisation of such land during the late 1980s and early 1990s.

The second rudimentary division apparent from Table 17.2 is that between large and small owners. Here a more even split is evident, with larger owners (54.2% of total) just outnumbering smaller ones (45.8%). While the research suggested that private individuals and trustees were almost insignificant as owners of large brownfield sites, private companies were the single most important owners identified. Moreover, as already suggested, private companies may well be under-represented in the research analysis, as they were often hard to contact, especially in Dundee. However, the simple division of companies into size categories on the basis of whether or not they are registered as PLCs is fraught with difficulties. Indeed, as the research showed, some of the private companies investigated played quite a substantial role in local, if not national, land and property markets. One private developer/dealer operating in Dundee, for example, was instrumental in finding new retail, leisure and residential uses for three former industrial sites in the city investigated in the research.

Table 17.2: Rudimentary owner typology

		No	%
Small private-sector owners	Private individual/trustee	6	5.0
	Private company	49	40.8
Large private-sector owners	Always private-sector PLC	15	12.5
	PLC privatized	17	14.2
Public-sector owners	Government or public body	11	9.2
	Local authority	22	18.3
Total		120	100.0

To evaluate whether and how far these 120 owners of vacant urban land or obsolete urban property encouraged or impeded redevelopment, detailed information was gathered on their:

- strategies, marketing attempts and site valuations;
- knowledge of development constraints and any action taken to resolve them;
- network of contacts;
- awareness of, and potential influence over broader structural factors;
- reaction to possible policy changes;
- particular legal and personal characteristics.

A qualitative analysis was then undertaken, owner by owner, with the results compared, and if necessary adjusted, to ensure consistency. Since each redevelopment site is unique, what would most encourage redevelopment must vary from site to site. Owners were therefore considered to have significantly encouraged redevelopment if, at some time from 1991 to 1995, they had themselves sought to resolve at least the most important of the five established tests of development feasibility (Barrett et al., 1978) for their particular site[3]. The distinctiveness of each site ensured that, on one, for example, the most significant contribution an owner could make to redevelopment was the fervent pursuit of planning permission through recourse to appeal, while on another where planning permission was relatively simple to obtain, the owner's energies needed to be concentrated instead on securing development grants or subsidies.

The results of this analysis are shown in Table 17.3 (see appendix) by owner type. Overall, the table reveals that 58% of the 120 owners analysed encouraged or significantly encouraged redevelopment during the research period of 1991 to 1995. A further 28% had a neutral impact on redevelopment and only 14% discouraged or significantly discouraged redevelopment. These results are not significantly different from those for an extended sample of 155 owners reported in a separate paper (Adams *et al.*, 1999).

What is interesting from Table 17.3 is the differential pattern of owner impact on redevelopment prospects between the various owner types. Overall, for example, 73% of public sector owners encouraged or significantly encouraged redevelopment in comparison with 53% of those in the private sector. Local authorities and other public bodies such as local enterprise companies, were usually highly active in pushing redevelopment forward. Within the private sector, combined scores for significant encouragement and encouragement were recorded of 67% for PLCs that had always been in the private sector, 53% for private companies, 47% for privatised PLCs and only 33% for the few trustees/private individuals investigated.

To highlight the overall importance of brownfield sites within ownership portfolios, owners were asked to estimate the value of the each site as a proportion of their overall assets and identify the number of other potential development they owned within a five mile distance. Those who assessed the relative value of the particular site at 5% or less of their total assets accounted for two-thirds of the 110 landowners prepared to answer this question. This proportion rose to 80% of owners for PLCs that had always been in the private sector and to 100% for both local authorities and privatised PLCs. In contrast, it fell to 42% for private companies and to zero for private individuals and trustees. However, while all the private individuals and trustees who responded to this question valued the particular site at between 6% and 25% of their total assets, private companies displayed the most diversified response of any group, with 11% suggesting that the particular site was worth 26% to 50% of their total assets and a further 13%

reporting that the site to be worth more than half of their total assets.

As might be expected, all the local authority respondents owned at least 10 more potential redevelopment sites within a five mile distance. Conversely, for all other groups, 80% or more of respondents held three or fewer potential redevelopment sites within five mile of the particular site investigated in the research. Private individuals and trustees were least likely to hold another redevelopment site within this distance. These findings suggest that local authorities, as landowners, were likely to be far more experienced in dealing with brownfield redevelopment than any other landowner type.

In summary, this section demonstrates that, although the private sector (including privatised PLCs) dominates the ownership of large brownfield sites, individual private-sector owners have significantly less experience of dealing with such sites than local authorities. In addition, public-sector owners as a whole appear keener to promote brownfield redevelopment than those in the private sector. However, as a general rule within the private-sector, the larger the organisation, the better experienced and more well-disposed owners appear to be to redevelopment.

For example, of the PLCs questioned that had always been in the private sector, 54% held at least 2 other potential redevelopment sites within a five mile distance, 67% encouraged or significantly encouraged redevelopment and 90% considered the particular site investigated to be worth 5% or less of their asset base. In contrast, the few private individuals and trustees encountered appeared more likely to regard the particular site as financially important within their asset base and as a whole, to be less encouraging towards redevelopment than the private sector as a whole. Privatised PLCs, however, remain an exception to this general rule since they normally had greater experience of dealing with brownfield land than the private sector as a whole and held sites that they regarded as financially inconsequential in relation to their overall asset base, but were less likely than the average private-sector owner to encourage or significantly encourage its redevelopment.

Owner Attitudes

To help understand their attitudes towards the broader context for land management and development, the 120 brownfield landowners were first asked whether or not it really mattered, either to themselves or others, that the particular site owned had not recently been used to its full potential. Of the 93 respondent owners, 59% felt that it mattered and 41% that it did not. However, this overall response disguised a marked variation between owner types since the proportion of respondents who considered that under-use mattered was recorded at 81% for public-sector owners, 57% for small private-sector owners and only 46% for large private-sector owners. Within the latter category, the figure for privatised PLCs fell to 39%.

Of course, debates on the importance of brownfield land redevelopment are much related to wider notions of environmental sustainability and to the appropriate balance between private initiative and state intervention in land management and development. The 120 owners were therefore asked to state whether they agreed or disagreed with eight "value statements" deliberately constructed to enable their attitudes towards these broader matters to be identified. The results are shown in Table 17.4.

As Table 17.4 shows, the value statement recording the highest level of agreement among respondent owners (at 94%) was "Customer satisfaction in new development depends very much on the skill and flair of the developer." Only four negative responses were recorded to this statement, all from privatised PLCs. Otherwise, there was no significant difference in owner responses between public-sector and private-sector owners. Indeed, public-sector respondents were almost all in agreement with this statement, perhaps indicating their lack of confidence in the ability of public policy to assist customer satisfaction in new development.

The second highest level of agreement (at 86%) concerned the statement that "Those whose cause dereliction should pay for cleaning it up", reflecting the increasingly widespread acceptance of the "polluter pays" principle. The highest level of agreement for this statement came from small private-sectors owners with it securing unanimous support among private individual/trustee respondents and virtually unanimous support among private companies. However, the level of support among public-sector and large-private sector owners fell to 80% and 74% respectively, perhaps indicating their fuller awareness of some of the practical difficulties faced in implementing the "polluter pays" principle.

The statement that "Compulsory purchase should always be regarded as a last resort" received the third highest level of agreement among respondents at 80%. There was very little difference between public- and private-sector responses to this statement. Even local authority respondents recorded a 75% rate of support for such limited use of compulsory purchase.

Table 17.4: Owner response to attitude statements

	Agree		No Opinion		Disagree		Total		No Response
	No	%	No	%	No	%	No	%	
Customer satisfaction depends on developer	108	94	2	2	5	4	115	100	5
Those who cause dereliction should pay for it	99	86	4	3	12	10	115	100	5
Compulsory purchase should always be regarded as last resort	93	80	6	5	17	15	116	100	4
What happens on private land is of public interest	76	66	14	12	25	22	115	100	5
State interference on land should be minimal	73	63	6	5	36	31	115	100	5
No greenfield development while urban land vacant	60	54	10	9	42	38	112	100	8
Green belts one of great planning successes	63	54	34	29	19	16	116	100	4
Public concern for environment gone too far	29	25	18	16	69	59	116	100	4

Support levels then started to fall, with the statement that "What happens on private land is a matter of legitimate public interest" recording only a 66% level of agreement. Here, however, the statement received much stronger levels of support among public-sector respondents (at 81%) than among small private-sector ones (at 66%). The least level of support for this statement (at 52%) was found among large private-sector respondents, irrespective of whether they had been privatised or not.

However, the related statement that "State interference in land management and development should be minimal" drew the agreement of

64% among respondents, reflecting the widespread belief among brownfield land owners that while state intervention in land management may be legitimate in principle, it should in practice be kept to a minimum. The highest level of hostility to state interference was recorded among small private-owners, with 73% of the private companies and 83% of the private individuals and trustees who responded to the statement supporting the position of minimal state intervention. This may well reflect a perception among such smaller private-sector owners that state interference is rarely to their advantage. In contrast, support for minimal state intervention fell to 58% for public-sector respondents and 50% for those large private-sector ones.

Turning to broader issues of environmental sustainability, the statement that "No development should take place on greenfield sites, while there is vacant land within cities" was supported by 54% of respondent owners. However, while this outcome was matched exactly by that for large private-sector respondents, smaller private-sector respondents were even more opposed to greenfield development, recording a figure of 63% in support of the statement. In contrast, almost two-thirds of public-sector respondents disagreed with the statement, with the figure for local authority opposition reaching 70%. This may well reflect the widespread awareness within public decision-makers that, much as such a policy might prove favourable to public opinion, entire reliance on brownfield sites to meet future land demands would be unrealistic.

Overall, the statement that "Green belts have been one of the great successes of postwar British planning" generated a similar mixed response. Although the average level of agreement was again 54%, smaller private-sector respondents (63% in favour of statement) appeared considerably more impressed by the success of green belts than larger private-sector respondents (52% in favour) or those in the public sector (only 42% in favour). Local authority owners, in particular, appeared undecided about the success of green belt policy, with 55% of these respondents having no particular opinion on the matter!

Finally, the statement that "Public concern for the environment has gone too far" was widely rejected, with only 25% of respondents in favour. However, significant variations in response were recorded between owner types. While no central government or public body agreed with this statement, 30% of local authority respondents did so. This was exceeded only by private companies (40% in favour) and private individuals and trustees (50% in favour).

It is evident from the above analysis that no clear distinction emerged between public and private-sector owners of brownfield land in their attitudes towards the broader context for land management and development. Both appeared to set limits on state intervention in land management and development and subscribe instead to notions of owner responsibility and developer initiative. Indeed, differences of attitude between large and small private-sector owners were often more interesting than those between the public- and private-sectors. For example, smaller private-sector owners were more "hawkish" than larger private-sectors ones in their support of the "polluter pays" principle, in their belief that no development should take place on greenfield sites while there is vacant land within cities and in the success they attributed to green belts. In each of these cases, the attitude of large private-sector respondents was closer to that of local authorities.

However, the position was reversed in relation to more general public concern for the environment, where public-sector and larger private respondents took the more "hawkish" stance. It is perhaps ironic that, despite the majority view among all brownfield owners investigated that compulsory purchase should be regarded as a last resort and that state interference should be minimal, the development prospects of many of the sites investigated were dependent on state support, either financially or in the form of land assembly.

Owner Networks

To discover how extensive were their networks of contacts, the 120 brownfield owners were asked to identify the extent and productivity of their contacts with 23 different categories of people and organisations. As Table 17.5 shows, a clear distinction emerged in the results between what might be considered a close-knit property development network into which the respondents landowners appeared well connected and a much broader policy and economic development network in which their connections were far looser.

The property development network appeared to comprise seven key contacts, with whom at least half of all the respondent brownfield owners had at some stage been in contact. The most frequent link proved to be the local planning department, which was contacted by 82% of all respondent owners, with 61% describing the contact as productive or highly productive. In most cases, local planning authorities are keen to promote brownfield redevelopment and their attitude was therefore likely to be viewed favourably by the particular owners interviewed for this research. However, on occasions, considerable conflict arose between respondents owners and planning authorities on the perceived potential for retail redevelopment, with the authorities seeking to concentrate such development in town and city centres against the will of site owners. It is likely that such policies accounted for most of the occasions where contact with the planning department was viewed as unfavourable by respondent landowners.

Professional consultants formed the next three members of the property development network with whom landowners were regularly in contact. Chartered surveyors and property agents play a diverse role in the development process, including offering advice on land valuation, marketing, development finance, and negotiation on price and rental levels. They were contacted by 74% of all respondent owners, with 60% describing the contact as productive or highly productive.

It is perhaps surprising to find that the next most frequently contacted profession by landowners were architects or planning consultants but it should be remembered that the research definition of owners including property development companies, who would use the services of architects in testing development feasibility. Planning consultants provided valuable support for owners in challenging unfavourable planning decisions, especially in relation to retail redevelopment. Overall, architects and planning consultants were contacted by 65% of all respondent owners, with 57% describing the contact as productive or highly productive.

An almost similar figure was recorded for lawyers who offered important advice to brownfield owners in relation to purchase and sale, planning difficulties and increasing on the technicalities of contaminated land legislation. Elsewhere in the research, the detrimental impact of ownership constraints to urban redevelopment was noted (Adams *et al.*, 1999). In this context, lawyers played an important role in clarifying unknown ownership, overcoming partial title and drawing together the bundle of ownership rights by ensuring, for example, that any leases and licenses on site were effectively terminated.

In this context, it is interesting to note that neighbouring landowners were contacted by 60% of all respondent owners, but that only 38% described such contact as productive or highly productive. This reflects the problematic nature of boundary disputes and the difficulties often inherent in assembling a realistic modern development site from the patchwork quilt of historic ownership within an area. In the wider study, it was apparent that such multiple ownership of land, in particular, proved hard to resolve without the prospect of lucrative commercial development and/or state intervention.

Table 17.5: Extent and productivity of owner contracts

	Productive		Unproduct		No Contact		Total		No Response
	No	%	No	%	No	%	No	%	
Local Planning Department	71	61	25	21	21	18	117	100	3
Chartered Surveyors/Property Agents	70	60	16	14	31	26	117	100	3
Architects/Planning Consultants	65	57	9	8	41	36	115	100	5
Lawyers	65	57	8	7	42	37	115	100	5
Private Commercial/Industrial Development Companies	49	43	13	11	52	46	114	100	6
Other Local Authority Departments	49	43	18	16	48	42	115	100	5
Neighbouring Landowners	44	38	25	22	46	40	115	100	5
Private Housebuilding Companies	29	25	10	9	75	66	114	100	6
DOE/Scottish Office	25	22	9	8	80	70	114	100	6
Local TEC or LEC	22	19	3	3	90	78	115	100	5
Banks and Building Societies	21	18	5	4	88	77	114	100	6
Housing Associations	16	14	10	9	90	78	116	100	4
Other Government Departments	15	13	4	4	94	83	113	100	7
English Partnerships/Scottish Enterprise	14	12	7	6	93	82	114	100	6
Pension Funds and Insurance Companies	14	12	6	5	94	82	114	100	6
Local Amenity and Environmental Groups	11	10	6	5	97	85	114	100	6
Venture Capitalists	7	6	6	5	101	89	114	100	6
National Amenity Environmental Groups	5	4	2	2	107	94	114	100	6
Member of Parliament	3	3	5	4	106	93	114	100	6
Minister of the Crown	3	3	2	2	109	96	114	100	6
Conservative Party	2	2	1	1	110	97	113	100	7
Labour Party	3	3	3	3	108	95	114	100	6
Other Political Party	0	0	2	2	112	98	114	100	6

Other local authority departments, apart from the planning office, were contacted by 58% of all respondent owners, with 43% describing the contact as productive or highly productive. It is likely that most of these discussions concerned technical issues such as access and drainage that had the potential significantly to affect development feasibility. Private industrial and commercial developers formed the final group within this close-knit property development network. It is apparent that brownfield landowners were in much greater contact with industrial and commercial developers than housebuilders (54% as against 34%) and that more respondent owners described contact with the former as productive or highly productive than with the latter (43% as against 25%). This probably again reflects the perception that retail and office development is likely to be commercially more lucrative in brownfield locations than residential development.

All those other people and organisations listed with Table 17.5 who might be considered to form a much broader policy and economic development network were contacted by less than half of the respondent owners and in most cases by less than a fifth. It is especially noticeable how little direct contact existed between large brownfield owners and the political community. However, it may well be that such important forms of contact were mediated through the property development network, with lawyers and property agents in particular acting as the conduit through which the concerns of landowners were drawn to the attention of leading politicians.

Although this possibility was not formally investigated in the research, it became apparent that such linkages existed, especially in Nottingham. Here, for example, one of the most influential chartered surveyors in the city had been drawn into a wider policy role and now serves alongside leading city councillors on the board on a public-private development partnership that is seeking to promote the regeneration of the city. If this pattern were to be replicated more widely, it could be argued that brownfield owners do not require direct contact with politicians, since their interests are well served by indirect contact through their professional representatives.

The research also sought to compare the extent of contact achieved by the different categories of owner introduced in the rudimentary typology set out earlier in Table 17.2, with the seven key contacts who comprised the property and development network. These were the local planning department, chartered surveyors/property agents, architects/planning consultants, lawyers, private commercial and industrial companies, other local authority departments and neighbouring landowners. As Table 17.6 shows, a simple multiplication of the 120 owners by these seven key contacts gave a total number of potential contacts of 840. Of course, not all these contacts will need to be made for every redevelopment. In each case, this will depend on the size and complexity of the proposed scheme and its relationship both to public policy and private interests. In the research, 65% of the 840 potential contacts were made, of which 51% were considered productive and 14% unproductive.

Table 17.6: Extent and productivity of key owner contacts

	Number of Owners	Number of Potential Contacts	Actual Contacts (%)		
			Productive	Unproductive	No Contact
Government or public body	11	77	67	11	22
Local authority	22	154	65	7	28
Always private-sector PLC	15	105	58	19	23
PLC privatized	17	119	54	11	35
Private company	49	343	41	18	41
Private individual/trustee	6	42	31	14	55
All Owners	120	840	51	14	35

This overall picture masks some interesting contrasts between the six different owner types. As Table 17.6 clearly shows, larger owners were more frequently in touch with key contacts than smaller owners, with the highest level of productive contact enjoyed by the public sector. Indeed, Table 17.6 would appear to suggest that private companies and individuals were more at the fringes than at the centre of these key development networks in urban regeneration. Private individuals and trustees, for example, enjoyed less than half the rate of productive contact experienced by government or public bodies. In the end, then, a tight nexus of critical contacts in urban redevelopment appears to exist, with the richness of that network most intensely experienced by larger, rather than smaller owners.

Conclusions

This chapter has shown that brownfield landowners form an important part of local property development networks in which they are most closely linked to local planning departments, chartered surveyors/property agents, architects/planning consultants, lawyers, private commercial and industrial companies, other local authority departments and their neighbouring landowners. However, public-sector owners appear slightly better connected to such networks than large private-sector owners and significantly better-connected than small private-sector owners.

These results are relevant in assessing the robustness of growth coalitions and associated networks within the context of selected cities in the United Kingdom. From its inception out of experience in the United States, the concept of growth coalition emphasises the efforts of landowners in contributing to local economic growth. However, there appears to be significant differences between the balance of land ownership interests in the United Kingdom and the United States.

In the United Kingdom out of historical coincidence and political influence, public sector ownership interests are more dominant than in the United States where the private sector interests are the key players. The research results indicate that public sector interests are more critical to redevelopment and local economic growth. Although local authorities, for example, have widespread ownership of redevelopment land in their areas, the research suggests that they have recently maintained significantly encouragement towards redevelopment. Harding (1991) contends that growth coalitions take different forms, but retain their substance. The research findings suggest possible coalition-building around public sector interests. The objective is consistently the achievement of local economic regeneration.

Although, growth coalitions stand to benefit from inward investment, it may be the case that some coalitions with limited local development capacity will have to wait for their turn in attracting mobile capital. However, the onus is on such local interests to create and maintain a necessary atmosphere through stable political and social systems in order that inward investment will reach them sooner rather than later.

Acknowledgements

The research on which this chapter is based was funded by the Economic and Social Research Council (Award Reference Number R 000 23 6081). The support of the Council is gratefully acknowledged. We would also wish to express our thanks to all those who facilitated the research through their provision of valuable advice and information.

Notes

1 Throughout the research, redevelopment was defined to include the re-use of property through substantial refurbishment.

2 It should be emphasised that the definition of owners, here adopted for the purposes of the research, does not distinguish between those who hold land expressly for the purpose of development and those who do not. We therefore include within our definition of owners those whose other characteristics might prompt them to be termed "developers" in everyday language.

3 The five tests specified by Barrett et al. (1978) relate to ownership, public procedures, project viability, physical conditions and market conditions.

References

Adams, D. and May, H. (1991), 'Active and passive behaviour in land ownership', *Urban Studies*, vol. 28, pp. 687-705.

Adams, D.; Disberry, A.; Hutchison, N. and Munjoma, T. (1999), 'Do Landowners Constrain Urban Redevelopment?', Aberdeen Papers in Land Economy 99-01, Department of Land Economy, University of Aberdeen, Aberdeen.

Barrett, S., Stewart, M. and Underwood, J. (1978), 'The Land Market and the Development Process', Occasional Paper No. 2, School for Advanced Urban Studies, University of Bristol.

Department of the Environment, Transport and the Regions (2000), *Our Towns and Cities: the Future - Delivering an Urban Renaissance*, Cm 4911, The Stationary Office, London.

Harding, A. (1991), 'The rise of urban growth coalitions, UK-style?', *Environment and Planning C*, vol. 9, pp. 295-317.

Kivell, P. (1993), *Land and the City*, Routledge, London.

Lloyd, M. G. and Newlands, D. A. (1988), 'The growth coalition and urban economic development', *Local Economy*, vol. 3, pp. 31-39.

Lowndes, V.; Nanton, P.; McCabe, A. and Skelcher, C. (1997), 'Networks, partnerships and urban regeneration', *Local Economy*, vol. 12, pp. 333-342.

Molotch, H. (1976), 'The city as growth machine', *American Journal of Sociology*, vol. 82, pp. 309-332.

Urban Task Force (1999), *Towards an Urban Renaissance*, E & F N Spon, London.

Appendix

Table 17.3: Owner typology by impact of strategies and actions on redevelopment prospects

	Private ind/trustee		Private Company		Always Private-Sector PLC		PLC privatised		Gov't/ public body		Local authority		Total	
	No	%	No	%	No	%	No	%	No	%	No	%	No	%
Significant Encouragement			12	24.5	4	26.7	1	5.9	1	9.1	4	18.2	22	18
Encouragement	2	33.3	14	28.6	6	40.0	7	41.2	7	63.6	12	54.5	48	40
Neutral Impact	2	33.3	17	34.7	5	33.3	4	23.5	1	9.1	4	18.2	33	28
Discouragement	2	33.3	5	10.2			3	17.6	2	18.2	1	4.5	13	11
Significant Discouragement			1	2.0			2	11.8			1	4.5	4	3
Total	6	100.0	49	100.0	15	100.0	17	100.0	11	100.0	22	100.0	120	100

18 Mixed Use, Densification and Public Choice

NIA BLANK, MARTYN SENIOR AND CHRIS WEBSTER

Introduction

We report on an empirical study designed to cast light on the question of individuals' willingness to live at *higher densities* and in neighbourhoods with a *greater mix of uses* than is customary in most parts of British cities. Much of the debate about sustainable city policy implies that the property market will follow strategic policy guidance. Households, it is assumed, will be happy to live in more compact and diverse urban environments; developers and builders will be willing to deliver them; and financiers will be prepared to invest in them. The experience to date suggests that these assumptions are not necessarily well founded. Developers and financiers, like producers in any other sector, are naturally cautious of innovative products because of uncertainty about demand. The biggest barrier to the successful implementation of compact city policy, however, might not be *uncertainty*, but the *certainty* that for decades many British householders have clearly revealed a preference for *higher per capita land consumption* not less, and for *less land use heterogeneity in neighbourhoods* not more. Sustainable land use policy in this respect works against the traditional market at a very fundamental level. That is not to say that it will be unsuccessful in changing the market; only that policy makers should have a realistic view of the markets they are trying to influence. Our research aims to make a statement about the demand-side of the compact city question with the belief that urban policy that disregards or glosses over the preferences of citizens will fail, either in the sense of being ignored or in the sense of creating social problems for the future. The study aims to measure householders' demand for innovative and possibly contentious forms of urban living, as well as for more traditional sustainable living environments such as dense inner suburbs. The next section summarises the background to the issue. Subsequent sections present: a theoretical model of locational choice in which the effects of density and land-use mix are made explicit; the statistical choice model; the experimental and survey designs; initial findings; and conclusions.

Background

The notion of the compact city with mixed-use neighbourhoods has gained support at all levels of political decision making since the Brundtland Report first popularised the sustainable cities debate in the mid 1980s. The Commission of the European Communities' (CEC) Green Paper on the Urban Environment (1990), criticises "the almost clinical separation of land-uses" (p48) in modern cities and proposes that "strategies which emphasise mixed use and denser development are more likely to result in people living close to work places and the services they require from everyday life"(p60). Since the Green Paper, European, national and local policy and discussion documents have repeated the dual theme: greater densities and mixing of land-uses will significantly contribute to emissions and energy consumption reduction via shorter and possibly less frequent intra-urban car-based trips and will raise the vitality, viability and safety of urban areas. At the European level, the message has been developed strongly in the Commission-funded report "Mixed uses in buildings, blocks and quarters" (CEC 1993) and the European Sustainable Cities Report and Good Practice Guide (CEC 1995a, 1995b). At a national level, the same theme is promoted in the UK Government's Strategy for Sustainable Development (DoE and FCo, 1994) and the Quality in Town and Country discussion document (DoE 1994). It is translated into formal guidance through planning policy guidance notes, such as PPG13 (DoE and DoT 1994), PPG6 (DoE 1996) and PPG3 (DETR, 2000). At a local level, development briefs and strategy documents throughout the country contain pro mixed-use and densification policies.

There is an implementation gap however. Schemes on the ground that seek to develop new, or re-develop old, urban areas following the new sustainability ethic are fewer than policy-makers might wish for and this says something about the poor understanding of the processes involved. In particular it demonstrates the need to have more understanding about citizens' preferences for sustainable living space and the land market's understanding of and attitude towards those preferences. Ironically, in cities where the city centre living market seems to have taken off, it is developers (of a certain type) who complain at the lack of potential schemes – lack of land with appropriate permission. In such cases it may be landowners (holding out for higher value uses with unrealistic hope values) and sometimes local planning authorities (holding out for a type of development that is uneconomic) who fail to understand the market. Because the driving concern behind sustainable city policy has been global ecological relationships, much of the published empirical research on sustainable cities focuses on relationships between emission and energy costs and urban form (Rickaby 1987; Breheny 1992, 1995; ECOTEC 1993; Bannister and Watson 1994; and others). The total, average and marginal energy and emissions-related costs of alternative development patterns are, however, only one component of the sustainability equation. The focus of such studies is principally on costs that are borne by society at large and in the end by the global community. The consumer demand question concerns externality costs and benefits borne more locally. *In many ways, the local costs and benefits have a more immediate relevance to the sustainable city policy debate than the bigger issues that drive the debate*. This is because, in economic terms, the marginal disutility of bad-neighbour externalities (crime, noise, local traffic problems, litter, poor environmental design) are arguably greater than the marginal disutility of more city-wide or globally-consumed externalities. In other words *local public bads are likely to be stronger components of a household's utility function than the more ubiquitous public bads of energy over-consumption and emissions over-production*. Evidence for this is easy to demonstrate, for example, suburbanites' willingness to endure the cost of long commutes for the benefits of avoiding some of the local externalities mentioned. There have been many studies of the determinants of residential location decisions but few if any that specifically embed the analysis within the sustainable city debate by attempting to measure trade-offs between important policy variables of density and land use mix.

Urban policy makers made a catastrophic mistake once before when prescribing grand solutions for urban living in pursuit of grand objectives. Le Corbusier is quoted as having said of his modernist city project *"people will have to learn to live in these buildings"*, and many post-war planners and governments apparently took the same view. A similar attitude is discernible in the current policy debate with an unsubstantiated belief that *people will at best want to, and at worst have to, learn to live in more dense and heterogeneous neighbourhoods*. Experience of the past clearly indicates a dominant preference for relatively homogeneous neighbourhoods. Contemporary trends in urban crime and other nuisances and the threat to social cohesion of increasing income differentials indicate that citizens are likely to continue to seek safety and security in numbers.

There is little evidence to suggest that the majority of the better off will do anything other than continue to seek such security in the comfort of lower densities. If clustering of similar uses reduces the probability of spillover costs, then it can be viewed as a risk minimising strategy. It may be conjectured that the inducements to experiment with inherently more risky living solutions have to be particularly high in a time of rising social and economic insecurity.

In this respect our endeavour relates to another body of literature - that relating to the global phenomenon of gated communities (Blakely and Snyder 1997; Kennedy 1995; Smith 1997; Clingermayer 1996). Citizen's preferences for homogeneous neighbourhoods are both a threat and an opportunity for compact cities. American suburban gated communities have the effect of institutionalising low-density sprawl, even incorporating suburbs as separate cities. Developments with a certain degree of enclosure, on the other hand, may well be an effective means, possibly the only means, of making city living an attractive alternative to a greater number of movers or new homeowners. However, high profile security measures may inadvertently send the wrong signal deterring potential residents.

Sustainable city policies relating to densification, use-mixing and use-zoning therefore have to be informed by a far more searching understanding of citizens aspirations, preferences and trade-offs. As with all urban planning in largely private-sector led development markets, successful policy is one that sensitively *manages* markets on the basis of detailed market intelligence. Our study conducts, in effect, a market demand survey of citizens, quantifying the contributions to household utility made by alternative features of development schemes and asking the question: how can we achieve a product mix that citizens (of various types) will "buy" and that is in line with governments' overall goals towards the sustainable city?

Theory

Assume that households seeking locations at which to live, maximise the following utility function:

$$U^h([X^h, N^h]) \qquad [1]$$

where:

$X^h = \{x_1, x_2, x_3 \ldots x_n\} = \{q^h, P^h, \Gamma^h, Z^h\}$ = a set of goods consumed by household h

q^h = land consumed by h

$P^h = \{p_1, p_2, p_3 \ldots p_n\}$ = a set of property attributes

$\Gamma^h = \{\gamma_1, \gamma_2, \gamma_3 \ldots \gamma_n\}$ = a set of locational attributes

$Z^h = \{z_1, z_2, z_3 \ldots z_n\}$ = a set of other goods consumed by h

N^h = a set $\{n_1, n_2, n_3 \ldots n_n\}$ containing the number of joint-consumers of each good in X^h

This is an adaptation of a model first proposed by Buchanan (1965) and is ideally suited to representing locational choice behaviour by virtue of its elegant handling of density and external neighbourhood effects (crucial in analysing land use mix). Buchanan's was a model of congestible joint consumption and was a qualification and generalisation of Samuelson's (1954) model of pure public goods (infinite joint consumption). Where N=1, a good is privately consumed (in our version, by a single household). Where N>1, a good is jointly consumed by more than one household. The number of joint-consumers N appears in the utility function because all shared neighbourhood goods (local public goods) are congestible after some point (club goods in Buchanan's terms). $[q\ n_q]$ is the density of land consumed by h (measured at, for example, plot or building scale). $[\gamma_1\ n_{\gamma 1}]$ is the level of congestion in a household's consumption of neighbourhood good γ_1, for example, a local park, a school or the pavement outside a property. Neighbourhoods with different land use mixes are distinguished by different profiles of values in Γ^h and households are assumed to be able to express preferences over these distinct *Gamma profiles*. In the stated preference model described below, home-mover respondents are asked to rank a series of cards showing various combinations of values for P^h and Γ^h in order to estimate the size and sign of utility function parameters.

Residents are assumed to seek to maximise [1] subject to budget constraint [2]

$$Y^h - \tau^h - B^h . X^h = 0 \qquad [2]$$

where:

Y^h is the income of household h

τ^h is the tax payment of h to governments that supply it with neighbourhood and civic goods and services

B^h is a vector of prices $\{b_1, b_2, b_3 \ldots b_n\}$ associated with goods in X^h

As well as providing a framework for conceptualising locational behaviour, utility function [1] provides a theory of classification that avoids the need to pre-classify urban goods and services as public or private (joint or individually consumed). Table 18.1 illustrates this.

Unpriced goods (externalities) may add to or diminish household welfare, entering into the utility function with a positive or negative sign. Buchanan demonstrated the conditions necessary for optimal solution of a model with this structure. In the context of locational choice, this amounts to finding the optimal mix of property and locational attributes (the last three rows of the table) taking into consideration the density of consumption (density of households at a location, density of land uses of different kinds, and density of positive and negative externalities – given by the n_is). These ideas are developed further by Webster (2001a and 2001b). The next section articulates the theory in an empirical choice model.

Table 18.1: Classification of private/public goods in locational preferences via a density of consumption term

x_i	n_I	b_I	
>0	1	0	Privately consumed unpriced good
>0	>1	0	Unpriced locational externalities (+ or -)
>0	>1	>0	Priced goods including private goods generating positive externalities, collective goods paid for via club arrangements and locational externalities reflected in property prices
>0	1	>0	Pure private goods

A Model for Stated Residential Preference

Residential movers are assumed to be rational decision-makers within their own individual frames of reference. They choose their most preferred (perceived highest utility) option from the set of residential alternatives available to them. Various attributes of these alternatives and of the movers will define their utility functions. However, it is very unlikely that we as researchers will be able to observe all influences affecting movers' choices. Hence, random utility theory is commonly invoked, whereby movers' utility functions are assumed to be composed of a systematic part, comprising influences on choices observable by the researcher, and a random part subsuming all unobserved effects.

Commonly, in travel demand and marketing research (Hensher and Johnson, 1981), a multinomial logit model (MNL) is derived from this random utility framework, because it provides a statistically tractable expression for the choice probabilities, as follows:

$$p_{ik} = \frac{\exp(V_{ik})}{\sum_{j}^{J} \exp(V_{ij})} \qquad j=1,2\ldots,k,\ldots J$$

where ***exp*** denotes the exponential function, p_{ik} is the probability of mover i choosing residential alternative k from all those (j=1,2....k...J) available, and V_{ik} is the observable component of the mover's utility function for residential alternative k. The right hand side of this model represents the perceived utility, or attractiveness, of residential package k, relative to the combined perceived utility or attractiveness of all residential alternatives available to mover i.

Data on respondents' revealed or stated choices can be used directly in this model. However, stated preference (SP) surveys ask respondents either to rate or to rank alternatives (Hensher, 1994). An important paper by Chapman and Staelin (1982) showed how rank-ordered SP data could be used to estimate the MNL choice model. Essentially, each respondent's ranking of, say 6, residential alternatives is "exploded" or decomposed into 5 choices.

First the most preferred residential alternative is chosen out of the six available. If the most preferred alternative were not available, the second most preferred is assumed to be chosen from the remaining five, and so on, until the fifth ranked alternative is chosen from the two remaining.

Stated Preference Experiment

A stated preference design is used because the intention is to explore movers' responses to both actual and potential mixes of attributes of the residential environment, especially in terms of mixed land use and density. Another advantage is that SP designs allow the researcher greater control over influences on residential choice, which then facilitates the estimation of the significance and strength of those influences. A number of studies (e.g. Louviere and Timmermans, 1990; van de Vyvere et. al., 1998) have used this SP approach to analyse residential choice. In particular, Molin et al. (1996) have examined housing density, among other factors. However, we are not aware of any previous SP study specifically examining the effects of mixed land use policies, although Hunt et al (1995) have modelled citizens' preferences for alternative urban development strategies that include density variations.

Eight dimensions are used to define the residential alternatives (Tables 18.2 and 18.3). These eight dimensions, along with the characteristics of the respondents (e.g. household composition, employment, car ownership), form the variables in the observable component, V_{ik}, of each respondent's utility function. These variables and their constituent parameters are selected to represent both the existing market possibilities and future development patterns as a result of policy implementation. For example, locational variables largely represent existing types of neighbourhood. However, the characteristics of a "major suburban centre" are designed to reflect potential densification and diversification of uses around a transport node. Dwelling type, amenity space and parking provision are employed as measurements of development density while the frequency of passers-by provides an indicator of intensity of use. By including a variable on the socio-economic characteristics of other residents, observations can be made concerning current policies promoting the development of neighbourhoods with a more diverse mix of residents.

These categories on the eight dimensions produce 10368 possible combinations of attributes defining the residential alternatives to be presented to respondents. This is known as a "full factorial" design, from which a very small, manageable "fractional factorial" design is chosen, using both intuition and the principles of experimental design. Certain combinations of attributes are not allowed. Thus it is not realistic to suggest that detached and semi-detached properties will have only on-road parking and/or no garden space of their own; also such dwellings are not likely to be provided in city centre locations.

Our fractional design comprises 36 residential scenarios. These are split into blocks of 6 as previous experience with a pilot study in Sheffield (Blank, 2000) suggested that more than six alternatives could lead to respondent fatigue and confusion. Each respondent is asked to rank the 6 scenarios in a block in order of preference. They are then requested to exclude the scenarios that they would not consider given their current circumstances. Finally respondents are asked to estimate a maximum price that they would be willing to pay to purchase each property. Respondents are required to repeat the process with a maximum of 3 blocks. A series of attributes are held constant throughout the experiment as the literature review indicated that these factors are likely to be important in housing choices. Thus all properties are in the respondent's price range, have the number of bedrooms they are looking for; and are in the catchment area of a good secondary school. Following this SP experiment, a brief interview is undertaken to investigate the decision making process in a more qualitative manner.

The parameters of the MNL model calibrated on the overall preference ordering in this data set can be interpreted as the part-worth utilities of individual attributes. The coefficients indicate the elasticity of citizen satisfaction with respect to changes in density, use mix and other urban system performance measures. The ratio of the coefficients of two attributes (for example, density and parking facilities; mixed land use and open space; social mix and dwelling type) give the trade-offs between those attributes.

Table 18.2: Dimensions and categories in the SP experiment

Location • City centre • Redeveloped dockland area • Inner-urban • Major suburban centre • Typical suburban area • Small town	Land Use Mix • Non-residential land uses shown on card X* within 2-5 minutes walk • Non-residential land uses shown on card X within 6-10 minutes walk • Non-residential land uses shown on card X not within reasonable walking distance
Dwelling Type • Detached house • Semi-detached house • Terrace house • Apartment	Dwelling Character • Dwelling with distinctive character (old or new) • Dwelling with non-distinctive character (old or new)
Open Space • Your own garden • Shared garden with access to immediate neighbours only • No garden but easy access to public open space	Residents' Socio-economic Characteristics • Most residents of higher socio-economic background than yourself • Most residents of similar socio-economic background as yourself • Most residents of lower socio-economic background than yourself • Residents with a mixture of different socio-economic backgrounds including those higher and lower than your own
Parking • Parking in secure garage • Parking in off-road parking space • Parking on-road	Frequency of Passers-by • Infrequent passers-by on the pavement in front of the property • Frequent passers-by on the pavement in front of the property

* card X refers to one of the six location-specific land use mixes in Table 18.3.

Table 18.3: Public transport and non-residential land uses present at each location

Location	Public Transport	Non-Residential Land Uses
City centre	Frequent bus, train and coach services to local and national destinations	• Wide range of major high street stores, post office, banks, small supermarket • Offices of all sizes • Cinemas, theatre, sports centre, cafes/bars, takeaways • Night clubs, library
Redeveloped dockland area	Bus and train services to city centre every 10 minutes	• Limited range of local shops, large supermarket, several high street stores • Medium and large sized offices • Hotels • Sports centre, cafes/restaurant • Bars, multi-screen cinema
Inner urban	Bus and train services to city centre every 10 minutes	• Wide range of local shops, post office, banks, small supermarket and some smaller high street stores • Small and medium sized offices • Small workshops • Takeaway, restaurant/café, pub • Drs, dentist, primary school, church • Sports centre, library
Major suburban centre	Bus and train services to city centre every 15 minutes	• As for Inner urban
Typical suburban area	Bus to city centre ever 30 minutes	• Small local shop or newsagents, pub • Drs, primary school
Small town	Bus to closest major urban area every hour	• As for Inner urban

The experiment was administered to a sample of home movers in Cardiff, selected from four types of origin neighbourhood: new suburban estate, traditional outer suburb, traditional inner urban ring, redeveloped dockland. This permits an analysis of subgroups including testing hypotheses about preferences of householders moving from homes in traditional and innovative sustainable neighbourhoods and those moving from traditional homogenous suburbs. From our interviews with developers, we expect to find signs of market fragmentation.

Initial Findings

Pilot and Household Surveys

The householder survey has not been fully analysed at the time of writing and as such results are limited to those of the Sheffield pilot study (61 individuals, all moving from terraced properties) and 19 respondents in Cardiff. Detailed model specifications and discussions will be reported in other papers. The Sheffield pilot study considered four variables via a stated preference experiment supplemented by discussion. Variables comprised: dwelling type; land use mix; the provision of additional security measures; and frequency of passers-by. The study had a particular focus on attitudes to land use diversity and the potential relationship between land use mix, street activity and householders' perception of security, be that personal or property related.

Both the statistical analysis of the stated preference experiment and the qualitative interviews found that respondents were trading variables and that dwelling type, security measures and frequency of passers-by were significant determinants in at least some of their choices. The statistical analysis suggested that individuals were not significantly influenced by land use mix in their selection of residential scenarios. However, further consideration of the experimental findings and the interviews showed that this was a reflection of the analysis of the respondents' choices simultaneously. In other words those preferring mixed use cancelled out those preferring single use in the statistical model and this is a good illustration of the dangers of taking the statistical results at face value without deeper analysis. In fact only one third of respondents were **not** using land use mix in their decisions. The remaining two thirds were equally split between those who preferred single-use areas and those who were actively seeking a mixed-use environment.

If the decision-making process is considered in more detail we find that dwelling type, as might be expected, was the strongest determinant and had a highly significant and large effect on respondent choices. Indeed 98%

of respondents stated that it was the most important determinant in their selection, with a pronounced aversion to apartments often being the first factor that influenced preference. Most individuals would consider terraced properties (only 11 people excluded them) and, although 77% of respondents preferred semi-detached properties, approximately one third of these were willing to trade a semi-detached scenario for a terraced scenario to gain or avoid other factors. 11% of respondents preferred terraces. The influence of security measures and frequency of passers-by was less significant in many householders' decisions. However, security measures particularly produced an interesting range of reactions. Only 39% of respondents used frequency of passers-by as a determinant in their choices with the remaining individuals stating that they did not mind. Of those who used it as a determinant the vast majority preferred few passers-by, usually for reasons for privacy and noise. In terms of land use mix, those who preferred mixed-use neighbourhoods principally did so for reasons of amenity. However, it was also suggested that they were: *more lively and attractive areas* (1 respondent); *organic areas which also create a demand for public transport and help property prices* (2 respondents); *and have a sense of community and are safer after dark* (2 respondents). Twenty-one respondents preferred single residential use areas, most giving the reason that these were "quieter". Three respondents felt that it was a safer environment for children as there would be less traffic. Despite these preferences all respondents stated that they would consider properties in either area and were more likely to trade this attribute off against other variables. This willingness to trade may help explain why people live in mono-use suburban areas and estates - because this is how the market often offers semi-detached and detached living scenarios and house type is a much stronger determinant.

Further discussion concerning attitudes to land use diversity found that perceptions of mixed use areas are complex and diverse and are commonly a composite reaction to the amenity value, potential externalities and type and number of people associated with specific uses, the latter being intimately linked to perceptions of the presence of people in general. Results clearly showed that noise, parking problems and traffic are the three major externalities which respondents were trying to avoid in their location decisions. These reasons were given for a wide variety of land use types, as were issues of privacy and for some a dislike of the general "business" associated with places frequented by a significant number of people. Individuals varied substantially in the proximity they would consider living to specific uses and one could observe a marked social class effect with those of "lower social class" being more tolerant of non-residential uses.

In terms of mixed use and crime/anti-social behaviour the survey and literature review both suggest that the perceived relationship is much more complex than many theorists and policy-makers imply. A significant number of respondents associate certain uses and the presence of more people with an increase in certain types of crime/anti-social behaviour rather than a reduction, as many suggest. However, other research and some findings of the pilot survey support the theorised associations that mixed use, and particularly the presence of people, engenders feelings of greater safety for a significant proportion of the population, largely as a result of "natural surveillance" and a "safety in numbers" attitude. As some have already recognised, the identity and perceived intent of people is a key factor in the formation of attitudes and it is tentatively concluded that the survey provides some support for writers such as Jacobs (1961) who suggest that feelings of safety are often linked to sentiments of trust. However, it is not clear whether trust is facilitated through the numerous casual interactions as a result of mixed use as she suggests. Another interesting observation is that dwelling type had a significant influence on perceptions of crime, independent of land use type, with the higher density housing (terrace) largely engendering feelings of safety in comparison to the lower density housing (semi-detached). Nevertheless, the highest density option (apartments) was often considered to be less safe due to poor natural surveillance in communal areas.

We add a note of caution in generalising these findings to a wider population as the survey respondents were representative of only one sector of the population - terrace movers who tended to live in the older, relatively built-up areas of Sheffield and who often had a mixture of land uses present at the neighbourhood scale. Residents in different cities and city centre and suburban dwellers, for example, may have quite different views. While results of the Cardiff study are not ready for dissemination we can make some initial observations. Perhaps the most striking aspect of the findings at the time of writing is the malleability of householders' choices. The majority of respondents state that they would consider a relatively wide variety of living scenarios if they were at the "right price". Indeed, their current search for a new dwelling has often led them to consider various areas as they try to balance price against product and neighbourhood. There are choice attributes that respondents are not willing, or at least are reticent, to compromise on, such as trading a house for an apartment or living in a city centre location rather than a small town in some cases. However, householders were willing to "trade" many aspects of the housing scenarios, such as a suburban location for a house in a redeveloped dockland environment or an apartment with a garden and off-road parking for a house without. The attitudes to new housing and redeveloped dockland environments were also very interesting. A marked division existed between those who were attracted to them by their "newness", their "attractive, well kept" surroundings and what they perceived to be their investment value as "up and coming areas", and those who viewed them as having little sense of community and often as being areas which were of unknown investment potential. A "sense of community" was cited as an attractive feature of certain neighbourhoods by many of the respondents and was often associated with small towns and inner urban areas. In our full analysis we shall pay attention to the relationship between these preferences and present locational characteristics.

Developers

Interviews were undertaken with a small selection of development interests with the initial aim of informing the household survey design. However, they also provided a wealth of other insights into the sustainable settlements debate, highlights of which are reported below. Interviewees included individuals from:

- 5 major house-building companies with a range of housing markets;
- a small, innovative development company specialising in inner city housing and mixed use projects;
- a family run property investment business which undertakes residential and mixed use developments; and
- a major financial institution with redundant space in many of its properties and which became involved in the Living Over the Shops Scheme (LOTS).

Discussion tended to focus on housing in mixed-use schemes and environments, particularly city centres. Initially the interviews explored the companies' roles and development in the housing market and this exposed a fascinating but highly complex set of factors governing their activities. The large house-building companies' "traditional" market can be typically characterised as new-build, private, largely greenfield, relatively low density, "suburban" type housing. However, all of the companies have *diversified* considerably. Some have become involved in social housing or mixed-use developments and increasingly they are all working in the city centre housing market and with brownfield sites. The reasons for diversification relate, among other things, to policy frameworks, market conditions and organisational characteristics.

The first observation to highlight is the *influence of policy and politics* on development activities. All of the house-builders were finding that the supply of greenfield land is increasingly being limited by restraint policies, whilst at the same time the shifting political climate means that there is greater policy support for urban housing projects. Both of these factors were considered to be influential in encouraging house-builders to look "inwards" for development opportunities, particularly towards city centres. Pressure from planning restraint policies was less influential on the 3 other interviewees, and perceived market opportunities largely stimulated entrance into inner urban and city centre housing markets. The specialist city centre company was set up as a result of individuals seeing a niche market, while the property development company saw an opportunity to expand its housing activity and found housing a more attractive alternative to secondary office space. The slump in the secondary office market was named as one of the influences on the financial institution as it sought alternative uses for redundant office space. However, the key influence in this case was the availability of government grants through partnership with housing associations. However, policy and local authority attitudes were still important. All interviewees stressed their significance in progressing planning applications and conditions and made the points that time delays can reduce the viability of schemes. They were all of the opinion that policy and attitude significantly affect the density, design, parking and mix of uses in a development. Some expressed frustration and questioned the market realism of certain local authorities. However, there was also recognition that planning activities can be beneficial and sometimes crucial to the success of a scheme. In particular a supportive planning framework and holistic planned approach to the improvement of an area was felt to reduce developer risk, and public realm improvements were often cited as an essential foundation for developing a successful city centre market. Thus policy, partnership and the activities of planners were means of managing developer risk by reducing uncertainty.

Uncertainty was a theme raised repeatedly in reference to city centre housing, mixed-use developments and refurbishment. The nature and degree of uncertainty varied depending on the company but often focused around a need to develop new skills, working practices and partnerships and a lack of information. One area that most companies expressed uncertainty over however was the city centre housing market. The market was seen to be immature in most areas and to be very geographically specific both in terms of particular cities and the areas within them. There was considerable uncertainty regarding the depth and sustainability of the market and, to an extent, the nature of the market. Different sectors of the population were considered as potential residents: students; short term professional tenants; people requiring social housing; childless professionals; and retired people whose children have left home. Generally developers saw the mixed-use environment of the city centre to be attractive to these sectors of the market and considered them relatively tolerant of potential externalities associated with mixed use in comparison to, say, the suburban consumer. Indeed, several interviewees implied that *the market was quite fragmented* and that those attracted to city centre living had quite different attitudes to more "traditional" and possibly conservative consumers. None of the interviewees saw city living as an option that would be attractive to families without further development of city centre facilities and schools. Some companies expressed a good degree of market knowledge pertaining to their specific area of activity partly based on market experience and in a few cases market research, but the majority appeared to base their perceptions on observations of others' activity, especially in cities such as Manchester where the market is more mature, and on a "gut feeling" of what would work. In a limited number of cases developers suggested that their city centre activity was limited by a lack of opportunities.

In terms of a *product*, most companies were providing apartments or, in some cases, townhouses in city centre locations at relatively high densities in order to make sites with high market values competitive for housing. Some developers were also incorporating other land uses at ground and first floor level to make schemes more economically attractive and possibly to comply with development briefs. In terms of the major house-builders most schemes were new build. However, some larger conversion or shell schemes were being progressed. The financial institution and specialist city centre developer were predominantly active in the refurbishment market, while the property investment company took on both types of scheme.

Overall the impression of the large house-builders was that some were not comfortable looking at the city centre market but considered it a necessary step given the economic and policy climate. Others appeared to be embracing the city centre market, viewing it as a major opportunity, which they may have pursued without the pressure of planning policy. The market strategies of the companies also differed considerably with some acting as pioneers, being the first or one of the first to develop in an area, taking risks that other companies do not take. The interviewees from these companies had a vision and the people involved had a good personal knowledge of the areas they choose to develop – speaking of a "gut feeling" that a market was waiting to be developed. Other companies were more cautious in their approach, acting as "settlers", coming into areas once the market was maturer. There was an observable diffusion of ideas both within and between the individual companies, with experiences in places such as Manchester, Liverpool and Leeds particularly inspiring market confidence. The reasons for these differences in attitudes and activities can only be guessed at, but appeared to be linked to several factors including the risk-taking culture of the company, its flexibility and ability to adapt to new partnerships and working practices, and its in-house skills and experience in this market.

Conclusion

The stated preference technique provides great potential for studying the consumer choices that create and sustain new and evolving housing markets. Beyond the benefits of quantification, the technique provides a focus from which to stimulate penetrating and semi-structured discussion.

Initial results indicate that there is an unexpected degree of flexibility among movers. This is heartening as it means that many preferences may not be as entrenched as some suggest and may be open to manipulation. It may also suggest that, as some developers claimed, the market for less traditional living solutions (particularly city centre) is beginning to take off. Some choice constraints however are likely to be inflexible. It is unlikely that any private developer will be able to convince a family of 4 to live in a gardenless apartment in the city centre for example. Such a household, might however, be persuaded to consider a large ground floor apartment with its own garden in a dockland or similar development. Restricted access facilities such as communal gardens, play areas and parking may strengthen the persuasion.

Some uses are likely to create such a degree of externalities that they are better separated from residential locations. Others need to be considered very carefully given the strength of aversion that some respondents showed. This has significant implications for planners and others alike as it emphasises the importance of addressing these externalities through design and management. What is clear is that planners have a key role in making mixed use more acceptable to residents. They need to be particularly careful that in promoting certain practices that are considered desirable, such as the Government's policy of reducing car parking requirements for offices, they do not create other problems. They must not fall into the trap that mixed use is "good" whatever the uses.

A further barrier may prove to be preferences towards lower density dwelling types. This is significant as non-residential land uses require a certain catchment to be viable and this is less attainable within reasonable walking distances at lower densities.

Individuals in our surveys varied substantially in the distance they would consider living from specific uses. The marked social class gradient in tolerance of different land uses also has interesting implications. There may be potential to advance the Government's policies for more mixed communities and reduced social exclusion through land use diversification - with a distance decay effect in social class within a neighbourhood. Potentially, if demand for dwellings in very close proximity of non-residential uses is lower, property price gradients could also result, providing a greater diversity in prices and thus also facilitating a more mixed neighbourhood of household types.

There needs to be a variety of dwelling types and parking and open space arrangements on brownfield sites to encourage those who perhaps prefer suburban locations to come back into more central urban locations. However, there are possible conflicts with market requirements to build at "high" densities to secure land from other uses.

What is clear is that, although certain perceptions regarding aspects of the compact city are common to respondents, there is no general interpretation, which can apply to everyone - attitudes are diverse. The implication of this is that no single living scenario will fit all, a portfolio of settlement patterns are needed to attract a wide variety of people to urban living. The market does seem to be maturing for city centre, dockland and other brownfield housing but the process is one of market fragmentation as different groups respond in their own way to opportunities created by innovative developers. The fact that a second, possibly even third wave of developers and house builders are following the pioneers attests to this maturing. Our full SP analysis will test this claim in a regional city with a dynamic economy and development sector. It will help us to interpret the underlying market demand which the more bullish developers sense; to explore the dimensions of market segmentation; to understand the trade-offs made by different types of household in evaluating traditional and non-traditional locations; and to speculate about the effectiveness and appropriateness of policy.

Acknowledgements

The research reported in this paper was supported by the ESRC (grant R000222878). We wish to thank the developers and householders who participated in our study. Views expressed by developers are personal opinions and do not necessarily reflect those of their employers.

References

Bannister, D. and Watson, S. (1994), 'Energy use in transport and city structure', Working paper 7, Planning and Development, UCL, London.

Blakely, E. and Snyder, M. G. (1997), *Fortress America*, Brookings Institute, Washington.

Blank, N. E. (2000), 'Preferences for mixed land use neighbourhoods: Empirical evidence from residential movers in Sheffield', unpublished paper presented to the Annual Conference of The Royal Geographical Society, University of Sussex, January 2000.

Breheny, M. (1992) (ed.), *Sustainable development and urban form*, Pion, London.

Breheny, M. (1995), 'Counterurbanisation and sustainable urban forms', in Brothchie, J.; Batty, M.; Blakely, E.; Hall, P. and Newton, P. (eds.), *Cities in Competition*, Longman, Melbourne.

Buchanan, J. M. (1965), 'An economic theory of clubs', *Economica*, vol. 32, pp. 1-14.

Chapman, R. G. and Staelin, R. (1982), 'Exploiting rank ordered choice set data within the stochastic utility model', *Journal of Marketing Research*, vol. 19, pp. 288-301.

Clingermayer, J. (1996), 'Quasi-judicial decision making and exclusionary zoning', *Urban Affairs Review*, vol. 31, pp. 544-553.

Commission of the European Communities (1990), *Green Paper on the Urban Environment*, CEC, Brussels.

Commission of the European Communities (1993), *Mixed uses in buildings, blocks and quarters*, Shankland Cox, London.

Commission of the European Communities (1995a), *European Sustainable Cities Report*, CEC, Brussels.

Commission of the European Communities (1995b), *Good Practice Guide*, CEC, Brussels.

Department of the Environment (1994), *Quality in Town and Country*, HMSO, London.

Department of the Environment (1996), *PPG6*, DoE, London.

Department of the Environment and Department of Transport (1994), *PPG13*, HMSO, London.

Department of the Environment and Foreign and Commonwealth Office (1994), *Sustainable development: the UK strategy*, HMSO, London.

Department of the Environment, Transport and the Regions (2000), *PPG3*, Stationery Office, London.

ECOTEC (1993), *Reducing transport emissions through planning*, HMSO, London.

Hensher, D. A. (1994), 'Stated preference analysis of travel choices: the state of practice', *Transportation*, vol. 21, pp. 107-133.

Hensher, D. A. and Johnson, L. W. (1981), *Applied Discrete Choice Modelling*, Croom Helm, London.

Hunt, J. D.; Abraham, J. E. and Patterson, D. M. (1995), 'Computer generated conjoint analysis surveys for investigating citizen preferences', in Wyatt, R. and Hemayett, H. (eds.), *Proceedings of the 4th International Conference on Computers in Urban Planning and Urban Management*, Melbourne.

Jacobs, J. (1961), *The Death and Life of Great American Cities*, John Dickens and Conner Ltd, London.

Kennedy, D. (1995), 'Residential Associations as State Actors: regulating the impact of gated communities on non-members', *The Yale Law Journal*, vol. 105, pp. 761-793.

Louviere, J. and Timmermans, H. (1990), 'Hierarchical information integration applied to residential choice behaviour', *Geographical Analysis*, vol. 22, pp. 127-144.

Molin, E.; Oppewal, H. and Timmermans, H. (1996), 'Predicting consumer response to new housing: a stated choice experiment', *Netherlands Journal of Housing and the Built Environment*, vol.11, pp. 297-311.

Rickaby, P. (1987), 'Six settlement patterns compared', *Environment and Planning B*, vol. 14, pp. 192-223.
Samuelson, P. A. (1954), 'The pure theory of public expenditure', *Review of Economics and Statistics*, vol. 36, pp. 387-389.
Smith, L. J. (1997), 'Gated Communities: private solution- public dilemma', *Urban Lawyer*, vol. 29, pp. 413-426.
Van de Vyvere, Y.; Oppewal, H. and Timmermans, H. (1998), 'The validity of hierarchical information integration choice experiments to model residential preference and choice', *Geographical Analysis*, vol. 30, pp. 254-272.
Webster, C. J. (2001a), 'Property rights and the public realm: gates, green belts and gemeinshaft', *Environment and Planning B* (forthcoming).
Webster, C. J. (2001b), 'Gated cities of to-morrow', *Town Planning Review* (forthcoming).

19 Brownfield Sites: Problems of Definition, Identification and the Evaluation of Potential

PETER ROBERTS, VICTORIA JOY AND GLYN JONES

Introduction

In the past brownfield sites have frequently been regarded as liabilities rather than assets. Many brownfield sites are contaminated, derelict, remote, inaccessible or, perhaps more importantly, poorly regarded by both the development industry and the public policy community. It is this latter characteristic of brownfield sites which has in the past frequently determined attitudes towards, and the market status of, brownfield land and premises.

However, during the past decade brownfield has undergone a transformation, due in part to the results achieved by the successful urban regeneration experiments of the 1990s (Roberts and Sykes, 2000), but also as a consequence of a change in attitude towards brownfield in both the public and private sectors (Urban Task Force, 1999). This change in attitude reflects both the gradual adoption of a more positive approach to the potential displayed by many brownfield sites and, conversely, the urgent need to find additional sites to accommodate development in regions experiencing excessive pressure for growth. As a consequence these drivers of change have generated a series of debates about the definition, identification, remediation and methods for the development of brownfield sites. Further debates have considered appropriate end uses and the need to adjust planning and other policies.

This chapter is set within the context of the new debates on brownfield potential which have emerged during the past decade. Much of the evidence presented in the chapter is drawn from research undertaken by Urban Mines and its partners, which sought to examine the definition and identification of brownfield sites at both UK level and in a number of selected case study areas (Alker et al, 2000). The findings from this research – the National Brownfield Sites Project – have wide-ranging implications for both theory and practice, including a number of insights which can be used to guide the design of policies aimed at encouraging the more extensive use of brownfield land and premises.

In the second section of this chapter, a review is presented of a number of the difficulties that are encountered in attempting to define brownfield and some suggestions are made with regard to ways in which brownfield can be classified and modeled. Section three provides a synopsis of the views of the stakeholders who are involved in brownfield redevelopment and illustrates at local level some of the difficulties and development potentials associated with brownfield sites. The final section of the chapter offers some conclusions and suggests a number of changes to current policy that would help to support the extended productive use of brownfield sites.

Defining and Classifying Brownfield

Key Issues

One of the major problems associated with understanding and developing brownfield sites is a consequence of the considerable range of types of site referred to as brownfield. In one sense this imprecision in terminology is to be expected, especially given the important role played by local and regional contextual factors in the determination of the viability and market potential of an individual brownfield site. Put simply, what is considered as a brownfield site in one area may not be seen to present typical brownfield characteristics elsewhere. This difficulty is illustrative of the need for a more precise and all-encompassing definition of brownfield, on the one hand, and indicative of the importance of recognising and accepting the presence of spatial and sectoral variations in the application of a definition, on the other hand.

However, even accepting that there are considerable sectoral and spatial differences in the specification and definition of brownfield, there are also a number of common factors, many of which are now reflected in legislation and statutory processes. These common factors relate, for example, to sites such as those defined as derelict land or land which is considered to be contaminated under Section 57 of the Environment Act 1995. Such factors and their accompanying definitions, which are quite satisfactory in the context of the purposes for which they were originally devised, can also be seen to represent elements or sub-sets of a more inclusive definition of brownfield. Reflecting this discussion, the Parliamentary Office of Science and Technology (POST) has observed that, given the adoption of a wider, rather than narrower, definition of brownfield, the "total brownfield resource may be so large that the geographical variations become insignificant" (POST, 1998: 24). This has certainly been the case in some regions in the past, especially in localities where the extent of brownfield has been hidden within the total areas of land that is defined as operational land; an example of this can be seen in an early survey of industrial land availability in the Black Country which identified a considerable area of unused brownfield land within the boundaries of operational industrial sites (Roberts et al, 1988).

A further issue that should be considered from the outset is the difficulty encountered in the regeneration of brownfield as a result of the different perceptions, conventions and aspirations of the various stakeholders who are involved with brownfield. The absence of an agreed overall definition of brownfield, and of a typology for the classification of the various categories or types of site contained within the definition, can be seen as impediments to regeneration. Three important dimensions associated with this issue are; first, the potential which exists for confusion or false interpretation with regard to the extent, physical condition, market status and value of a brownfield site; second, the difficulty of conducting a physical assessment, risk analysis or evaluation of alternative remediation methods in relation to an individual site if its characteristics are imperfectly known or understood; and, third, the problem experienced by public sector planners and developers alike of devising, negotiating and implementing a suitable redevelopment programme in a situation in which the strategic significance, infrastructure requirements and development potential of a site cannot easily be assessed.

A number of governmental and academic studies undertaken in the UK and elsewhere have reflected the difficulties associated with the absence of a common definition of, and taxonomy for, brownfield sites. In discussing urban housing capacity, the Department of the Environment, Transport and the Regions observes that "Large vacant and derelict sites can be a blind spot for capacity studies" and that "several capacity studies appear to have excluded large sites by default due to the use of typical urban area techniques" (DETR, 2000: 13). Syms (1999: 3) in discussing the problems of defining brownfield and identifying the potential of previously used land for development, has noted that "Taken altogether, therefore, these problems contribute to the tortuous path of redeveloping brownfields". Gibbons et al (1998: 152) in examining the situation in the USA observe that, among other issues, brownfield development is hindered due to "uncertainties arising from inadequate site information and competing redevelopment objectives", whilst the establishment of the CLARINET programme by the European Union reflects the need "to develop technical recommendations for social decision making in the rehabilitation of contaminated sites" (Bardos et al, 1999: 237).

Defining Brownfield

The term "brownfield" would appear to have two main derivations. The first is that it signifies the opposite of greenfield and, although there is no official recorded definition in the UK of either greenfield or brownfield (Raynsford, 1998), the use of the term greenfield is generally taken to mean land which has not previously been developed (POST, 1998). Therefore, the term brownfield has been adopted to describe land which has previously been subject to development. Development is defined in the Town and Country Planning Act 1990 as follows: "The carrying out of building, engineering, mining or other operations, in, on, over or under land, or the making of any material change in the use of any buildings or other land".

The use of the term "development" in this definition could be interpreted as a simple change of use. However, as types of development vary, and in some interpretations could include development as recreational space where no permanent buildings are erected, the use of the term brownfield is generally related to some former hard end-use that has involved physical construction, or to a situation where an industrial process, such as mineral extraction or sewage farming, takes place.

A second source of the term "brownfield" is the United States Environmental Protection Agency (US EPA), which has adopted the following definition: "Abandoned, idled, or under-used industrial and commercial facilities where expansion or redevelopment is complicated by real or perceived environmental contamination" (US EPA, 1996: 1). This definition can cause confusion in the UK because it incorporates contamination of land and, therefore, it implies that all brownfield is contaminated. Indeed, in the USA, land that is labeled as brownfield is almost always contaminated. As a result, the definition has led to problems through the generation of stigma and blight associated with brownfield sites, to the extent that many states have now introduced brownfield redevelopment programmes. These programs address both environmental and economic constraints through the provision of tax incentives that are underpinned by state-based legislation (Dennison, 1997). It should be noted that the US EPA definition also incorporates the terms "idled" and "abandoned", which in the UK can best be interpreted as vacant land, although some authors equate abandonment with dereliction (Meyer et al, 1995). However, the use of the term "under-used" implies that a partially occupied site can be classified as brownfield.

The term "brownland" has also been used in the UK since the early 1990s. This term has emerged from the activities of a group of landowners, many of whom own land that is in need of some form of intermediate or final treatment. The treatment of such land may include the removal of derelict structures and the remediation of contamination prior to its redevelopment or sale. This forum of landowners is known as the Brownlands Group.

In a document that was sent to local authorities by the DETR in connection with the first phase of the National Land Use Database (NLUD) survey, the purpose of the exercise was stated to be the collection of "information about previously developed land that may be available for new housing and other development" (DETR, 1998: 2). Within the NLUD exercise, brownfield is taken to mean previously developed land and buildings. This definition incorporates:

- previously developed land which is now vacant;
- vacant buildings (excluding single residential dwellings);
- derelict land or buildings (as defined in the 1993 Derelict Land Survey);
- derelict land or buildings allocated for any development in an
- adopted plan or having planning permission for housing; and
- other previously developed land or buildings where it is known
- there is a potential for redevelopment.

Comparing the above definition with the 51 land use categories contained in the NLUD classification, most sites which are within the definition fall within category 12: Vacant Land and Buildings. However, there are also a number of overlaps between categories and examples of activities that are difficult to categorise. To take a trivial example, should derelict hen sheds, which are usually categorised as 11.6: Agricultural Buildings, be classified as brownfield? Such land uses may exist in areas that, otherwise, have not previously been developed and which a local authority would not wish to see developed in future. Therefore, although land or property may be classified as brownfield, this should not be taken to imply that it is available for redevelopment. The issue of whether sites are actually available for redevelopment at a specific moment in time has also influenced the category to which they are assigned within the NLUD exercise. This may have affected the NLUD final results.

All of the above issues and considerations have to be taken into account in attempting to produce a definitive and widely-applicable definition of the term "brownfield". In order to illustrate the difficulties encountered in attempting to move towards consensus, it is appropriate to list a number of the recent definitions that have been developed and applied by organisations and individuals in the UK:

- Brownfield sites are buildings and land either now vacant or that could become vacant or suitable for development, during a relevant (development) plan period (Gwilliam, 1997: 23).
- Any areas of land which have previously been the subject of man-made or non-agricultural use of any type. This would include industrial uses such as chemical works, heavy engineering, ship building and textile processing, together with unfit housing clearance sites and docklands, both inland and coastal, as well as mineral extraction and those sites used for landfill purposes (Syms, 1994: 63).
- Brownfield sites are those sites within urban areas which have previously been used by industry, such as docks or steelworks, or by institutions, such as former schools or hospitals. The category includes both redevelopment and infill sites, as well as conversion of existing buildings. A redevelopment site is one which has been specifically cleared for house building; an infill site is vacant or derelict land within an urban area (Strathclyde Regional Council, 1995: 12).
- A parcel of land which has been previously built on or despoiled by mineral extraction or waste disposal, and is now derelict, vacant or unused (Cheshire Housing Land Monitoring System, 1998: 20).

In addition to these UK definitions, the term "brownfield" has been defined in many different ways elsewhere. The American standard definition has already been referred to, and in this case the US EPA has associated brownfield with the actual or suspected presence of contamination. Other examples of non-UK usage are given below:

- Abandoned pieces of land, mainly in inner cities, which are often blocked for economic development due to their ecological and economic risks (Freier, 1998: 4).
- A brownfield site is one which has been urbanised or used industrially, subsequently vacated and available for re-urbanisation (Duany Plater-Zyberk and Co, 1998).
- Brownfields were developed sites with buildings and facilities from earlier industrial periods that often could not serve new types of businesses. Other things being equal, they were economically inferior to their greenfield competition (Meyer et al, 1995: 4).
- Brownfields, abandoned or underused industrial or commercial property, are universal. These properties generally consist of derelict, dilapidated buildings with rotting machinery … brownfields can also present environmental and health hazards, and negatively impact the social and economic health of the surrounding communities (Gibbons et al, 1998: 151).

The above definitions drawn from theory and practice in the UK and elsewhere provide, first, evidence of the significant potential for confusion and conflict which is present in any discussion of the characteristics, problems and opportunities associated with brownfield sites, and second, a number of building blocks that can be used to assemble a more complete and more generally acceptable definition of brownfield. The latter point is of particular interest in the context of this chapter, because it was the search for a more satisfactory definition of brownfield that provided the initial impetus for the establishment of the National Brownfield Sites Project. Most of the definitions referred to so far in this chapter can be seen to display four principal characteristics:

- they make reference to the status of brownfield sites – dilapidated, vacant, unused, disused, idled and abandoned are terms frequently used to describe the status of sites;
- they describe a stage or point in the life cycle of a site – now vacant, partially occupied, could become vacant and redevelopment site are typical descriptions;
- they indicate the physical state of site – contaminated, despoiled, derelict and ecological risk are terms frequently used to indicate the state of a brownfield site;
- they refer to the socio-economic and institutional situation within which a site can be assessed – the situation can be seen as both an evaluation of the consequences of brownfield (blight, negative image, etc) and a description or statement of the actions required in order to bring brownfield back into productive use (plans, strategies, investment programs, financial inducements, special planning regimes, etc).

Having isolated the fundamental characteristics of brownfield, there are four other issues to be addressed before a satisfactory definition can be proposed. The first of these concerns the image, as against the reality, of brownfield; this is an issue that has been discussed for many years. The Hunt Report (1969) on the Intermediate Areas pointed to the negative image of economic potential associated with the presence of brownfield, whilst the Select Committee on Scottish Affairs (1972) suggested that, given the contribution to national or regional wealth that was made in the past by economic activities located on sites that have become brownfields, it is reasonable to assume that "the nation as a whole has a clear obligation to assist in rapid reclamation" (Select Committee on Scottish Affairs, 1972: 46 and 47). These two examples demonstrate, first, that such issues have long been of interest to policymakers, and, second, that the negative effects of brownfields frequently extend beyond the boundary of an individual despoiled site. However, whilst image and reality are coincident in some cases, in other situations the preconception and the reality diverge. This may be the case, for example, when brownfield is used by a local community as an informal recreational resource or in instances where brownfield sites provide valuable habitats (Kivell and Hatfield, 1998).

A second issue, which is also a matter that has been discussed for many years, relates to the importance of analysing and understanding the process of brownfield formation within the context provided by the overall framework of planning and development activity. Central to this issue is the question of when an active site becomes a brownfield. In considering this question it is essential to distinguish between those cases whereby brownfield results from a chance or random action, which may be considered unavoidable, and other cases in which the creation of brownfield is regarded as an externality and is associated with a deliberate action. It is the latter category of brownfield, which can frequently be predicted, that should be avoided, and to this end a number of recent strategic plans have attempted to identify potential brownfield sites and then to plan their transition from one productive use to another, thereby avoiding such sites becoming brownfield (Glasgow and the Clyde Valley Structure Plan Joint Committee, 1999). This notion of the early identification of the potential redundancy of land or buildings in order that an alternative future use can be planned and implemented in a seamless manner, has also been advocated by the Urban Task Force (1999) and by some sections of the development industry (House Builders Federation, 1998).

The third issue to be considered relates to the socio-economic dimension of brownfield. This issue is associated, on the one hand, with the blighting, deprivation and poor local environment that is part of what could be described as the brownfield syndrome; Handley (1996: 18) has described brownfield as "not just a technical problem – there is an important human dimension". On the other hand, brownfield can also be viewed as an important economic resource; the fact that a site is brownfield may have prevented its use in the past and, if it is located in a prime position, such a site may now be seen as a valuable asset rather than a liability. Canal and railway lands frequently fall into this category and some such sites have proved to offer major locational advantages as the emphasis in development policy and practice has swung away from the outer city back to the inner city urban core (Latham and

Swenarton, 1999).

A fourth issue relates to the desirability of providing an overall assessment of the costs and benefits that are associated with the presence and/or regeneration of brownfield. Accepting that not all future urban development can or should be accommodated on brownfield sites, and acknowledging that there is a need to develop a balanced portfolio of greenfield and brownfield in order to accommodate future urbanisation, it is clear that the costs of not mobilising brownfield are considerable and should be set against the costs attached to remediation and restoration. By adopting a total costing approach, which has been advocated by both North American and European researchers (see, for example, De Sousa, 2000 and Fulford, 1998), it is possible to construct a definition of brownfield which avoids the preconception that they are inherently a problem.

Having explored previous attempts to define brownfield, and having examined a number of characteristics and issues that help to frame the brownfield debate, it is possible to offer a definition that both provides an objective description of brownfield and attempts to encapsulate most eventualities. The definition generated by the National Brownfield Sites Project is:

> A brownfield site is any land or premises which has previously been used or developed and is not currently fully in use, although it may be partially occupied or utilised. It may also be vacant, derelict or contaminated. Therefore a brownfield site is not necessarily available for immediate use without intervention (Alker et al., 2000: 22).

This definition, which attempts to encapsulate all of the necessary characteristics and to avoid any ambiguity, consists of a number of important elements (see figure 19.1). Brownfield land is:

- land that has previously been developed;
- all land not in current use which presents actual or suspected land contamination;
- land which is currently wholly developed and not used;
- land in both rural and urban locations;
- land defined as contaminated under Section 78A(2) of the Environmental Protection Act 1990;
- land classified as derelict;
- land classified as vacant.

The definition also provides a solid foundation upon which to base a clearer and more consistent approach to brownfield regeneration. It is important to note that the definition does not indicate availability for redevelopment in terms of planning status – for example, a brownfield site in a environmentally protected area would not always be considered as suitable for future development – and it does not imply that a landowner is willing to sell, lease or redevelop a site. The most important element of the definition is contained in the final sentence – not necessarily available for immediate use without intervention. Whilst this condition is implied in some other definitions (for example, Syms, 1994, or the definition of statutory derelict land), the scope of intervention envisaged under the proposed definition is much wider and includes the removal of physical, legal, ownership, planning status, regulatory, infrastructure, insurance and other barriers to the redevelopment of brownfield sites.

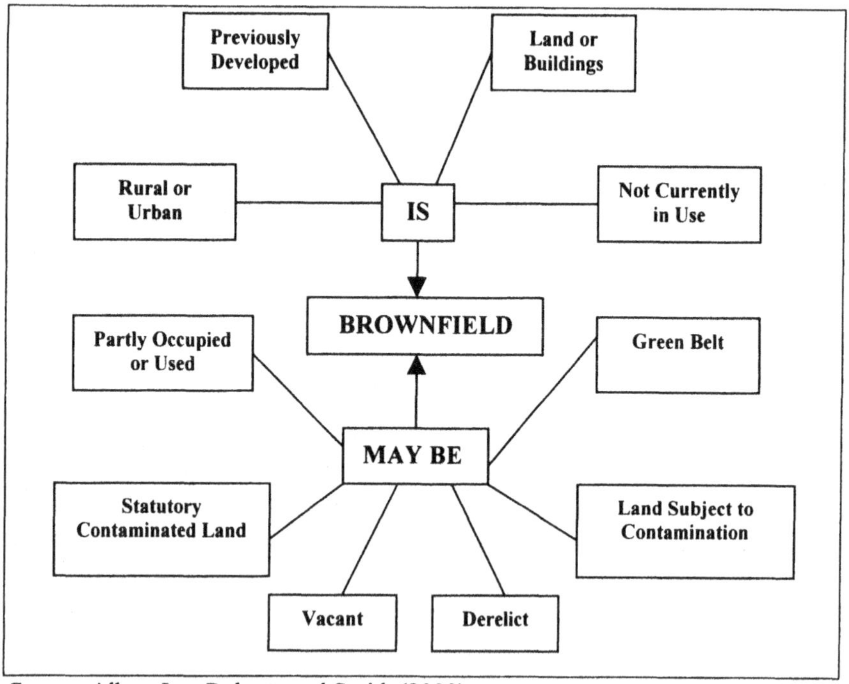

Source: Alker, Joy, Roberts and Smith (2000).

Figure 19.1: Elements in the definition of brownfield

Classifying and Modelling Brownfield

Having arrived at a more comprehensive and satisfactory definition of brownfield, the next stage in the development of an enhanced capability to understand and manage brownfield is the production of a system of classification (Roberts et al, 1998). Most previous attempts to classify brownfield have used one or more of the following methods and models:

- checklists,
- set models,
- matrix models.

Checklists are the most commonly used method for the classification of brownfield. Standard checklists have been developed for various categories of brownfield. Syms (1997), for example, has developed checklists for use in relation to contaminated sites. The strengths of this approach are that all reasonable steps can be taken to investigate the conditions and factors evident on a particular site at a specific moment in time and that the development and use of the methodology does not require extensive equipment. However, checklists also display a number of weaknesses, including the static nature of the method, the difficulties encountered in attempting to compare sites at different locations or at different moments in time, and the problems encountered in attempting to add extra factors to an original checklist.

Set models provide a more flexible and complete method for the classification of brownfield. The POST (1998) report on brownfield, "A Brown and Pleasant Land", advocated the use of a set approach in order to demonstrate the relationships between the various types of brownfield. This model includes sites classified as derelict or vacant, suffering from land contamination, urban and rural land, previously developed land, sites with poor ground conditions, land falling within the statutory definition of contaminated land, and under-used sites, including those that are within the boundaries of existing active enterprises. The strengths of set models are that they allow the inter-relationships between the various characteristics of brownfield to be mapped and understood, and they also provide a dynamic capacity that can be used to plot changes over time in terms of the features and status of sites. A major weakness is that set methods cannot easily be applied spatially.

The third group of methods utilise a matrix approach. A two-dimensional matrix, such as the method developed for the purposes of conducting environmental impact assessment (Glasson et al, 1994), offers considerable potential because it allows the required elements – the characteristics and the various sites – to be brought together in a single display. A three-dimensional version of the model, which also incorporates temporal variations, offers even greater potential. However, few matrix models of brownfield have been developed. In part this reflects the technical complexity of such models, but is also illustrates the difficulty of obtaining sufficient reliable information.

The Urban Mines research programme attempted to address the problems associated with all three types of model. Building upon the characteristics developed in the definition, a matrix approach was eventually adopted, principally because it allowed the characteristics evident at an individual site to be recorded and compared with the characteristics evident at other sites. This method also allows for the tracking of the changing nature and status of an individual site and of the entire inventory of sites at local, regional or national levels.

The following section develops the theoretical material discussed in this section and also encapsulates the lessons from previous practice. Emphasis is placed on the importance of attempting to incorporate the view of the various stakeholders who are involved in the redevelopment of brownfield and of identifying the information requirements of these stakeholders.

The Views and Requirements of Stakeholders

Characteristics and Types of Brownfield

In order to progress from a definition and theoretical model of brownfield, to an operational method of analysis and classification which meets the requirements of stakeholders, the research attempted to isolate the characteristics that should be incorporated in any typology. The most important characteristics to be incorporated included those related to:

- the current status, stage, state and situation of a site;
- the possible end uses for a site;
- the technical options for remediation and/or regeneration;
- the likely costs involved in remediation and/or regeneration.

In addition, individual stakeholder groups expressed a number of specialist requirements, including, for example:

- insurers may wish to know more about the extent and severity of liabilities and risks;
- landowners may want to be reassured about market demand in relation to alternative end uses;
- planners may require additional information about the presence and adequacy of on-site infrastructure provision.

Combining an extensive review of the strengths and weaknesses of existing methods of analysis and classification with the results of an initial survey of a sample of stakeholders, a list of over 120 characteristics or indicators was initially isolated. A shorter list of 38 characteristics was devised by combining two or more characteristics or indicators from the original list. These characteristics were grouped under four headings: environmental, economic, social and overlapping (see table 19.2).

A second stage in the development of a classification system incorporating the views of stakeholders, required the research team to produce a typology of brownfield sites. At this stage the main elements from the definition were isolated by reference to the requirements of stakeholders. Five key terms in the definition were used to establish a broad typology of sites: vacant, derelict, contaminated, partially occupied or utilised, and requiring intervention. Combining these terms provided sixteen types of brownfield. The full typology is shown at table 19.2.

Table 19.1: Environmental, economic, social and overlapping characteristics of brownfield sites

Environmental	**Economic**	**Social**
Ecological	Economic outlook	Community interaction
Geological	Economics of Infrastructure	Crime
Geotechnical	Economics of remediation	Demographic information
Pollution	Economics of site development	Education and training
Renov. and reclamation	Investment supply	Employment opportunities
Topology	Monetary incentives	Heritage
Water qual. and supply	Site economics	Recreation and leisure
		Urban capacity
Overlapping		
Benefits	Locational	Risk
Costs	Perception	Regional variations
Development issues	Planning instruments	Site availability
End use(s) of site	Policy instruments	Statutory/reg. Controls
Infrastructure	Post-development impacts	
Liabilities	Pre-development impacts	

Table 19.2: A typology of brownfield sites

I	Vacant, available for immediate use.
II	Vacant, partially occupied or utilised, available for immediate use.
III	Vacant, requiring intervention.
IV	Derelict, requiring intervention.
V	Contaminated, requiring intervention.
VI	Vacant and derelict, requiring intervention.
VII	Vacant and contaminated, requiring intervention.
VIII	Vacant, derelict and contaminated, requiring intervention.
IX	Derelict and contaminated, requiring intervention.
X	Vacant, partially occupied or utlised, requiring intervention.
XI	Derelict, partially occupied or utilised, requiring intervention.
XII	Contaminated, partially occupied or utilised, requiring intervention.
XIII	Vacant and derelict, partially occupied or utilised, requiring intervention.
XIV	Vacant and contaminated, partially occupied or utilised, requiring intervention.
XV	Vacant, derelict and contaminated, partially occupied or utilised, requiring intervention.
XVI	Derelict and contaminated, partially occupied or utilised, requiring intervention.

Each of the sixteen types of brownfield site was further elaborated by reference to the 38 characteristics or, in individual cases, to site-specific or activity-related characteristics. For the purposes of the following stages of the research programme, the 38 characteristic model was used.

Stakeholder Information Sources and Perceptions

Moving beyond the overall classification of the characteristics and types of brownfield, the research sought to investigate the views and types of information sources used by stakeholders, and the extent to which stakeholders considered these sources to be reliable. This stage of the research involved the participation of some 230 stakeholders in 22 regional seminars. An important component part of the identification of stakeholder views was the use of a structured ranking exercise through which stakeholders were asked to identify which of the 38 characteristics (and, by definition, categories of information) they took into consideration when assessing the development potential of a brownfield site. Stakeholders were also asked to rate the reliability of the information available in each of the 38 categories. For the purpose of these exercises stakeholders were grouped under the following headings:

- developers (A)
- landowners (B)

- local authority – environment (C)
- local authority – planning (D)
- professional advisers (E)
- environmental consultants and engineers (F)
- regulators (G)
- community interests – social (H)
- community interests – environmental (I)
- others (J).

Tables 19.3 and 19.4 summarise the results from the two exercises. In each exercise stakeholders were asked to score each category from 0 to 100. A critical category of information was defined as one for which over 90 per cent of the stakeholder group had given a score and the average score was over 70. For a category of information to be considered reliable, over 90 per cent of the respondents had to give it a score of over 70.

Comparing the results from these exercises with other research evidence is difficult, because either the other studies examined different aspects of brownfield development, or they considered all stakeholders as a single group. However, there are a number of points of validation available from other research evidence, including the importance placed on:

- the specification of a suitable end use or uses;
- the issue of cost;
- the need to address contamination and pollution issues;
- the extent and control of liabilities;
- the estimation of risk;
- the need to identify and assess suitable methods for renovation and reclamation;
- the need to operate in accord with relevant statutory and regulatory guidance and controls;
- the desirability of considering the overall economics of redevelopment; and, as ever,
- location, especially vis a vis other economic opportunities.

Table 19.3: Critical categories of information for each stakeholder group

	Stakeholder Group									
Factor	A	B	C	D	E	F	G	H	I	J
Benefits		X		X				X	X	X
Community interaction								X		
Costs		X		X		X	X			X
Crime										
Demographic information								X		
Development issues	X	X		X						
Ecological				X					X	
Economic outlook										
Economics of infrastructure	X									
Economics of remediation	X	X				X	X			X
Economics of site development	X	X				X				
Education and training								X		
Employment opportunities								X		
End uses(s) of site	X	X	X	X	X	X	X	X	X	X
Geological	X		X			X	X			
Geotechnical	X					X				
Heritage										
Infrastructure				X						
Investment supply		X								
Liabilities	X				X	X	X			
Locational	X	X		X				X		X
Monetary incentives	X	X						X		
Perception		X								
Planning instruments		X	X	X					X	
Policy instruments				X			X			
Pollution	X		X	X	X	X	X			X
Post-development impact		X		X						X
Pre-development impact		X		X						X
Recreation and leisure										
Regional variation										
Renovation and reclamation	X	X	X	X	X	X	X			
Risk	X	X	X		X	X	X			
Site availability	X	X								
Site economics	X	X								
Statutory and reg. controls	X		X	X	X	X	X		X	
Topology				X		X				
Urban capacity				X						
Water supply and quality			X			X				

Table 19.4: Reliable categories of information for each stakeholder group

	Stakeholder Group									
Factor	A	B	C	D	E	F	G	H	I	J
Benefits										
Community interaction										
Costs										
Crime										
Demographic information								X		
Development issues										
Ecological										
Economic outlook										
Economics of infrastructure										
Economics of remediation		X								
Economics of site development	X	X								
Education and training										
Employment opportunities										
End uses(s) of site			X	X			X	X		
Geological		X								
Geotechnical										
Heritage										
Infrastructure		X		X						X
Investment supply		X								
Liabilities		X								
Locational	X	X						X	X	X
Monetary incentives										
Perception										
Planning instruments	X	X	X	X	X		X		X	
Policy instruments										
Pollution		X								
Post-development impact										
Pre-development impact										
Recreation and leisure										
Regional variations										
Renovation and reclamation		X								
Risk										
Site availability		X								
Site economics										
Statutory and reg. controls	X	X	X	X	X	X	X		X	
Topology	X	X	X	X		X	X		X	
Urban capacity				X						
Water supply and quality		X								

Work undertaken by Fulford (1998) indicates that developers pay attention to many of the above issues, with particular emphasis placed on the assessment of the extent and severity of contamination and on the question of cost. Syms (1999) reports the presence of similar factors in a study which involved a more extensive group of stakeholders, with especially high scores recorded for risk assessment, the actual or potential presence of contamination and pollution, and a category described as "further considerations" which included regulatory and statutory guidance and controls. Further substantiation for the evidence produced during the course of the Urban Mines research program can be seen in the work of De Sousa (2000) in Toronto and of Gibbons et al (1998) in the USA.

Given the scale of the consultation exercise which was conducted, together with the spread and selection of participants in the survey and the presence of substantiation from other empirical research, it is considered that the views represented in figures 4 and 5 can be considered to be reliable.

Three further issues emerge from this aspect of the research, First, there is considerable variation between stakeholder groups in terms of the number of categories of information that they take into consideration when examining a brownfield site. Some groups consider a large number of categories of information – developers, landowners, planners, environmental consultants and regulators typically consider ten or more categories of information to be important. This group has a requirement for multiple sources of information and tends to take an integrated approach to the assessment of brownfield. Other groups consider a smaller number of categories of information to be important – professional advisers and community groups tend to focus on a shorter list of factors. Professional advisers, such as lawyers and insurers, frequently focus attention on those issues that are directly relevant to their work, whilst community groups are often topic or issue-based.

A second issue reflects the generally poor quality and reliability of much of the information available regarding the condition and potential of brownfield. Over half of the factors listed in figure 5 are not considered to be reliable by any of the stakeholder groups. Five or more of the ten stakeholder groups considered that reliable information was available with regard to:

- locational factors,
- planning instruments,
- statutory and regulatory controls,
- topology.

The third issue which emerges is of greater concern. Comparing the results of the two surveys that are summarised in figures 4 and 5, it is evident that in the case of four of the five most important categories of information, the information which is seen as essential is also considered to be unreliable. Table 19.5 compares the two sets of results.

Table 19.5: Critical categories of information and reliability

Factor	Score on Importance	Score on Reliability
End use of site	10	4
Pollution	7	1
Renovation and reclamation	7	1
Statutory and regulatory controls	7	8
Risk	6	0

It can be surmised from this discussion that, in addition to the general confusion which surrounds the definition and identification of brownfield and which as a consequence inhibits the regeneration of sites, the absence of reliable information on a number of key factors further adversely affects the progress, scale and cost of regeneration.

In addition to the specific answers provided by stakeholders, they also made a number of informal observations regarding the definition, typology and classification of brownfield and the brownfield site characteristics that were proposed by the research team. The creation of a common "language" of brownfield terminology was one of the intended outcomes of the project, and it would appear that stakeholders place a value on this contribution. Other favourable comments were received with regard to the classification method and typology.

Illustrations of the Application of the Typology

The definition, typology and classification system developed by the project were tested in eight case study areas. Using a combination of primary data collection and secondary data sources, all brownfield sites in an area were identified, assessed and classified. For each case study area the available information, on a site-by-site basis, was recorded using Arcview GIS software. A summary of the finding of the surveys is presented at table 19.6.

Table 19.6: Statistics for case study areas

Case Study Area	Total area surveyed (km^2)	Number of brownfield sites	Total area of brownfield (km^2)	Percentage of area that is brownfield
Bishop's Stortford	6.0	10	0.05	0.9
Blaenau Gwent	20.7	121	0.91	4.4
Bradford	5.0	183	0.46	9.1
Crewe	6.0	23	0.25	4.2
Hounslow	5.2	61	0.32	6.0
Sheffield	14.0	142	0.60	4.3
Sunderland	5.1	37	0.57	11.1
Wolverhampton	10.0	89	1.50	15.0

Three illustrations are sufficient to demonstrate the analytical capabilities of the typology and the method of site assessment:

- Blaenau Gwent is a semi-rural area in South Wales with many abandoned coal and metallurgical sites. Key barriers to brownfield development identified in the survey included economic obstacles (economic outlook and investment supply), social factors (employment opportunities, community interaction, education and training, demographic information), environmental problems (topology and geotechnical) and overlapping factors (end-use and development issues). Few positive features were observed or recorded.
- Bradford is a traditional industrial centre; the textile industry was once dominant, but this has declined and services are now more important. Key barriers to brownfield redevelopment that were recorded included economic factors (economics of site development and investment supply), social issues (most factors), environmental factors (renovation and reclamation, topology), and overlapping barriers (post development impact). Positive factors include economic issues (economic outlook, monetary incentives), social factors (recreation and leisure, urban capacity), environmental advantages (most) and overlapping factors (end-use, development issues).
- Hounslow is located to the west of London; a previous history of industrial activity has been replaced by service activities. Positive elements include all of the listed economic factors, most social issues and most overlapping factors. Barriers to redevelopment include social factors (demographic information, crime), most of the listed environmental barriers (and especially renovation and reclamation,

geological and geotechnical) and some overlapping issues (end-use, risk).

Conclusions and Policy Implications

This chapter has demonstrated the importance of providing a clear and generally-accepted definition of brownfield, and of developing this definition in order to provide a method of classification that can be used to identify and assess brownfield sites. In addition, the chapter has illustrated the differing perceptions and experiences of the various stakeholder groups involved with brownfield development.

A number of points have emerged which are of considerable importance for policy. For public policy makers the study has demonstrated that much of the brownfield resource is "hidden" due to problems of identification and definition. Across the eight case study areas, some 6.4 per cent of the total area surveyed was identified as brownfield, suggesting that an average level of 5 per cent of the total land area for urban areas requiring regeneration in England and Wales can be classified as brownfield. Even in those areas in which little brownfield is evident, it is suggested that between 1 and 2 per cent of the land area can be classified as brownfield, although many of the sites may be very small. The implication for the public policy community is that detailed survey using a consistent definition, which is wide enough to cover most types of site, is essential. In addition, it is important to up-date survey information on a regular basis, to ensure that potential brownfield is identified in advance and to bring together a brownfield database with land-use and economic planning processes.

The need for regular review and updating can be illustrated by reference to a recent resurvey of one of the Union Mines case study areas. In Bradford, the original survey in 2000 identified 183 sites covering an area of $0.46km^2$. A resurvey of the same area in 2001 revealed that 71 sites have been developed, but that 90 new sites have been added to the brownfield "stock". The total area of brownfield ($0.46km^2$ in 2000) has been reduced, with some $0.39km^2$ of brownfield present in 2001. It would appear that the larger sites have been developed first, but that none of the contaminated sites have been brought back into use (Urban Mines, 2001).

For the development community, the major implications of the research are related to the many difficulties currently encountered due to the lack of an accepted system of classification which can allow individual sites to be identified, assessed and registered. The provision of a standard listing procedure, whereby sites of a given type and specified size can be recorded by location, would aid redevelopment, especially if the list or register was in the public domain.

For both the public and private sectors, the research also demonstrates the lack of reliable, detailed and consistent site information, and the barriers to redevelopment which result from the absence of such indispensable management information. This suggests that there is an urgent need to refine and extend the coverage of registers such as that provided by NLUD, and to make the information collected available to the general public. Identifying brownfield sites is not enough, it is also essential to provide the information which is required in order to speed the processes of decision making and redevelopment.

References

Alker, S.; Barrett, P.; Clayton, D.; Jones, G.; Joy, V. and Roberts, P. (2000), *Delivering Regeneration: A Brownfield Renaissance*, Urban Mines, Norwood Green.

Alker, S.; Joy, V.; Roberts, P. and Smith, N. (2000), ' The Definition of Brownfield', *Journal of Environmental Planning and Management*, vol. 43, no. 1, pp. 49-69.

Bardos, P.; Lowe, M. and Baverstock, A. (1999), 'UK Participation in CLARINET', *Land Contamination and Remediation*, vol. 4, no. 4, pp. 237-243.

Cheshire Housing Land Monitoring System (1998), *User Handbook*, Cheshire County Council, Cheshire.

Dennison, M.S. (1997), *Brownfields Redevelopment*, Government Institutes, Rockville Maryland.

Department of the Environment, Transport and the Regions (1998), *Letter to Chief Executives and Chief Planning Officers of Local Authorities in England, re: National Land Use Database*, DETR, London.

Department of the Environment, Transport and the Regions (2000), *Tapping the Potential*, DETR, London.

De Sousa, C. (2000), 'Brownfield Redevelopment versus Greenfield Development', *Journal of Environmental Planning and Management*, vol. 43, no. 6, pp. 831-853.

Duany Plater-Zyberk & Co. (1998), *Architects and Town Planners: Glossary of Terms*, www.dpz.architects.com/terms.htm

Freier, K. (1998), *CLARINET Working Group1: Brownfield Redevelopment Workplan*, German Environment Agency, Berlin.

Fulford, C. (1998), *Urban Housing Capacity and the Sustainable City*, Town and Country Planning Association, London.

Gibbons, J. S.; Attoh-Okine, N.O. and Laha, S. (1998), 'Brownfields Redevelopment Issues Revisited', *International Journal of Environment and Pollution*, vol. 10, no. 1, pp. 151-162.

Glasgow and the Clyde Valley Structure Plan Joint Committee (1999), *Consultation Draft Structure Plan*, GCVSPJC, Glasgow.

Glasson, J.; Therivel, R. and Chadwick, A. (1994), *Introduction to Environmental Impact Assessment*, UCL Press, London.

Gwilliam, M. (1997), 'Something Old, Something New', *Planning*, no. 1233, p. 23.

Handley, J. (1996), *The Post-Industrial Landscape*, Groundwork Foundation, Birmingham.

House Builders Federation (1998), *Urban Life*, HBF, London.

Hunt, Sir Joseph (Chairman) (1969), *The Intermediate Areas – Report of a Committee under the Chairmanship of Sir Joseph Hunt*, Cm.3998, HMSO, London.

Kivell, P. and Hatfield, S. (1998), 'Derelict Land – Some Positive Perspectives', in Kivell, P.; Roberts, P. and Walker, G. (eds.), *Environment, Planning and Land Use*, Ashgate, Aldershot, pp.118-129.

Latham, I. and Swenarton, M. (1999), *Brindleylace: A Model for Urban Regeneration*, Right Angle Publishing, London.

Meyer, P.B.; Williams, R.H. and Yount, K.R. (1995), *Contaminated Land*, Edward Elgar, Aldershot.

Parliamentary Office of Science and Technology (1998), *A Brown and Pleasant Land*, POST, London.

Raynsford, N. (1998), *Hansard Written Answers to Parliamentary Questions*, 8 April, col.290, Stationary Office, London.

Roberts, P.; Collis, C.; Healey, M.; Noon, D.; Bozeat, N.; Johnson, C. and Medhurst, J. (1988), *A Study of the Performance of the Dudley Local Economy*, ECOTEC, Birmingham.

Roberts, P.; Joy, V. and Alker, S. (1998), 'Towards a Brownfield Sites Taxonomy: Issues in the Definition and Classification of Problems and Potentials', Unpublished Paper Presented at the *Remediation of Brownfield Sites for Housing Conference*, October 28, University of Ashton.

Roberts, P. and Sykes, H. (2000), *Urban Regeneration*, Sage, London.

Select Committee on Scottish Affairs (1972), *Land Resource Use in Scotland*, vol. 1, Cm. 5428, HMSO, London.

Strathclyde Regional Council (1995), *Strathclyde Sustainability Indicators*, Strathclyde Regional Council, Glasgow.

Syms, P.M. (1994), 'The Funding of Developments on Derelict and Contaminated Sites', in Ball, R. and Pratt, A.C. (eds.), *Industrial Property Policy and Economic Development*, Routledge, London.

Syms, P.M. (1997), *Contaminated Land*, Blackwell, Oxford.

Syms, P.M. (1999), *Redeveloping Brownfield Land: The Decision Making Process*, Paul Syms Associates, Macclesfield.

United States Environmental Protection Agency (1996), *www.epa.gov/swerosps/bf/html-doc/brinit.htm*

Urban Mines (2001), *Bradford Update*, Urban Mines, Norwood Green.

Urban Task Force (1999), *Towards an Urban Renaissance*, E & F N Spon, London.

20 Who Needs Housing in the South East?

CHRISTINE M. E. WHITEHEAD

Introduction

The Government's housing policy, enunciated in the Housing Green Paper (DETR, 2000), states that "demand for new housing will continue to grow: household projections indicate that somewhere close to 3.8 million households may form between 1996 and 2021 - equivalent to around 150,000 new households each year". It further states that patterns of supply and demand vary greatly at regional and local level and over time - making it difficult to match housing provision to needs.

The Green Paper then makes it clear that policies about the provision of additional housing are fundamentally matters for the planning system: "Our policies for future housing requirements are set out in planning guidance", and that responsibilities for these decisions on housing provision - how much, where, and how much should be affordable- lie with local authorities as long as these approaches are consistent with overall totals agreed by regional planning bodies. It is thus planners who have the responsibility to take the household projections and other estimates of future housing need and relate these to information on the use of the existing stock of dwellings and residential land and expected densities in order to determine how much additional land to release for housing purposes. It is also clear that they are charged with ensuring adequate provision - it is their role to allocate totals to their most appropriate locations and densities. It is not the role of the planning system to constrain that additional supply to less than the anticipated demand (see Crook and Whitehead, forthcoming, for a more detailed analysis of this process).

The objective of this chapter is to examine how this approach may pan out in the context of the most pressurised regions of England - London and the South East.[1] It draws on three reports which analysed in detail the basis of the household projections in these areas, the linkages between these projections and the demand for labour, the limitations on supply and the implications of the potential mismatch between demand and supply on the regional housing market and the wider economy (Whitehead *et al*,

1999; Travers *et al*, 1999; Holmans *et al*, 2000).

The Demographic Base

There is little doubt that the population of London and the South East is set to increase over the next two decades. How rapidly will depend on two main pressures -the growth of the indigenous population and continuing international inmigration. The first is almost certain to occur; the second depends on underlying economic pressures and on government policies with respect to general immigration and particularly to refugee and asylum seekers and where they are housed.

These numbers are not easy to predict - as is clear from the evidence of past decades. Patterns in London and the South East (table 20.1) show that population in London continued to decline, particularly in inner London, until well into the 1980s, before starting to increase quite rapidly in the late 1980s and 1990s. At the same time population in the South East continued to rise, initially not by enough to offset the decline in Greater London but from the early 1980s generating significant net increases in the 1980s and 1990s.

Table 20.1: Total resident population (London and South East) and official 1996 projection (000s)

	Inner London	Outer London	Greater London	South East
1961	3,481	4,496	7,977	--
1971	3,060	4,470	7,529	6,830
1981	2,550	4,255	6,806	7,245
1991	2,627	4,263	6,890	7,679
1996	2,708	4,366	7,074	7,895
2001	2,765	4,450	7,215	8,134
2011	2,863	4,607	7,470	8,534
2021	2,963	4,773	7,736	8,905

Source: Whitehead et al. (1999).

The official projections from 1996 onwards suggest very significant continuing increases in all parts of the region - by 9.3% in London and 12.8% in the South East up to 2021 (DETR 1999/ONS 1999). The 1998 projections add almost another million to the projected population to 2016 (Shaw, 2000).

Within this total some 450,000 are expected to be in London and perhaps 180,000 in the South East (Holmans, 2000; Holmans, 2001). Thus not far short of one half of the upward revision to the projected population of England is expected to be in London with a further twenty percent in the South East.

Table 20.2 shows how these population projections translate into households. They reflect the continuing decline in average household size throughout the period and the growing importance of people living alone. Over the twenty-five years from 1996 when official projections suggest an increase of 3.8 million nationally, household numbers are projected to go up by 21% in London and 26% in the South East. However, unofficial household projections based on the 1998 figures (Holmans, 2001) suggest that the national increase may be as much as 4.3 million between 1996 and 2021. Within this total over about 840,000 are projected to be in London and perhaps a further 850,000 in the South East - around 40% of the national increase is thus expected to be concentrated in these two regions (Holmans, 2001).

Table 20.2: Households in London and South East since 1971 with projections to 2021 (000s)

	Greater London	South East
1971	2,705	2,312
1981	2,635	2,644
1991	2,841	3,035
1996	3,002	3,225
2001	3,128	3,403
2011	3,377	3,735
2016	3,520	3,905
2021	3,645	4,060

Source: DETR (1999).

Behind these projections (not forecasts) are important changes in the ways that both the growth of the indigenous population and inmigration are affecting expected population and household change. First, and across the country, people are living longer, particularly men - adding to the overall population. Secondly, the South East and especially London, have relatively young populations, with a higher proportion therefore within child bearing ages - and this is particularly true of international inmigrants. Thirdly, international inmigration is now expected to add almost 100,000 to the UK population each year. The majority of that increase will initially come to London and to a lesser extent to the South East.

Table 20.3 shows how this has played out in London during the early part of the 1990s. Over the period 1991-1997, the population increase in London was some 230,000, very little more than measured natural change in the indigenous population. Net international inmigration however accounted for over 250,000 people but this was more than offset by outmigration mostly to the South East and other Southern regions.

Table 20.3: The make-up of population change in London in the 1990s (000s) 1991–1997

Starting Population 1991 (mid-year)	6889.9
Natural Change	229.1
Net Migration	- 195.0
(of which international migration	+ 87.8)
Asylum seekers and visitor switchers	178.5
Total migration	- 16.5
Other changes	+ 19.6
Final Population	7,122.2

Source: Travers et al. (1999).

In the later part of the decade the number of international inmigrants has almost so that by mid 1999 the mid-year population was estimated as 7,285 - with the total change of 395,000 over the eight-year period made up of 316,000 natural change and 80,000 net inmigration. Two of the most important questions with respect to the accuracy of the population projections are therefore what are the determinants of this figure and how volatile is it likely to be in the future. Economic migration is strongly related to the buoyancy of the economy, particularly that relating to financial services. The numbers of refugees and asylum seekers depend on world wide crises outside the UK's control but also on government attitudes - including the success or otherwise of policies to relocate people outside London. If the pressures of the late 1990s were to continue, the projections above would be significant underestimates - as in 1998 alone net international inmigration was estimated at 180,000 as against the 95,000 assumed in the projections for the medium term.

These figures show that the fundamental processes of migration are very different from popular understanding, which emphasises movement from north to south. This is clearly refuted by the evidence (Bate *et al*, 2000). It is international migration which is the engine of change. This then leads to increased outmigration from London to other regions, generating further pressures for outmigration to other regions. Table 20.4 shows that both in 1991 and 1997 net internal migration is negative for both London and the wider South East - although by 1997 the overall Southern subtotal has become positive as a result of growth in the East and South West regions.

Table 20.4: Net migration within UK (selected regions) (000s)

	1991	1997	1998	1999
London	-53	-55	-48	-65
South East	+13	+24	+20	+20
Eastern (GOER)	+20	+20	+19	+22
Subtotal SE	-31	-10	-8	-23
South West	+22	+32	+27	+33
Subtotal South	-9	+21	+19	+10
North	-10	-24	-21	-16
England	-16	-4	+2	0

Source: Population Trends, Table 8.1, Winter (2000).

Longer Term Employment Pressures

An important question, which will help determine the accuracy of longer term projections, is the expected demand for labour and the make up of that demand in London and the South East. This must be compared to the supply of labour implicit in the population and household projections, taking account of the extent and variation in commuting levels both within the region and from other regions.

The make up of employment has been changing rapidly throughout the country. Over the last decades the decline in manufacturing industry and the continuing deindustrialisation of the economy has been probably the most important force in generating population movement out of the urban areas (Urban Task Force, 1999). These pressures have been particularly strong in London and the South East (table 20.5). Engineering employment, for instance, has fallen in London by three times and in the South East by twice as much as in the country as a whole. Similar patterns are found in many of the traditional industries.

Table 20.5: Types of employment changes in employment 1987–1997 (%)

	London	SE	Great Britain
Engineering	-54	-33	-16
Other manufacturing	-40	-26	-14
Utilities	-73	-53	-42
Construction	-20	-6	-5
Catering	+29	+27	+22
Other business services	+38	+53	+46
IT, R&D, telecoms	+14	+39	+34
Retail	+5	+16	+18

Source: Census of Employment, NOMIS (1987 and 1997).

The growth in employment comes mainly from business services, telecoms, and in personal services such as catering and retailing. All of these sectors have grown, but only catering has grown proportionately more in London than in the country as a whole. The South East on the other hand has additionally done better than average in both telecoms and other business services. London's economy is concentrated in high paid, productive financial services, but also depends heavily on the personal services necessary for London to maintain its advantages in these high tech/high value added sectors. The South East is taking on many of the same attributes. What is clear is that, without adequate housing to accommodate both highly paid households looking for good quality/easy access housing and those on much lower incomes who provide necessary services, the economy cannot be expected to maintain its impetus into the longer term (Greater London Authority, 2000 and 2001).

More general evidence on labour markets suggest that while across the country demand and supply are roughly in balance until 2005, thereafter the demographics generate an imbalance with demand significantly outstripping supply as the population ages (Whitehead *et al*, 1999). This problem is being recognised across the (post) industrialised world and is leading to changes in working practices, notably with respect to the compulsory retirement age, and to far more positive attitudes to the immigration of skilled workers. These pressures suggest that, even if the economy were to grow less rapidly than in the late 1990s, inmigration is likely to be maintained at at least current levels. They also suggest that attitudes to refugees might become more generous as it becomes more obvious that many can rapidly join the labour force (Greater London Authority, 2001).

Thus the picture that emerges is one in which, if the national economy is to remain buoyant, housing must facilitate the continuing growth of the economy in London and the South East. Unless older and non-participant households are to be excluded from the region this implies much greater housing provision. It also requires that a significant proportion of the additional housing provided must be affordable to employees in personal, transport and public services. In the public sector in particular this raises important concerns about affordability. National pay scales make it difficult to address housing problems through the wage packet, particularly where promotion is linked to experience and qualifications as, for instance, in the case in the police force (Greater London Authority, 20001). In all sectors it generates problems of how to maintain economic competitiveness (Monk and Whitehead, 2000).

There is also a more fundamental concern in that, while London and the South East is undoubtedly the engine of the national economy, it also shows considerable signs of relative economic fragility. Both labour and housing markets appear to have become more volatile than the rest of the country, while only in a narrow range of internationally competitive sectors has employment grown as rapidly as elsewhere. Equally house prices both fell further in the early 1990s and have risen faster in the latest cycle. While the longer term trend is almost certainly upward - unless there is worldwide long term recession - the ride may well be very bumpy, especially if the Government's macroeconomic forecasts turn out to be optimistic (Travers *et al*, 1999).

Supply Capacity

Translating the household projections into a requirement for new housing involves additional analysis of the potential for reusing the existing stock of housing and land, the capacity to reduce vacancies and the likely extent of demolition and conversion.

The balance between households and dwellings, especially in London but also in the South East, is much tighter than in the rest of the country. In the country as a whole there are probably around 600,000 more dwellings than households. In London, on the other hand, the tightness of the market has increased to the point where the number of households exceeds the number of dwellings, while in the South East the situation is almost exactly balanced (table 20.6). This implies that there is already a large shortfall in provision as well-operating housing markets require a reasonable vacancy rate. This is borne out by the latest figures for the backlog in the provision of affordable housing estimated by Alan Holmans which suggests that over one third of the estimated 650,000 shortfall is located in London (Holmans, 2001; Holmans *et al*, 2000).

Table 20.6: Dwellings and households (thousand)

London

	Dwellings	Households
1971	2,555	2,705
1981	2,661	2,635
1991	2,912	2,841
1996	3,003	3,002

South East

	Dwellings	Households
1981	2,750	2,644
1991	3,120	3,035
1996	3,245 (est)	3,225

Source: Whitehead et al. (1999).

All the evidence suggests that very little of the requirement for additional housing can be found from the existing stock. Similarly, although there are possibilities of transferring commercial and other buildings into residential use, especially in London, the total numbers obtainable are relatively small -and will be less the greater the future success of the economy in the South East (Kleinman *et al*, 1999).

The latest estimates of London's capacity to provide for the projected increase in housing requirements suggested that perhaps 580,000 additional homes can be added in the capital over the twenty five year period 1992 - 2016 (London Planning Advisory Committee, 1999). Even this figure is based on "saving" almost 100,000 units by reducing the vacancy rate and increasing the number of households living in non self-contained accommodation. It also takes account of the potential for change of use of non-residential buildings and building new dwellings at higher densities on available brownfield sites.

This capacity suggests a very significant shortfall in meeting demands from additional households, in the capital alone, of at least 200,000 units over the next fifteen years. Moreover even this capacity implies an output of over 23,000 dwellings per annum, well above current output levels. The Mayor's Commission's estimates are much higher at 43,000 per annum, because they envisage providing for both the backlog of unmet need and for additional households (Greater London Authority 2000).

This tension between requirement and capacity can be alleviated in three main ways:

- housing standards could fall -although it is difficult to imagine that, as a nation, we will be prepared to consider lower standards than presently exist in two decades time when average incomes can be expected to be at least 50% higher;
- London's capacity can be increased by changing land use from non-residential to residential, recycling brownfield sites more effectively and increasing densities. This policy is being carefully examined at the present time, but is unlikely to achieve the very great increases required, especially given the tightness of the original estimates and the increasing standards likely to be demanded in the private sector -let alone the constraints on construction capacity in the capital; and/or,
- more households could move out of London and find accommodation in the wider South East, putting further pressure on the rest of the region -where increased requirements are already far greater than the land supply that districts feel able to identify.

The outcome is likely to be a mixture of all three possibilities - although the third will probably dominate. Thus, allowing for some overspill from London, the required output in the South East Government Office Region alone is probably of the order of 50,000 dwellings per annum - a million more homes over the next two decades. This is far more than either regional or local plans currently contemplate. This conclusion would still hold even if the increase in population and households in London were somewhat slower than the projections suggest.

Within these totals there are further complexities, associated with the Government's changing policy agenda, which planners must take into account. First the greater emphasis on brownfield sites makes it difficult to expand construction output rapidly. At the same time it increases the costs of that production. In London almost all output will be on brownfield sites - but as the easier ones are taken up it will be hard to maintain output levels, let alone expand them. In the wider South East, brownfield sites are in short supply across much of the region. Where they do exist there may be other, perhaps higher valued, uses available. Most identified brownfield sites are in the Thames Gateway area or on the coast. However, demand is concentrated to the West and South of London – placing additional pressure on already overstretched infrastructure.

The second major policy initiative is to increase densities. The density of population in London, especially inner London, is very low by international standards (Travers *et al*, 1999). Yet, there is no impetus for higher rise residential building as can be found in most other capital cities. Instead planners are expected to increase densities using traditional residential patterns of development. The main concern here is the potential difference between densities of occupation and planning densities. As incomes rise people require more space, both within their home and in terms of the site on which the dwelling is built. The challenge is to provide what is demanded at the same time as increasing planning densities – possible with greatly improved urban design, but not yet observed to any great extent in developments on the ground (Urban Task Force, 199; Travers *et al*, 1999).

In much of the South East densities are the lowest in the country, reflecting both high average incomes and accepted planning requirements with respect to parking and other services. Moreover in some areas these densities have continued to fall even in the changed planning environment (Whitehead *et al*, 1999; Monk and Whitehead, 2000). Planning advice has undoubtedly become much more directed at increasing these densities in the last two years – and quite small increases could reduce the land take significantly – but it will require a total reversal of longer-term trends if densities suggested in national guidance are to be achieved.

Implications for Housing and the Economy

If the national economy is to maintain its potential, the emphasis in London and the South East almost certainly has to be one of expansion. Underlying economics and demographics suggest that there will be significant increases in population and households in both London and the South East. Moreover that population will be relatively young and racially diverse. The greater the success of the economy the larger will be the net international inmigration and therefore the greater the pressure on the housing system.

The overall balance between dwellings and households in London and the South East has undoubtedly been tightening in the 1990s. This is reflected in increasing house prices and rents across the region as well as higher relative prices in London. This, in turn, generates increasing problems of affordability, especially among key workers in the public sector as well as higher levels of commuting as more workers move out of central areas.

A further economic pressure arises from the growing inequalities of incomes among employed households. This reflects the increasing importance of high valued added business and financial services on the one hand and the relatively greater expansion of low paid, often part time, service employment on the other. It is likely to result in even higher prices in market housing as well as more desirable areas becoming less accessible to poorer households. All these factors suggest a larger proportion of households needing some help with housing costs.

The pressures on the housing market are likely to continue to grow as supply finds it difficult to respond to both market and social sector requirements. If the urban renaissance transforms demand so that single person households in particular are prepared to live in higher density urban developments, whether in London or the wider South East, urban capacity can be increased and the main constraints will lie with the construction industry and ensuring greater planning flexibility. This scenario would also be consistent with the sustainability agenda of less travel and greater dependence on public transport. However it depends on major changes in attitudes so that planning and demand pressures are complementary rather than fighting against one another.

If existing trends towards larger homes, lower density living and greater suburbanisation cannot be reversed – even if they are significantly reduced – the tensions between planning and housing requirements will continue to grow. Many Londoners will want to move out at the same time as indigenous demand increases in the South East. Both lower paid workers and better off households will move further out increasing the need for commuting and generating further pressures in the wider region. If planners see their role as more to accommodate local requirements than to meet incoming demand, and, more fundamentally, if the move towards greater local and regional responsibility leads to a tightening of planning constraints, much of the pressure will have to come out in higher house prices and greater problems of affordability for those on lower incomes. This will reduce the competitiveness of the regions with the greatest economic opportunities in the UK and will adversely impact on our capacity to operate effectively within the Europe market.

The provision of large scale additional residential development on both brownfield and greenfield sites, the better utilisation of the existing stock and the development of housing markets in areas which are currently seen as relatively undesirable, such as the Thames Gateway and coastal towns, are all fundamental to achieving sustained economic growth in the early twenty-first century. It will take a very sophisticated, and consumer oriented, approach by planners and developers alike to achieve this agenda.

Note

[1] The chapter draws heavily on a number of research projects undertaken at the Cambridge Centre of Housing and Planning Research and particularly on the work of Alan Holmans (see bibliography for details).

References

Bate, R.; Best, R. and Holmans, A. (2000), *On the Move: the Housing Consequences of Migration*, York Publishing Services, York.

Crook, A. and Whitehead, C. (forthcoming), *Social Housing and Housing Subsidies, Planning Gain and Betterment Tax: A Critical Assessment of Using the Planning System to Provide Affordable Housing.*

Department of Environment, Transport and the Regions (1999), *Projections of Households in England to 2021*, TSO, London.

Department of the Environment, Transport and the Regions/Department of Social Security (2000), *Quality and Choice: a decent home for all*, DETR, London.

Greater London Authority (2000), *Homes for a World City*, the Authority, London.

Greater London Authority (2001), *Key Issues for Key Workers*, the Authority, London.

Holmans, A. (2001), *Housing Demand and Need in England, 1996 - 2016*, Royal Town Planning Institute, London.

Holmans, A., Kiddle, C. and Whitehead, C. (2000), *Housing Needs in the South East and London: An Update*, Research Report III, Cambridge Housing and Planning Research, University of Cambridge, the Centre, Cambridge.

Kleinman, M.; Aulakh, S.; Holmans, A.; Morrison, N.; Whitehead, C. and Woodrow, J. (1999), *No Excuse Not to Build*, Shelter, London.

London Planning Advisory Committee (1999), *London Housing Capacity Guidelines 1992 - 2016*, Report 85/99, LPAC, London.

Monk, S. and Whitehead, C. (2000), *Housing Key Workers in Surrey*, Cambridge Housing and Planning Research Report No II, the Centre, Cambridge.

Office for National Statistics (1999), *1996 based Subnational Population Projections*, Series PP3 No 10, DETR, London.

Office for National Statistics (2000), *Population Trends*, Winter.

Shaw, C. (2000), '1998 based national population projections for the United Kingdom and constituent countries', *Population Trends*, Spring.

Travers, T.; Whitehead, C.; Holmans, A. and Gordon, I. (1999), *Housing in London: Future Perspectives*, LSE London Discussion Paper No 4, London School of Economics and Political Science, London.

Urban Task Force (2000), *Towards an Urban Renaissance*, E and F.N. Spon, London.

Whitehead, C.; Holmans, A.; Marshall, D.; Royce Porter, C. and Gordon, I. (1999), *Housing Needs in the South East*, Property Research Unit Discussion Paper No 112, Department of Land Economy, Cambridge.

For Product Safety Concerns and Information please contact our EU representative GPSR@taylorandfrancis.com Taylor & Francis Verlag GmbH, Kaufingerstraße 24, 80331 München, Germany

Batch number: 10726013

Printed by Printforce, the Netherlands